Lecture Notes in Bioinformati

Edited by S. Istrail, P. Pevzner, and M. W

Subseries of Lecture Notes in Computer Science

Corrado Priami Rainer Breitling
David Gilbert Monika Heiner
Adelinde M. Uhrmacher (Eds.)

Transactions on Computational Systems Biology XII

Special Issue on Modeling Methodologies

Springer

Series Editors

Sorin Istrail, Brown University, Providence, RI, USA
Pavel Pevzner, University of California, San Diego, CA, USA
Michael Waterman, University of Southern California, Los Angeles, CA, USA

Editor-in-Chief

Corrado Priami
The Microsoft Research - University of Trento
Centre for Computational and Systems Biology
Piazza Manci, 17, 38050 Povo (TN), Italy
E-mail: priami@dit.unitn.it

Guest Editors

Rainer Breitling
University of Groningen, Groningen Bioinformatics Centre
E-mail: r.breitling@rug.nl

David Gilbert
Brunel University, School of Information Systems
E-mail: david.gilbert@brunel.ac.uk

Monika Heiner
Brandenburg University of Technology, Computer Science Institute
E-mail: monika.heiner@informatik.tu-cottbus.de

Adelinde M. Uhrmacher
University of Rostock, Institute of Computer Science
E-mail: lin@informatik.uni-rostock.de

Library of Congress Control Number: 2009943838

CR Subject Classification (1998): J.3, F.1-2, F.4, I.6, I.2, C.1.3

ISSN	0302-9743 (Lecture Notes in Computer Science)
ISSN	1861-2075 (Transactions on Computational Systems Biology)
ISBN-10	3-642-11711-2 Springer Berlin Heidelberg New York
ISBN-13	978-3-642-11711-4 Springer Berlin Heidelberg New York

springer.com

Printed in Germany

Typesetting: Camera-ready by author, data conversion by Scientific Publishing Services, Chennai, India
Printed on acid-free paper SPIN: 12831975 06/3180 5 4 3 2 1 0

Preface

Inspired by the interesting contributions to the 6th Conference on Computational Methods in Systems Biology (CMSB 2008) and the Dagstuhl Seminar 09091 on "Formal Methods in Molecular Biology" in February 2009, papers have been selected for this special issue of the journal Transactions on Computational Systems Biology, under the title Modeling Methodologies.

The special issue starts with a position paper on "Biomodel Engineering – from Structure to Behavior", which discusses the potential that concepts from traditional computing science hold for creating more powerful models of biological systems and identifies venues for challenging future research.

The technical contributions of the special issue cover a broad range of modeling methodologies that have recently been developed in computational systems biology.

First, two new modeling languages are presented. The "Attributed π-Calculus with Priorities" presents a combination of a concurrent process language, i.e., the π-calculus, and a sequential core language, i.e., the λ-calculus; being equipped with priorities the language enables the description of diverse spatial phenomena, different kinetics, as well as an individual-based and population-based modeling of biological systems. It is shown how the different dialects of the π calculus, including the pi@-calculus, can be expressed. Its stochastic semantic is firmly rooted in CMTC's and is reflected in the stochastic simulator.

"A Language for Biochemical Systems" (LBS) focuses on supporting a modular modeling of biological systems. Modules are parameterized processes and are defined with a clear interface to facilitate the construction of complex models and a reuse of model components. Processes are systems of reactions between modified complexes, and reactions are modeled as rules, combined into systems using a parallel combinator. LBS takes its core inspirations from Petri Nets and process calculus, and supports a variety of semantics including colored Petri nets, ODE's and CTMC's. Unlike the π-calculus, and related formalisms, processes model systems of reactions rather than individual molecules.

The following case studies illustrate the use of formal modeling languages. The paper "Mechanistic Insights into Metabolic Disturbance during Type-II Diabetes and Obesity Using Qualitative Networks" presents the qualitative network approach and demonstrates its value in a case study based on wet-lab experiments. It enables new insights into the function of a biological system, particularly when no or only little quantitative data is available and a large number of complex biochemical interactions are involved.

With the remaining contributions, the special issue leaves the area of qualitative modeling, to move toward the quantitative, in particular discrete-event, realm.

In "Modelling Self-assembly in BlenX", the use of BlenX, a programming language based on the Beta-binders process calculus, is shown in its application to modeling molecular self-assembly. Filaments, trees and rings form the building blocks that, by being equipped with typed and constrained interfaces, can be used to describe phenomena like actin polymerization in a bottom-up manner.

The paper "Rule-Based Modeling of Transcriptional Attenuation at the Tryptophan Operon" introduces a rule-based model of gene regulation. A compact modeling language is applied to the transcriptional attenuation showing that rule-based modeling allows a concise and elegant description, and thereby analysis, of fine-grained mechanisms that have before been neglected. The more fine-grained model leads to questioning the results of earlier simulation models.

The paper "Modelling and Analysis of the NF-κB Pathway in Bio-PEPA" illuminates the strength of Bio-PEPA as a formal intermediate representation of biological signaling pathways on which various kinds of analysis can be performed. Recently added features of the language, such as explicit representation of locations and time-dependent events, play a crucial role in representing compartments and the activation of NF-κB by an external stimulus. The analysis which is conducted on the model offers new insights into the robustness of the NF-κB oscillations between the cytosol and the nucleus, and the nature of the sensitivity of some parameters.

Throughout the above technical contributions certain features emerge as central for current state-of-the-art modeling approaches in computational systems biology. These include the support of n-ary reactions, different kinetic models, different levels of description, spatial phenomena, distinguished interfaces, and modular model design. We hope to see more research on compact and easy-to-use languages, that integrate many of those features in a seamless manner. These efforts need to be complemented by applications that extend and question existing models and urge a close interaction with wet-lab experimenters as discussed above, and thereby demonstrate how formal languages help to push the envelope of modeling, simulation, and analysis in systems biology.

November 2009

Rainer Breitling
David Gilbert
Monika Heiner
Corrado Priami
Adelinde Uhrmacher

LNCS Transactions on Computational Systems Biology – Editorial Board

Table of Contents

Biomodel Engineering – From Structure to Behavior

Rainer Breitling[1], Robin A. Donaldson[2], David R. Gilbert[2,3], and Monika Heiner[4]

[1] Groningen Bioinformatics Centre, University of Groningen, Kerklaan 30, 9751 NN Haren, The Netherlands
r.breitling@rug.nl
[2] Bioinformatics Research Centre, University of Glasgow, Glasgow G12 8QQ, UK
[3] School of Information Systems, Computing and Mathematics, Brunel University, Uxbridge, UB8 3PH, UK
[4] Brandenburg University of Technology at Cottbus, Dept. of Computer Science, Postbox 10 13 44, D 03013 Cottbus, Germany

Abstract. Biomodel engineering is the science of designing, constructing and analyzing computational models of biological systems. It forms a systematic and powerful extension of earlier mathematical modeling approaches and has recently gained popularity in systems biology and synthetic biology. In this brief review for systems biologists and computational modelers, we introduce some of the basic concepts of successful biomodel engineering, illustrating them with examples from a variety of application domains, ranging from metabolic networks to cellular signaling cascades. We also present a more detailed outline of one of the major techniques of biomodel engineering – Petri net models – which provides a flexible and powerful tool for building, validating and exploring computational descriptions of biological systems.

Keywords: Systems biology, Petri nets, computational modeling, differential equations.

1 What Is Biomodel Engineering?

Biomodel engineering takes place at the interface of computing science, mathematics, engineering and biology. It provides a systematic approach for designing, constructing and analyzing computational models of biological systems. Some of its central concepts are inspired by efficient software engineering strategies. Biomodel engineering does not aim at engineering biological systems *per se,* but rather aims at describing their structure and behavior, in particular at the level of intracellular molecular processes, using computational tools and techniques in a principled way. The two major application areas of biomodel engineering are systems biology and synthetic biology. In the former, the aim is the design and construction of models of existing biological systems, which explain observed properties and predict the response to experimental interventions; in the latter, biomodel engineering is used as part of a general strategy for designing and constructing synthetic biological systems with novel functionalities.

C. Priami et al. (Eds.): Trans. on Comput. Syst. Biol. XII, LNBI 5945, pp. 1–12, 2010.

2 Building a Computational Model

In this paper we focus on computational models in the narrow sense, i.e. models that use formal computational descriptions or algorithms to describe and understand natural phenomena (these are also sometimes called "executable models" [1]). There are many strategies for building a predictive computational model of a biological system. All of them involve a high level of abstraction and need to start with a careful analysis of what the model is supposed to achieve (requirements capture analysis). In that context, it is most important to determine what kind of properties the biologists can observe (How do they measure it? What are the accuracy, scale and temporal resolution of the measurements?), and which components of the system are known to influence system behavior and how they can be manipulated (By knocking down genes? By inhibiting of enzymes? By changing environmental conditions?). These factors determine which type of model should be built and which variables have to be included in it. In general, models (of intracellular pathways) can be *qualitative*, describing the topology or wiring diagram of the biochemical entities, or *quantitative*, with the addition of kinetic information, e.g. equations and rates. The latter can be in general *continuous* or *stochastic* [2, 3].

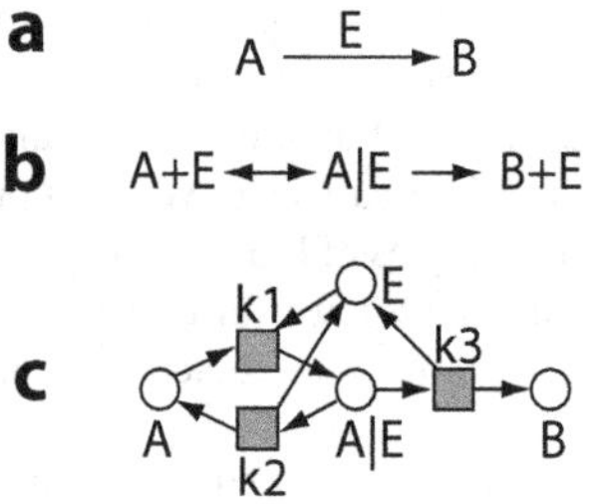

d

$$\frac{d[A]}{dt} = -k_1[A][E] + k_2[A|E]$$

$$\frac{d[E]}{dt} = -k_1[A][E] + k_2[A|E] + k_3[A|E]$$

$$\frac{d[A|E]}{dt} = k_1[A][E] - k_2[A|E] - k_3[A|E]$$

$$\frac{d[B]}{dt} = k_3[A|E]$$

Fig. 1. A simple enzyme-catalyzed reaction in four different representations. (**a**) As a biological cartoon, (**b**) as a set of chemical equations, (**c**) as a computational Petri net model, and (**d**) in the form of the corresponding system of differential equations. The equivalence of the last two representations can be clearly seen. The Petri net can be uniquely translated into differential equations (although the reverse is not generally the case).

If a model of a cellular system has to be built from scratch, the next step is to describe all relevant causal connections between the components in the system (chemical reactions, regulatory influences, physical interactions). This is most conveniently done in a graphical fashion, which comes as close as possible to the cartoon models of biological systems that are common in the biological literature (**Figure 1a**). The main challenge is a proper disambiguation of the informal descriptions provided in these cartoons. A popular software program for this purpose is, for example, Cell Illustrator [4-6], which uses a graphical user interface that employs an intuitive graphical notation to describe interactions between molecules. This is then translated into a Petri net, one of the most powerful and general types of computational models. More extensive reviews of this and alternative computational model types are provided in [1, 7, 8].

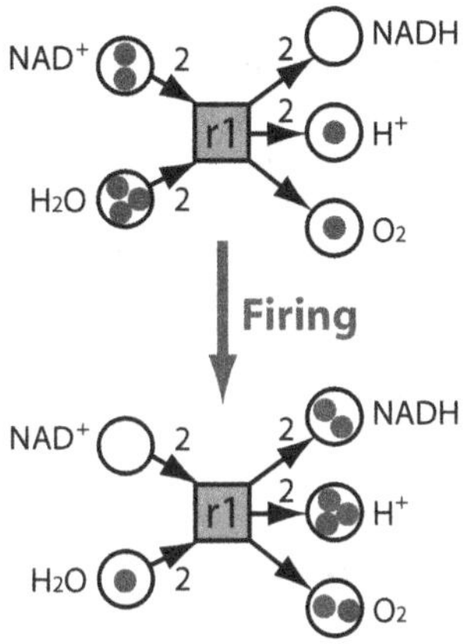

Fig. 2. Illustration of the token game of a simple chemical reaction represented as a Petri net

3 Playing the Token Game – An Example of Computational Modeling

To illustrate the properties of computational models, we will briefly describe one of them, Petri nets, in more detail. Petri nets are just one of the many formalisms that can be used in biomodel engineering. They are a graphical tool for the description of complex system behavior that was originally inspired by the description of chemical processes [7, 9, 10]. A Petri net is a graph containing two types of nodes, *places* and *transitions*. In a (bio-)chemical system, places correspond to chemical entities, and transitions to the reactions that convert them. The state of a system (e.g., the current concentration of molecules) is indicated by the amount of *tokens* that are present at each place. Simulating the behavior of a biological system using Petri nets is done by "playing the token game" (**Figure 2**): Whenever there are sufficient tokens at the places upstream of a particular transition, the transition can *fire*, i.e it consumes tokens at upstream places and creates new tokens downstream. Weights on the transitions indicate the number of tokens that are "used" when a transition fires, i.e. they correspond to the stoichiometry of the chemical reaction.

The token game can be played computationally, or using a visualization tool like Snoopy (**Table 1**), which can show how tokens move through the system over time. Such a direct exploration of molecular and signaling fluxes in a system can provide useful insights into the validity of the model and the underlying biology.

Places, tokens and transitions in a Petri net can be interpreted in multiple ways, and different interpretations can be combined in a single model, which explains the flexibility of Petri net models in biology. For example, places can not only encode chemical compounds, but also different states of compounds, including the status of protein complexes. Tokens can represent individual molecules, but also discrete levels of concentrations (e.g., high–medium–low). Transitions can correspond to mass action chemical reactions, but also to complex enzymatic reactions, complex formation or the transmission of a cellular signal.

The Petri nets described above are classical, discrete nets, where the amount of tokens is always an integer and there is not explicit notion of time (it is only implicitly given in the causality). Continuous Petri nets are not limited in this way and can describe continuous reactions in the same way as differential equations. In fact, they can be directly translated into the corresponding differential equation system using simple rules. Additional extensions of the Petri net formalism include *stochastic modeling*, which makes use of the fact that the firing of transitions is non-deterministically delayed, so that multiple token games (simulations) of the system can result in quite different outcomes each time. All these Petri nets can employ *hierarchical modeling*, where certain subgraphs in the model can be defined independently and composed into larger models.

4 Automated Model Construction

However, as systems biology is advancing, it is becoming more likely that some form of quantitative system model already exists and initially just needs to be translated into a computational model.

The translation of quantitative systems models into computational models is particularly straightforward for systems of ordinary differential equations, which can be shown to be equivalent to the continuous Petri net model of the system (although it is important to note that only the translation from Petri nets to differential equations is unique, while the reverse is not guaranteed to be the case). For small systems, the translation algorithm is easy to implement manually. For larger systems, it is more convenient to use automated translation tools, for example [11], which can directly convert Systems Biology Markup Language (SBML) descriptions of the differential equations into a Petri Net Markup Language (PNML) description of the corresponding Petri net. Automated translation has also been used on a larger scale to translate entire databases of biological knowledge into the corresponding Petri net description: Nagasaki et al. [12] demonstrated the feasibility of this approach by translating the TRANSPATH database of signaling pathways [13] into a Petri net model, stored in Cell System Markup Language (CSML), yet another systems biology file format.

Another obvious target for translation into a computational model are the stoichiometric matrices used for Flux Balance Analysis [14] and related popular approaches [15-17]. A stoichiometric matrix corresponds to the adjacency matrix of a Petri net graph. It has been shown that many of the key ideas of metabolite network analysis, such as elementary modes/extreme pathways [18] and metabolic pools [19] are equivalent to traditional concepts of Petri net theory, such a T(ransition)- and P(lace)-invariants [18, 20, 21]. Even extended models, which include Boolean descriptions of protein complexes, alternative splicing and isozymes [22], and transcriptional regulatory interactions [23], can be unified by translation into, for instance, a Petri net model to facilitate model management and exploration.

5 Managing Models in Biomodel Engineering

Biomodel engineering goes well beyond building a computational model. In particular it provides powerful ways for managing collections of models, including different versions of the same model. In biomodel engineering, computational models are considered in the same way as large computer programs in software engineering. For example, it is important to implement rules for version control, which document sources of information and each step in the model construction process. This makes it possible to explore alternative hypotheses about the system structure. Another important aspect is the systematic identification of building blocks and sub-models (modules), which can be described independently. It will be possible to reuse model modules across the model, in a similar fashion to objects in computer programming. They can be modified in a hierarchical manner, so that there may be one general module for enzymatic reactions, and specialized instances for reactions with different kinetics, which inherit the main structure and can in turn be modified to allow different mechanisms of enzyme inhibition.

Within a model management system, version control and modularity permit an efficient distributed multi-modeler approach to model building, which is essential for genome-scale modeling by large teams of curators. All of the modelers have access to the same database of building blocks, allowing for rapid composition of larger models.

6 Model Behavior Checking

Another large and very promising area of research in biomodel engineering is based on the concept of model behavior checking. Once a computational (or mathematical) model of a biological system has been created, it needs to be validated in a principled way. Does it produce reasonable predictions of system behavior? Sometimes this can be done 'by eye', but for large models this is not an option. Most importantly, it is often necessary to explore model behavior in a wide range of conditions. For this purpose, the technique of formal model behavior checking can be very useful. It is based on the principles of model checking in traditional computer engineering, an approach that uses formal logics to express certain properties of a computer system, for example whether it is safe against deadlock and other forms of systems failure [21, 24-28], which are then proved in a rigorous way. The approach we present here differs from this classical model checking in that it can also be used to check the behavior of real biological systems, e.g. to check whether the observed time series behavior in a wet laboratory experiment conforms to some formally defined properties.

Simulative model behavior checking comprises four main components:

Model. A model of the system of interest. This can be implemented in a wide range of formalisms.

Simulator. A program that produces simulated time courses from the system model. These can be stochastic or continuous simulations of a single model, or variant models that cover a range of possible parameters.

Property. A temporal logic formula describing a (desired or expected) biochemical behavior. Such a logic is essentially a specialized "mini-language" to make statements about model behavior, such as "the concentration of [P] is oscillating" or "the peak of [X] occurs at least 5 minutes before the peak of [Y]" or "an increase of activator [A] is followed by a transient repression of its target [B]".

Model Checker. A program which tests whether the model exhibits the specified temporal property, using a collection of simulated runs as input.

There are five major areas of application for model behavior checking in biology:

1. Model validation: Does the model behave in the way we expect?

2. Model comparison: How similar are two models, independent of their underlying formalisms?

3. Model searching: In a database of alternative models, which of them show a particular property? This can be used to select among competing descriptions of a system produced by various research groups. Given a large database of models (either a general collection or variants of the same model), one can use model behavior checking to perform systematic database queries, such as "find all models that show oscillatory behavior under some conditions" or "find all descriptions of this signaling pathway that are transiently active after growth factor stimulation"

4. Model analysis: In a collection of variants of a model (e.g., *in silico* gene knock-outs), which models show a certain behavior? E.g., how many knock-outs lead to a loss of oscillating behavior?

5. Model construction: Which modifications of a model lead to a desired property? Modifications can involve changes in kinetic parameters or initial concentrations, but they can also be more complex, for example changing the topology of the model by removing or even adding new components. How to do this efficiently is still an active area of research.

The model behavior checking approach can be combined with automated model modification (adding/removing/modulating edges in the network; or adding/removing molecular species in the network). Given a particular collection of behaviors that a correct model should exhibit, one can for example find the minimum number of modifications that are necessary for a given model that does not yet fulfill the conditions or detect the critical components that are most often added/removed to achieve a particular behavior [25].

Ideally, such analysis would be performed in a comprehensive biomodel engineering environment, which supports all of the above. First steps in this direction are currently being made, but no fully featured system is available yet. Eventually, biomodel engineering would be used as part of an integrated biosystem design and construction process, i.e. it will be performed in close interaction with synthetic biologists.

7 Computational versus Mathematical Models

The traditional approach of describing biological system dynamics is mathematical and is based on differential equations. They are familiar to many biologists, for example in the form of mass action equations, Lotka-Volterra equations or Michaelis-Menten enzyme kinetics. Mathematical and computational models are closely related. It can be shown that the graphical model description shown in **Figure 1c** can be directly translated into the differential equation system in **Figure 1d**, and vice versa. Both of them describe the same biological behavior. So, how do computational models differ from differential equation models?

The main difference is the systematic structure enforced by the graphical form of the computational model. Differential equations lack this rigid structure; they can be re-arranged and transformed in infinite ways. In our view, this leads to three major advantages of a computational model description:

Firstly, the major conceptual advantage is the gain in explanatory power, due to the formal structure of the computational models. In a computational model, each component corresponds to a biological entity that can be the target of an experiment (or at least a thought experiment). The model thus implies a well-defined set of *modified* systems, and predicts the response to a range of experimental interventions. Differential equation systems don't allow this straightforward mapping onto biological entities that could be manipulated experimentally. To be interpreted and to be useful as a predictive biological tool, mathematical modeling critically depends on such a mapping, but this is provided only implicitly in the analysis process, rather than as an integral part of the model engineering process as for computational models.

Secondly, their formal structure makes computational models eminently suitable for the storage of a diverse biological knowledge base: detailed quantitative information (e.g., kinetic rate laws), linear constraints (e.g., stoichiometric ratios), Boolean relationships (e.g., gene regulatory patterns), and qualitative data (e.g., protein-protein interactions) can all be stored and analyzed in one unified framework. For example, as described above, Petri nets can be both continuous and qualitative without a change in the overall formalism, and they can even incorporate stochastic elements, which are important for an accurate simulation of many biology processes [29-31]. This multiple nature also makes it possible to first validate that the structure of a model is reasonable (using topology-based analysis methods), before going on to quantitatively parameterize the description.

Finally, the graphical structure of computational models allows straightforward exploitation of the modularity of biological systems. The structure of many biological processes, ranging from enzymatic mechanisms to regulatory motifs, occurs repeatedly in the system. Once its basic structure has been described, it can be re-used as a building block across the system, with only some re-parameterization necessary. Similar plug-and-play approaches, which are essential for structured biomodel engineering, are not possible for pure mathematical models. This ability to define structural modules is a major advantage for the use of computational modeling in synthetic biology: for instance, one can envisage that each standard biological part (such as the "BioBricks" [32]) comes along with a computational model of its function, so that predictive models of new systems can rapidly be assembled. Even automatic model composition and exploration of model behavior would be possible, making biomodel engineering a key tool for the new generation of large-scale synthetic bioengineering.

8 Case Studies of Biomodel Engineering

Many groups have recently shown the power of computational models for understanding biological systems. In this brief review we can only present a few selected examples that highlight the range of systems biology applications.

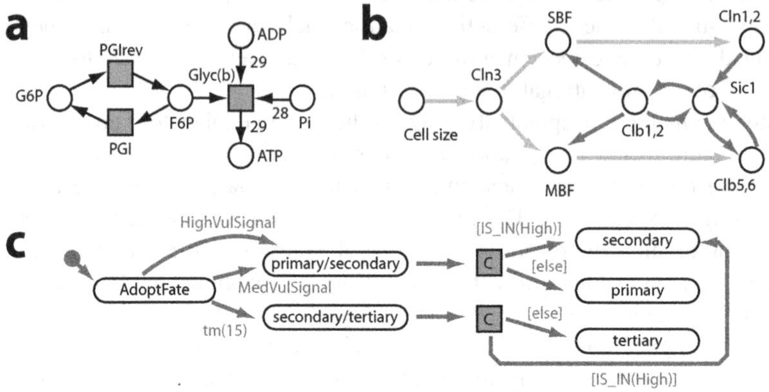

Fig. 3. Examples of computational models for different types of biological networks. (**a**) Part of the Petri net model of sucrose breakdown in the potato tuber discussed by Koch et al. [33]. (**b**) A subgraph of the simplified Boolean description of the yeast cell-cycle network presented by Li et al. [34]. (**c**) The main statechart of the vulval precursor cells in the model of *C. elegans* vulva development by Fisher et al. [35].

Metabolic networks were the first molecular biological systems to be described and analyzed by differential equations and related mathematical models, and they were also the first to be modeled by computational models such as Petri nets [36, 37]. Computational models have been shown to allow all the analytical methods that are used in the successful Flux Balance Analysis and related techniques [15]. For example, using a Petri net model of sucrose breakdown in the potato tuber, Koch et al. [33] use invariant analysis to identify sets of conserved co-factors (P-invariants) and independent sub-pathways (T-invariants) (**Figure 3a**).

Gene regulatory networks also were an early application area of biomodel engineering. Their topology can be determined experimentally on a large scale. The directionality of regulatory connections is also accessible, but the strength of the interaction is very hard to measure. Therefore, Boolean descriptions are a natural way of computationally modeling gene regulation. Li et al. [34] described the cell-cycle regulatory network of yeast as a dynamic Boolean network and identified its global dynamic properties, including a surprising level of robustness against network perturbations (**Figure 3b**).

Cellular signaling pathways are another popular target for computational modeling. For example, Ruths et al. [38] exploit the intermediate position of Petri nets between purely topological descriptions and fully kinetic descriptions to come up with a nonparametric strategy for characterizing the dynamics of cellular signaling. Their method is fast and does not require detailed kinetic information, while still providing testable predictions about the temporal change of signaling intermediates, based on repeatedly playing the token game.

Developmental differentiation processes are one of the main challenges for large-scale system modeling, because they combine gene regulation and cellular signaling across multiple temporal and spatial scales. So far, computational models have been applied mainly to small, well-characterized model systems. For example, vulval development in the worm *C. elegans* involves only three cell types and was successfully modeled using interacting state machines [35] (**Figure 3c**). The model included different modes of crosstalk between signaling pathways, and the analysis was carried out using model checking to predict novel feedback loops that are required to explain the system behavior and robust patterning observed in biological experiments.

9 Conclusions

In this paper we have shown how using concepts from traditional computing science can be used to create more powerful models of biological systems. We have demonstrated that computational models are more than just quantitative descriptions of the system. Their structure enables systematic predictions of model behavior, and model behavior checking can be used to explore options for updating a model after new experimental evidence becomes available. The modularity of computational models makes composition of genome-scale models based on standardized building blocks much more efficient. The intuitive graphical nature of computational models and their ability to integrate qualitative/structural and quantitative/kinetic information contributes to their raising popularity in many areas of biology. We predict that we will see a continued surge of biomodel engineering throughout systems biology and synthetic biology in the coming years.

References

1. Fisher, J., Henzinger, T.A.: Executable cell biology. Nat. Biotechnol. 25, 1239–1249 (2007)
2. Gilbert, D., Fuss, H., Gu, X., Orton, R., Robinson, S., Vyshemirsky, V., Kurth, M.J., Downes, C.S., Dubitzky, W.: Computational methodologies for modelling, analysis and simulation of signalling networks. Brief Bioinform. 7, 339–353 (2006)
3. Gilbert, D., Heiner, M., Lehrack, S.: A Unifying Framework for Modelling and Analysing Biochemical Pathways Using Petri Nets. In: Calder, M., Gilmore, S. (eds.) CMSB 2007. LNCS (LNBI), vol. 4695, pp. 200–216. Springer, Heidelberg (2007)
4. Nagasaki, M., Doi, A., Matsuno, H., Miyano, S.: Computational modeling of biological processes with Petri Net-based architecture. In: Chen, Y. (ed.) Bioinformatics Technologies, pp. 179–242. Springer, Heidelberg (2005)
5. Matsuno, H., Li, C., Miyano, S.: Petri net based descriptions for systematic understanding of biological pathways. IEICE Trans Fundam Electron Commun. Comput Sci. E89-A, 3166–3174 (2006)
6. Nagasaki, M., Doi, A., Matsuno, H., Miyano, S.: Genomic Object Net: a platform for modeling and simulating biopathways. Applied Bioinformatics 2, 181–184 (2003)

7. Chaouiya, C.: Petri net modelling of biological networks. Brief Bioinform. 8, 210–219 (2007)
8. Heiner, M., Gilbert, D., Donaldson, R.: Petri nets for Systems and Synthetic Biology. In: Bernardo, M., Degano, P., Zavattaro, G. (eds.) SFM 2008. LNCS, vol. 5016, pp. 215–264. Springer, Heidelberg (2008)
9. Petri, C.A., Reisig, W.: Petri nets. Scholarpedia 3, 6477 (2008)
10. Breitling, R., Gilbert, D., Heiner, M., Orton, R.: A structured approach for the engineering of biochemical network models, illustrated for signalling pathways. Brief Bioinform. 9, 404–421 (2008)
11. Shaw, O., Koelmans, A., Steggles, J., Wipat, A.: Applying Petri Nets to Systems Biology using XML Technologies. Technical Report Series. University of Newcastle upon Tyne (2004); CS-TR-827
12. Nagasaki, M., Saito, A., Li, C., Jeong, E., Miyano, S.: Systematic reconstruction of TRANSPATH data into cell system markup language. BMC Syst. Biol. 2, 53 (2008)
13. Krull, M., Pistor, S., Voss, N., Kel, A., Reuter, I., Kronenberg, D., Michael, H., Schwarzer, K., Potapov, A., Choi, C., et al.: TRANSPATH: an information resource for storing and visualizing signaling pathways and their pathological aberrations. Nucleic Acids Res. 34, D546–D551 (2006)
14. Edwards, J.S., Palsson, B.O.: Metabolic flux balance analysis and the in silico analysis of Escherichia coli K-12 gene deletions. BMC Bioinformatics 1, 1 (2000)
15. Reed, J.L., Famili, I., Thiele, I., Palsson, B.O.: Towards multidimensional genome annotation. Nat. Rev. Genet. 7, 130–141 (2006)
16. Becker, S.A., Feist, A.M., Mo, M.L., Hannum, G., Palsson, B.O., Herrgard, M.J.: Quantitative prediction of cellular metabolism with constraint-based models: the COBRA Toolbox. Nat. Protoc. 2, 727–738 (2007)
17. Palsson, B.O.: Systems Biology: Properties of Reconstructed Networks. Cambridge University Press, Cambridge (2006)
18. Papin, J.A., Stelling, J., Price, N.D., Klamt, S., Schuster, S., Palsson, B.O.: Comparison of network-based pathway analysis methods. Trends Biotechnol. 22, 400–405 (2004)
19. Famili, I., Palsson, B.O.: The convex basis of the left null space of the stoichiometric matrix leads to the definition of metabolically meaningful pools. Biophys. J. 85, 16–26 (2003)
20. Heiner, M., Koch, I., Will, J.: Model validation of biological pathways using Petri nets–demonstrated for apoptosis. Biosystems 75, 15–28 (2004)
21. Heiner, M., Koch, I.: Petri net based model validation in systems biology. In: Cortadella, J., Reisig, W. (eds.) ICATPN 2004. LNCS, vol. 3099, pp. 216–237. Springer, Heidelberg (2004)
22. Duarte, N.C., Becker, S.A., Jamshidi, N., Thiele, I., Mo, M.L., Vo, T.D., Srivas, R., Palsson, B.O.: Global reconstruction of the human metabolic network based on genomic and bibliomic data. Proc. Natl. Acad. Sci. U S A 104, 1777–1782 (2007)
23. Herrgard, M.J., Lee, B.S., Portnoy, V., Palsson, B.O.: Integrated analysis of regulatory and metabolic networks reveals novel regulatory mechanisms in Saccharomyces cerevisiae. Genome Res. 16, 627–635 (2006)
24. Heath, J., Kwiatkowska, M., Norman, G., Parker, D., Tymchyshyn, O.: Probabilistic model checking of complex biological pathways. Theoretical Computer Science 391, 239–257 (2008)
25. Kwiatkowska, M., Norman, G., Parker, D.: Using probabilistic model checking in systems biology. ACM SIGMETRICS Performance Evaluation Review 35, 14–21 (2008)

26. Calder, M., Vyshemirsky, V., Gilbert, D., Orton, R.: Analysis of Signalling Pathways using Continuous Time Markov Chains. In: Priami, C., Plotkin, G. (eds.) Transactions on Computational Systems Biology VI. LNCS (LNBI), vol. 4220, pp. 44–67. Springer, Heidelberg (2006)
27. Chabrier, N., Fages, F.: Symbolic model checking of biochemical networks. In: Priami, C. (ed.) CMSB 2003. LNCS, vol. 2602, pp. 149–162. Springer, Heidelberg (2003)
28. Clarke, E.M., Grumberg, O., Peled, D.A.: Model checking. MIT Press, Cambridge (1999)
29. Losick, R., Desplan, C.: Stochasticity and cell fate. Science 320, 65–68 (2008)
30. Gillespie, D.T.: Stochastic simulation of chemical kinetics. Annu. Rev. Phys. Chem. 58, 35–55 (2007)
31. Kaern, M., Elston, T.C., Blake, W.J., Collins, J.J.: Stochasticity in gene expression: from theories to phenotypes. Nat. Rev. Genet. 6, 451–464 (2005)
32. Shetty, R.P., Endy, D., Knight Jr., T.F.: Engineering BioBrick vectors from BioBrick parts. J. Biol. Eng. 2, 5 (2008)
33. Koch, I., Junker, B.H., Heiner, M.: Application of Petri net theory for modelling and validation of the sucrose breakdown pathway in the potato tuber. Bioinformatics 21, 1219–1226 (2005)
34. Li, F., Long, T., Lu, Y., Quyang, Q., Tang, C.: The yeast cell-cycle network is robustly designed. Proc. Natl. Acad. Sci. U S A 101, 4781–4786 (2004)
35. Fisher, J., Piterman, N., Hubbard, E.J., Stern, M.J., Harel, D.: Computational insights into Caenorhabditis elegans vulval development. Proc. Natl. Acad. Sci. U S A 102, 1951–1956 (2005)
36. Hofestädt, R.: A Petri Net Application of Metabolic Processes. Journal of System Analysis, Modelling and Simulation 16, 113–122 (1994)
37. Reddy, V.N., Mavrovouniotis, M.L., Liebman, M.N.: Petri Net Representation in Metabolic Pathways. In: First International Conference on Intelligent Systems for Molecular Biology 1993, pp. 328–336. AAAI Press, Menlo Park (1993)
38. Ruths, D., Muller, M., Tseng, J.T., Nakhleh, L., Ram, P.T.: The signaling Petri net-based simulator: a non-parametric strategy for characterizing the dynamics of cell-specific signaling networks. PLoS Comput. Biol. 4, e1000005 (2008)

Table 1. A small selection of software and websites for biomodel engineering

	URL
BioNessie. ODE-based pathway construction, simulation and analysis	http://www.bionessie.org
CellDesigner. Graphical editing and simulation of biochemical models	http://www.systems-biology.org/cd/
Cell Illustrator. Draw, model, elucidate and simulate biochemical models; Petri net	http://www.cellillustrator.com/
Charlie. Tool for analyzing Petri net models	http://www.informatik.tu-cottbus.de/~wwwdssz/software/charlie.html
CPN Tools. Editing, simulating and analyzing Coloured Petri Nets	http://wiki.daimi.au.dk/cpntools/cpntools.wiki
MC2. Probabilistic linear-time temporal logic checker	http://www.brc.dcs.gla.ac.uk/software/mc2/
Pathway Logic Assistant. Create, simulate and analyze based on rewriting logic	http://pl.csl.sri.com/software.html
Prism. Probabilistic branching-time temporal logic model checker	http://www.prismmodelchecker.org
Snoopy. Petri net editor, simulator, and analysis interface	http://www.informatik.tu-cottbus.de/~wwwdssz/software/snoopy.html
SPiM. Stochastic pi-calculus simulator	http://research.microsoft.com/~aphillip/spim/
Cell System Markup Language	http://www.csml.org/
Petri Net Markup Language	http://www2.informatik.hu-berlin.de/top/pnml/about.html
Systems Biology Markup Language	http://sbml.org/Main_Page

The Attributed Pi-Calculus with Priorities

Mathias John[1], Cédric Lhoussaine[2,4], Joachim Niehren[3,4], and Adelinde M. Uhrmacher[1]

[1] University of Rostock
Institute of Computer Science, Modeling and Simulation Group
[2] University of Lille 1
[3] INRIA, Lille
[4] BioComputing group, LIFL* & IRI**

Abstract. We present the attributed π-calculus for modeling concurrent systems with interaction constraints depending on the values of attributes of processes. The λ-calculus serves as a constraint language underlying the π-calculus. Interaction constraints subsume priorities, by which to express global aspects of populations. We present a non-deterministic and a stochastic semantics for the attributed π-calculus. We show how to encode the π-calculus with priorities and polyadic synchronization π@ and thus dynamic compartments, as well as the stochastic π-calculus with concurrent objects SPiCO.

We illustrate the usefulness of the attributed π-calculus for modeling biological systems at two particular examples: Euglena's spatial movement in phototaxis, and cooperative protein binding in gene regulation of bacteriophage lambda. Furthermore, population-based model is supported beside individual-based modeling. A stochastic simulation algorithm for the attributed π-calculus is derived from its stochastic semantics. We have implemented a simulator and present experimental results, that confirm the practical relevance of our approach.

1 Introduction

Biological systems are populations of molecules that evolve over time. Numeric modeling approaches based on differential equations assume that the evolution is continuous and deterministic. Stochastic approaches founded in continuous time Markov chains (CTMCs) consider the evolution of populations as discrete and non-deterministic. Transitions of a CTMC express state changes of the population. They can be obtained by modeling biochemical systems on population level or on individual level, for instance, by viewing chemical reactions as transformation rules for populations or respectively as rules for interactions between individuals. Deterministic models can be obtained from stochastic models by averaging over possible behaviors [1,2,3].

A plethora of formal concurrent programming languages for stochastic modeling of biological systems has been proposed, since the seminal work of Regev

* Laboratoire d'Informatique Fondamentale de Lille, CNRS UMR8022.
** Interdisciplinary Research Institute, CNRS USR3078., Lille

C. Priami et al. (Eds.): Trans. on Comput. Syst. Biol. XII, LNBI 5945, pp. 13–76, 2010.

and Shapiro [4,5] on the stochastic π-calculus. Semantically, programs in such languages define CTMCs. Operationally, they enable stochastic simulation without constructing the CTMC, so that only a (small) part of the (often huge) state space of the CTMC is inspected on the fly.

We distinguish rule-based and object-centered modeling languages. Note that this distinction is largely independent of a population-based or individual-based modeling style. Rule-based languages provide rewrite rules, by which to express chemical reactions and pathways being composed thereof. Multiset rewriting is the foundation of BIOCHAM [3] and Petri nets [6], while graph rewriting underlies the κ-calculus[7,8] and Bigraphs [9].

Object-centered modeling languages describe concurrent systems as collections of interacting objects. Various interaction concepts were investigated. Channel communication underlies the stochastic π-calculus [4,10,11,12], and its spatial relatives [13,14], while the participation of a species in a chemical reaction is the approach of Bio-PEPA [15], and communication through a global constraint store the principle of stochastic concurrent constraint programming (sCCP) [16]. Different interaction concepts may lead to different modeling perspectives. The π-calculus and its relatives support the modeling of molecules as objects (individual-based), Bio-PEPA features the modeling of species as objects, whose amounts are changed by chemical rules (species-based). The idea of sCCP is to model chemical reactions as objects that change the population residing in the constraint store (population-based).

Various kinds of local and global constraints on interactions were proposed for the π-calculus. Polyadic synchronization [17] can express local constraints, which impose equality of tuples of channels. Priorities yield global constraints that were added to the π-calculus [18] in order to model global aspects of populations in individual-based modeling, such as global state updates following local interactions. Stochastic rates provide another kind of global constraints, that were proposed for the π-calculus [19,20,12,21] in order to enable stochastic modeling and simulation. As shown in [18], dynamic compartments become expressible in the π-calculus with priorities and polyadic synchronization π@. There exists a variant called sπ@ that is equipped with a stochastic simulation algorithm [22]. It can express dynamic compartments with variable volumes, but does neither provide a stochastic semantics in terms of CTMCs nor any support for general purpose attributes beside compartment volumes.

Basic Ideas. In this paper, we propose the attributed π-calculus as a unifying framework, in order to express local and global interaction constraints for the π-calculus. We obtain it by extending the π-calculus with attributed processes and interaction constraints, that may depend on the attribute values.

Attribute values of various types are useful in order to define various properties of biological processes. Two examples on different levels of abstraction illustrate the basic idea: first, a cell *Cell(coord,vol)* is attributed by coordinates $coord \in \mathbb{R}^3$ and a volume $vol \in \mathbb{R}_+$. Second, protein *Prot(comp)* is attributed by the cellular compartment $comp \in \{\texttt{nucleus}, \ldots\}$ in which it is located.

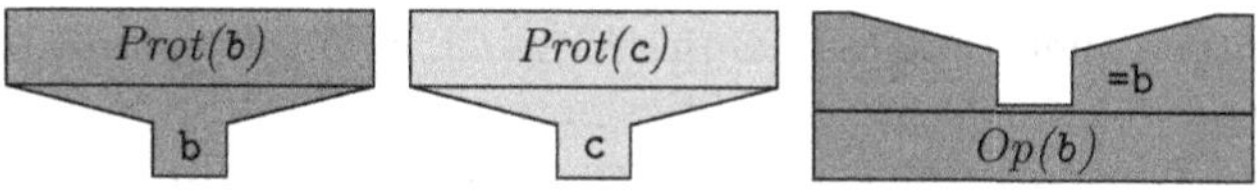

Fig. 1. Only the protein of sort b is permitted to bind to the operator of sort b, but not the protein of sort c. Equality of sorts is tested by the operator, once it sees the sort of the protein, by applying the test function $\lambda x.x{=}\mathtt{b}$.

Interaction constraints on attribute values may be imposed in order to define, whether or not two attributed processes are allowed to communicate. Consider e.g. proteins $Prot(x)$ of sort $x \in \{\mathtt{b}, \mathtt{c}\}$ and operators $Op(y)$ of sort $y \in \{\mathtt{b}, \mathtt{c}\}$ able to bind a protein of the same sort. For the binding, equality of their sorts $x{=}y$ is required. In the attributed π-calculus this can be expressed by the following definitions, which are valid for all possible values of the attributes x and y:

$$Prot(x) \triangleq bind[x]!().0$$
$$Op(y) \triangleq bind[\lambda x.x = y]?().OpBound(y)$$

This says that before enabling a binding action, $Prot(x)$ needs to provide its sort as specified by $bind[x]$. The operator receives the value of x and tests it for equality with its own sort, as imposed by $bind[\lambda x.x{=}y]$; the equality constraint is expressed by the Boolean valued function $\lambda x.x{=}y$. Consider for instance a system with two actors $Prot(\mathtt{b})$ and $Op(\mathtt{b})$:

$$Prot(\mathtt{b}) \mid Op(\mathtt{b}) \rightarrow OpBound(\mathtt{b})$$

The interaction constraint for these two actors is composed by function application $(\lambda x.x{=}\mathtt{b})\mathtt{b}$. It evaluates to the truth value of $\mathtt{b}{=}\mathtt{b}$ which is true. This enables the above binding action, by which the operator turns into its bound state. The slightly more complex system $Prot(\mathtt{b}) \mid Prot(\mathtt{c}) \mid Op(\mathtt{b})$ where only the first protein is permitted to bind to the operator is illustrated in Fig. 1.

In the above example, interaction constraints evaluate to Boolean values. If we permit constraints that evaluate to positive real numbers, we can express stochastic rates that depend on attribute values. Or else, if we generalize constraints such that they return values of an ordered set, we obtain a π-calculus where priorities may depend on attribute values.

Contributions. We introduce the attributed π-calculus $\pi(\mathcal{L})$ with priorities, in which $\mathcal{L}$ is a language providing attribute values and constraints. We propose to use a call-by-value λ-calculus as attribute language $\mathcal{L}$, in which we leave the choice of constants of $\mathcal{L}$ parametric. This enables the many reasonable choices, such as positive real numbers for stochastic rates or arbitrary ordered sets for priorities. Note that the role of the attribute language is analogous to the role of the constraint language in constraint logic programming CLP($\mathcal{L}$) [23] or concurrent constraint programming CCP($\mathcal{L}$) [24], except that we now permit higher-order languages $\mathcal{L}$.

We present two operational semantics for $\pi(\mathcal{L})$, a non-deterministic semantics with priorities and a stochastic semantics in terms of CTMCs, under the assumptions that successful constraints evaluate to stochastic rates in $\mathbb{R}_+^\infty$. We show that the stochastic semantics is a proper refinement of the non-deterministic semantics, in that it permits the same reduction steps (Proposition 5). Our stochastic semantics does not impose any syntactical restrictions such as the biochemical form. This is relevant, since it simplifies all technical results in contrast to most previous approaches (including the conference version [25] of the present article). We also provide a type system, that extends that of the simply typed λ-calculus to the process level, and prove type safety (Theorem 1). Types express invariants that are useful to detect program errors for higher-order languages, and thus when modeling biological systems in such languages.

We illustrate the usefulness of the attributed π-calculus for modeling and simulation in systems biology. We first describe the light dependent movement of Euglena, a single celled organism living in inland water. This illustrates the treatment of spatial aspects with location dependent attributes. The second example regards cooperative binding, a phenomenon, which is often observed in gene regulatory systems [26,27]. This is an example of an interaction between three actors that can be simplified by using attributes. We also discuss new opportunities for population based-modeling of chemical reactions in the attributed π-calculus. Here, populations are represented by attributed processes for multisets such as $P(4,5,3)$ and reactions as actors that transform populations. This modeling style was proposed for sCCP [16].

We present an encoding of the π-calculus with priorities in an ordered set into an attributed π-calculus with priorities as constants, and prove it correct for the non-deterministic semantics (Theorem 2). We show that the same translation can be used to encode the stochastic π-calculus into the attributed π-calculus with stochastic rates as constants and two levels of priorities and that the encoding is correct with respect to the stochastic semantics (Theorem 3). Types are preserved by the encoding.

We show how to encode the π-calculus with priorities and polyadic synchronization π@ into the attributed π-calculus with equality tests, and prove the correctness wrt. the non-deterministic semantics (Theorem 4). Thereby, we lift encodings of the spatial modeling languages BioAmbients [13] and Brane [14] to the attributed π-calculus (see [18,28]). Note that there exists neither a stochastic semantics in terms of CTMCs for π@ nor a type system.

Finally, we present a stochastic simulation algorithm for the attributed π-calculus, that is inferred directly from the stochastic semantics. We have implemented this algorithm on top of the modeling and simulation framework JAMES II [29]. We discuss the performance of our simulator at run-time experiments with the Euglena model, and show that it yields comparable results with the SPiM simulator for the stochastic π-calculus [11]. This confirms the practical relevance of our approach.

Compared to the conference version [25], we added priorities to the non-deterministic semantics of the attributed π-calculus, and extended the encoding of $\pi@$ in the attributed π-calculus, such that it accounts for priorities. The whole presentation has been changed, such that the attributed π-calculus becomes a proper generalization of the π-calculus with priorities and the stochastic π-calculus. Annotations have been moved from channels to communication prefixes, so that no global environments needed to be introduced. The changes have led to an important simplification of the stochastic semantics, which enables our correctness proofs. We also improved our implementation and extended the practical experiments.

Discussion. We argued that the attributed π-calculus supports both modeling styles, molecules as processes (individual-based) and reactions as processes (population-based). In the individual-based approach, priorities are needed to express global aspects of the population, while no priorities are not needed in the population-based style.

The stochastic semantics of the attributed π-calculus only slightly liberalizes the law of mass action, in that the propensity of an interaction remains the product of multiplicities of the interaction partners and the stochastic rate of the interaction. The only difference is that the stochastic rate may now be dependent on the values of the attributes of the interacting processes, so that other aspects than their masses may intervene. A weakness of the π-calculus approach presented here is that it does not support different kinetics such as Michaelis-Menten in individual-based modeling (while it does for population-based modeling).

Whereas an increased expressiveness of a modeling language typically will ease the development of models, it requires additional support for developing models, e.g. to ensure type consistency, and burdens model analysis and simulation. Here we followed a tradition in concurrent programming languages, to combine a process level and a sequential core language (expression level). Since only the λ-calculus with types of low (first or second) order are used in practice, we believe that our extension is justified. This holds in particular, when accepting the π-calculus as a starting point, since it is higher-order anyway.

Outline. In Section 2 we present the π-calculus with priorities and the stochastic π-calculus in a uniform manner. We start from an ordered set $(R, <)$ whose elements may be either priorities or stochastic rates. We provide a unified syntax for processes in both calculi, in which communication prefixes (rather than channels) are annotated by values of R. We then present two operational semantics for the same syntax, a non-deterministic semantics as for the π-calculus with priorities, and a stochastic semantics as for the stochastic π-calculus.

In Section 3, we present the attributed π-calculus with priorities. In the syntax, we permit λ-expressions instead of channels, priorities or stochastic rates. Again, we introduce two operational semantics, the one is non-deterministic and the other stochastic. We present a type system, which lifts the simple type system of the λ-calculus to attributed process, and prove type safety under the assumption that the attribute language is type safe.

Section 4 contains the examples for modeling biological systems in the attributed π-calculus, and a discussion about population- versus individual-based modeling. In Section 5, we encode the π-calculus with priorities and π@ into the attributed π-calculus with priorities, and prove correctness. We provide an encoding of SPiCO and discuss variants of the stochastic and attributed π-calculus. In Sections 6 and 7, we present the stochastic simulation algorithm and discuss the performance of our implementation.

Related work. The π-calculus with polyadic synchronization was first proposed in [17] but could be extended by more general data terms as in [30] without particular difficulties. An extended π-calculus with polyadic synchronization, data terms, and explicit substitutions motivated by security applications was proposed in [31]. Another instance of the idea of interaction constraints is present in BlenX[32,33], where interactions are allowed only between prefixes sharing a compatible type annotation.

Further modeling approaches based on the stochastic π-calculus can be found in [34,35]. All stochastic π models can be translated into the attributed π-calculus. This requires to encode the versions of the stochastic π-calculus used there into the attributed π-calculus. We show in Section 5.3, that we can express the version of the stochastic π-calculus where stochastic rates are annotated to channels.

The closest approach to ours is stochastic concurrent constraint programming (sCCP) [16]. There, one can indeed express attributed processes with stochastic rates depending on attribute values. This enables the modeling of "reactions as objects" and "populations as stores" (population-based), but not of "molecules as objects" (individual-based), since direct interactions between concurrent actors are not permitted. In sCCP they interact by indirection though a global constraint store. It should also be noticed that the constraints of the attributed π-calculus are higher-order while those of sCCP are first-order.

2 Pi-Calculus with Priorities

We start from the π-calculus with priorities, for which we provide a non-deterministic and a stochastic operational semantics. This latter leads us to a new version of the stochastic π-calculus, which in contrast to [11,25] does not impose any syntactic restrictions (such as biochemical forms).

2.1 Design Decisions

We search for a unified treatment of the π-calculus with priorities and the stochastic π-calculus. Both, priorities and stochastic rates both express global properties, that concern the whole population of objects. For selecting a communication step with highest priority, one has to inspect all potential communication steps. Similarly, the probability of a communication step in a stochastic setting depends globally on all possible communication steps. In both cases, the difficulty is thus to reason globally about all possible communication steps in a given population.

Our first design decision is to permit process definitions with recursion and parameters in the syntax (rather than replication). Definitions are very convenient for modeling and simulation in systems biology, and therefore supported by all current simulators for the stochastic π-calculus [12,21,10]. The difficulty in the presence of priorities is to discover all potential communication steps in a given process, since some of them may be hidden by definitions. In order to solve this problem, we propose to always apply process definitions exhaustively before selecting any communication step. Fortunately, the resulting operational semantics remains nicely simple, and can be generalized properly to the stochastic setting. The alternative approach elaborated in [18] consists in adding the replication operator to the syntax of the π-calculus, and to hide its application in the structural congruence of processes. Since applications of the structural congruence must be restricted when counting interaction opportunities as needed in the stochastic semantics, this approach cannot be applied here.

Our second choice is to annotate communication prefixes rather than channels by elements in an ordered set $(R, <)$ which may either contain priorities or stochastic rates. Note stochastic rates were annotated to channels in the conference version of the attributed π-calculus [25], and that priorities were not considered in the non-deterministic version there. The change toward prefix annotations simplifies our semantics considerably (and some proof obligations tremendously). Note also that we only annotate priorities to sender prefixes, rather than to sender and receiver prefixes, since otherwise one has to resolve conflicting priorities somehow [18]. See Section 5.3 for further discussions.

2.2 Syntax

Let $\texttt{Bool} = \{\texttt{true}, \texttt{false}\}$ be the set of Booleans, $\mathbb{N}$ the set of natural numbers starting from 1, $\mathbb{N}_0 = \mathbb{N} \cup \{0\}$, $\mathbb{R}_+$ the set of non-negative real numbers, and $\mathbb{R}_+^\infty = \mathbb{R}_+ \cup \{\infty\}$.

We start from a partially ordered set $(R, <)$ of priorities, an infinite set *Vars* of channel names $x, y \in \mathit{Vars}$, and an infinite set of process names $A \in \mathit{Proc}$, that have fixed arities in $\mathbb{N}_0$. Whenever we write a term $A(x_1, \ldots, x_n)$ we assume that n is the arity of A. We use tuple notation in many places, as for instance $\tilde{x}$ for tuples of channels. If $\tilde{x} = (x_1, \ldots, x_n)$ then we define the length of the tuple by $|\tilde{x}| = n$. Whenever we use terms $A(\tilde{x})$ we assume that the length of $\tilde{x}$ is equal to the arity of A. Substitutions replacing y by x are denoted by $[x/y]$. Substitutions $[\tilde{y}/\tilde{x}]$ apply to tuples of the same length $|\tilde{y}| = |\tilde{x}|$.

The syntax of the π-calculus with priorities is defined in Fig. 2. In addition to channel names $x \in \mathit{Vars}$ and priorities $r \in R$ there are 4 syntactic categories: prefixes π, processes P, sums M, and definitions D. A prefix is either a receiver $x?\tilde{y}$ or a sender $x{:}r!\tilde{z}$. We assume that all channel names in $\tilde{y}$ are pairwise distinct (since they are distinct formal parameters). A receiver is supposed to receive a tuple of values for $\tilde{y}$ on channel x, and a sender to send a tuple of values $\tilde{z}$ on channel x. The priority r of an interaction is determined by the

Prefixes	$\pi ::= x?\tilde{y}$	receiver
	$\mid\ x{:}r!\tilde{z}$	sender
Sums	$M ::= \pi.P$	prefixed process
	$\mid\ M_1 + M_2$	choice
Processes	$P ::= M$	sums
	$\mid\ A(\tilde{x})$	defined process
	$\mid\ P_1 \mid P_2$	parallel composition
	$\mid\ (\nu x)P$	channel creation
	$\mid\ \mathbf{0}$	idle process
Definitions	$D ::= A(\tilde{x}) \triangleq P$	parametric process definition

Fig. 2. Syntax of the π-calculus with channels $x, \tilde{x}, \tilde{y}, \tilde{z} \in \mathit{Vars}$ and priorities $r \in R$

$fv(M_1 + M_2) = fv(M_1) + fv(M_2)$	$fv(\mathbf{0}) = \emptyset$
$fv(x?\tilde{y}.P) = \{x\} \cup (fv(P) \setminus \{\tilde{y}\})$	$fv(P_1 \mid P_2) = fv(P_1) \cup fv(P_2)$
$fv(x{:}p!\tilde{z}.P) = \{x\} \cup fv(\{\tilde{z}\}) \cup fv(P)$	$fv(A(\tilde{x})) = \{\tilde{x}\}$
$fv((\nu x)P) = fv(P) \setminus \{x\}$	$fv(A(\tilde{x}) \triangleq P) = fv(P) \setminus \{\tilde{x}\}$

Fig. 3. Free channel names

sender. A term $\pi_1.P_1 + \ldots + \pi_n.P_n$ is a sum of guarded prefixes, that we denote by $\sum_{i=1}^{n} \pi_i.P_i$ equivalently. A process P may be either a defined process $A(\tilde{x})$, or a parallel composition $P_1|\ldots|P_n$ that we denote equivalently as $\prod_{i=0}^{n} P_i$, or a process $(\nu x)P$ creating a new channel x with scope P. If $\tilde{x} = (x_1, \ldots, x_n)$ then we write $(\nu\tilde{x})P$ instead of $(\nu x_1)\ldots(\nu x_n)P$. Note that our syntax provides empty products but not empty sums, i.e. if $n = 1$ then $\prod_{i=1}^{n} P_i = \mathbf{0}$ is the idle process, while $\sum_{i=1}^{n} P_i$ is undefined.

The free channel names $fv(P)$ are defined as usual in Fig. 3. The three variable binders are new binders $(\nu x).P$, formal parameters $\tilde{y}$ in input prefixes $x?\tilde{y}.P$, and formal parameters $\tilde{x}$ in definitions $A(\tilde{x}) \triangleq P$. We say that bound variables are renamed apart in P, if 1) no variable is bound twice in P, 2) no bound variable of P has a free occurrence in P, and 3) no bound variable of P has a free occurrence in some definition. We generally assume, that all processes P are renamed apart, before applying any interaction step to any subprocess of P.

The structural congruence on processes $\equiv$ remains the least congruence satisfying the axioms given in Fig. 4, i.e. consistent renaming of bound variables, associativity and commutativity of parallel composition and summation, the rule of the neutral element of $\mathbf{0}$ with respect to parallel composition, and scope intrusion and extrusion for ν-binders. Note that every process P is congruent to some process in *prenex form* $(\nu\tilde{x})\prod_{i=1}^{n} P_i$, where all processes P_i are either sums M or defined processes $A(\tilde{y})$, such that all all bound variables are renamed apart, also from the free variables in the definitions of the defined processes.

$$
\begin{array}{rcl}
(P_1 \mid P_2) \mid P_3 \equiv P_1 \mid (P_2 \mid P_3) & & P_1 \mid P_2 \equiv P_2 \mid P_1 \\
(M_1 + M_2) + M_3 \equiv M_1 + (M_2 + M_3) & & M_1 + M_2 \equiv M_2 + M_1 \\
P \mid \mathbf{0} \equiv P & & (\nu x)(P \mid Q) \equiv (\nu x)P \mid Q \text{ if } x \notin fv(Q) \\
P \equiv_\alpha Q \Rightarrow P \equiv Q & & (\nu x)(\nu y)P \equiv (\nu y)(\nu x)P
\end{array}
$$

Fig. 4. Axioms of structural congruence

As an example, we consider silent actions `delay`:$r.P$. A salient action becomes active with priority r without any communication partner and then behaves like P. In our syntax of the π-calculus, silent actions can be expressed by processes $(\nu\, delay)(delay?().P \mid delay{:}r!().\mathbf{0})$ where a dummy interaction partner sends the empty tuple on local channel *delay* with priority r and then disappears.

2.3 Operational Semantics

The operational semantics of the π-calculus with priorities is presented in Figure 6. It is defined by a reduction relation $\rightarrow$ that is based on three kinds of binary relations, reductions $\xrightarrow[nd]{r}$ with priority $r \in R$, reductions $\xrightarrow[nd]{err}$ leading to errors, and reductions $\xrightarrow[nd]{app}$ applying process definitions. The label *nd* distinguishes non-deterministic from stochastic reduction steps, *err* stands for error and *app* for application.

Fig. 5 contains the axioms of the binary relations on processes $\xrightarrow[nd]{\alpha}$ where $\alpha \in R \cup \{app, err\}$, which define local interaction opportunities on the level of individuals. A communication step (COM) applies to two parallel sums with matching prefixes, a sum with a receiver $x?\tilde{y}.P_1 + M_1$ and another with sender $x{:}r!\tilde{z}.P_2 + M_2$ for the same channel x and with the same number of arguments $|\tilde{y}| = |\tilde{z}|$. The sender hands over its arguments $\tilde{z}$ to the receiver and continues with P_2, while the receiver replaces its formal parameters $\tilde{y}$ by $\tilde{z}$ and continues with $P_1[\tilde{z}/\tilde{y}]$. All alternative choices in M_1 and M_2 are discarded. The whole step may be performed with priority r contributed by the sender. A communication error (E.COM) is raised, if two matching prefixes on the same channel x offer different arities $|\tilde{y}| \neq |\tilde{z}|$. Here we write $\perp$ for an arbitrary erroneous expression. A single application step (APP) replaces a defined process by its definition. We assume that there exists a unique definition for all defined processes.

Fig. 6 provides the closure rules for these relations, and defines the global reduction relation $\rightarrow$ between processes, which depend globally on the set of all potential interactions in the population. Communication and error steps are closed under structural congruence (STRUCT), and permitted under parallel composition (PAR) and new binders (NEW). Rule (PRIOR) states that only communication steps with highest available priority may be selected by final reduction relation $\rightarrow$. The set of all communication prefixes becomes apparent only after having applied definitions exhaustively (CONV). Application may not terminate such as for $A()$ if defined by $A() \triangleq A()$. Such nonterminating definitions block

Communication and application steps

$$(\text{COM})\ \frac{|\tilde{y}| = |\tilde{z}|}{x?\tilde{y}.P_1 + M_1 \mid x{:}r!\tilde{z}.P_2 + M_2 \xrightarrow[nd]{r} P_1[\tilde{z}/\tilde{y}] \mid P_2} \qquad (\text{APP})\ \frac{A(\tilde{x}) \triangleq P}{A(\tilde{y}) \xrightarrow[nd]{app} P[\tilde{y}/\tilde{x}]}$$

Program errors

$$(\text{E.COM})\ \frac{|\tilde{y}| \neq |\tilde{z}|}{x?\tilde{y}.P_1 + M_1 \mid x{:}r!\tilde{z}.P_2 + M_2 \xrightarrow[nd]{err} \bot}$$

Fig. 5. Axioms of operational semantics of π-calculus with priorities

Structural rules where $\alpha \in \{err, app\} \cup R$

$$(\text{PAR})\ \frac{P_1 \xrightarrow[nd]{\alpha} P_1'}{P_1 \mid P_2 \xrightarrow[nd]{\alpha} P_1' \mid P_2} \quad (\text{NEW})\ \frac{P \xrightarrow[nd]{\alpha} P'}{(\nu x)P \xrightarrow[nd]{\alpha} (\nu x)P'} \quad (\text{STRUC})\ \frac{P \equiv P_1 \quad P_1 \xrightarrow[nd]{\alpha} P_2 \quad P_2 \equiv P'}{P \xrightarrow[nd]{\alpha} P'}$$

Error-free convergence of application

$$(\text{CONV})\ \frac{P \xrightarrow[nd]{app}{}^{*} P' \quad P' \equiv (\nu\tilde{x}) \prod_{i=1}^{n} M_i \quad \neg P' \xrightarrow[nd]{err} \bot}{P \Downarrow P'}$$

Reduction $(r \in R)$

$$(\text{PRIOR})\ \frac{P \Downarrow P' \quad P' \xrightarrow[nd]{r} Q \quad \neg\exists r_1 \in R.\exists Q_1.\ r < r_1 \wedge P' \xrightarrow[nd]{r_1} Q_1}{P \to Q}$$

Fig. 6. Rules of operational semantics of π-calculus with priorities in $(R, <)$

all potential subsequent communication steps. Similarly communication errors $P \xrightarrow[nd]{err} \bot$ block all communication steps on P. Finally note, that we define the reflexive transitive closure $(\xrightarrow[nd]{app})^*$ used in rule (CONV) such that it contains the structural congruence $\equiv$.

Example 1. We consider the example of forwarders $Fwd(x, y)$ which receive some value on channel x and forward it to channel y. Forwarders can be used to let objects flow along lists, such as RNAP polymerases along DNA sequences. We assume two levels of priorities $low < high$ and give highest priority to forwarding actions.

$$Fwd(x, y) \triangleq x?(z).(y{:}high!(z).\mathbf{0} \mid Fwd(x, y))$$

We first use forwarders in order to define a list with two elements, which an object z traverses.

$$List_2() \triangleq x_1{:}low!(z).\mathbf{0} \mid Fwd(x_1, x_2) \mid Fwd(x_2, x_3)$$

Process $List_2()$ can be reduced as follows:

$$\begin{aligned} List_2() &\rightarrow Fwd(x_1, x_2) \mid x_2{:}high!(z).\mathbf{0} \mid Fwd(x_2, x_3) \\ &\rightarrow Fwd(x_1, x_2) \mid Fwd(x_2, x_3) \mid x_3{:}high!(z).\mathbf{0} \end{aligned}$$

Beside lists, we can construct rings or other cyclic data structures from forwarders:

$$Ring_3() \triangleq x_1{:}low!(z).\mathbf{0} \mid x_2{:}low!(z).\mathbf{0} \mid Fwd(x_1, x_2) \mid Fwd(x_2, x_3) \mid Fwd(x_3, x_1)$$

One of the two z objects is turning around in the ring forever, while the other can never enter the ring, since entering actions are given lower priority.

$$\begin{aligned} Ring_3() &\rightarrow x_1{:}low!(z).\mathbf{0} \mid Fwd(x_1, x_2) \mid Fwd(x_2, x_3) \mid x_3{:}high!(z).\mathbf{0} \mid Fwd(x_3, x_1) \\ &\rightarrow x_1{:}low!(z).\mathbf{0} \mid x_1{:}high!(z).\mathbf{0} \mid Fwd(x_1, x_2) \mid Fwd(x_2, x_3) \mid Fwd(x_3, x_1) \\ &\rightarrow x_1{:}low!(z).\mathbf{0} \mid Fwd(x_1, x_2) \mid x_2{:}high!(z).\mathbf{0} \mid Fwd(x_2, x_3) \mid Fwd(x_3, x_1) \\ &\rightarrow \ldots \end{aligned}$$

2.4 Convergence

We show that the order of application steps does not matter, so that the result of exhaustive application of definitions is always unique. Otherwise, the approach presented here would be computationally unfeasible and thus useless in practice.

Let R be a binary relation between processes. We define the reflexive transitive closure of R up to structural congruence by $R^* = \cup_{i=0}^{\infty} R^i$ where R^0 is the structural congruence $\equiv$ and $R^{i+1} = R \circ R^i$ for all $i \in \mathbb{N}$, and write R^{-1} for the inverse relation. We call R confluent modulo structural congruence, if $(R^*)^{-1} \circ R^* \subseteq R^* \circ (R^*)^{-1}$.

Lemma 1. *The rewrite relation* $\xrightarrow[nd]{app}$ *is confluent modulo structural congruence; thereby irreducible processes are congruent to processes* $(\nu\tilde{x}) \prod_{i=1}^{n} M_i$.

Proof. The lemma relies on the fact, that we assume a unique definition for every defined process, and that the order of application of these definitions does not matter. We start with a standard analysis of the structural congruence.

Claim. Let $P = (\nu\tilde{x}) \prod_{i=1}^{n} P_i$ be a prenex normal form in which all bound variables are renamed apart, and such that all P_i are sums or defined processes. In this case, $P \xrightarrow[nd]{app} P'$ if and only if the following rule applies:

$$\frac{1 \le j \le n \quad P_j = A_j(\tilde{z}_j) \quad A_j(\tilde{y}_j) \triangleq Q_j \quad P' \equiv (\nu\tilde{x})(\prod_{i=1, i\neq j}^{n} P_i \mid Q_j[\tilde{z}_j/\tilde{y}_j])}{P \xrightarrow[nd]{app} P'}$$

Application defines a relation on equivalence classes of processes modulo structural congruence, such that $[P]_\equiv \xrightarrow[nd]{app} [P']_\equiv$ if $P \xrightarrow[nd]{app} P'$. The above claim shows that application terminates on equivalence classes of processes of the form $[(\nu\tilde{x})\prod_{i=1}^{n} M_i]_{\equiv}$, since we assume that there exists at least one definition for every defined process. We next show that application on equivalence classes is uniformly confluent [36], i.e., if $P \xrightarrow[nd]{app} P_1'$ and $P \xrightarrow[nd]{app} P_2'$ then $P_1' \equiv P_2'$ or there exists P'' such that $P_1' \xrightarrow[nd]{app} P''$ and $P_2' \xrightarrow[nd]{app} P''$. Uniform confluence implies strong confluence and thus confluence [37]. To see uniform confluence, we assume that $P \xrightarrow[nd]{app} P_1'$ and $P \xrightarrow[nd]{app} P_2'$ and let j_1 and j_2 be the positions of the respective reduction step (according to the above rule). If $j_1 = j_2$ then $P_1' \equiv P_2'$, since we assume that there exists at most one definition for every defined process. Otherwise if $j_1 \neq j_2$, then we can set P'' to $(\nu\tilde{x})(\prod_{i=1, i\notin\{j_1,j_2\}}^{n} P_i \mid Q_{j_1}[\tilde{z}_{j_1}/\tilde{y}_{j_1}] \mid Q_{j_2}[\tilde{z}_{j_2}/\tilde{y}_{j_2}])$. □

There exists processes P that do not converge to any P' since the application of process definitions does not terminate. Our semantics ensures that such processes cannot be reduced any further, even though they might not contain an immediate error $P \xrightarrow[nd]{err} \bot$. For instance, consider the process $A()$ with the following definition $A() \triangleq A()$ that is not well-founded. An implementation may either run into an infinite loop unfolding the definition of A repeatedly, or report the erroneous cycle.

Proposition 1 (Convergence). *For all process P there exists at most one class $[P']_{\equiv}$ such that $P \Downarrow P'$.*

Proof. This follows immediately from the confluence result in Lemma 1.

This proposition states that the idea of exhaustive application of definitions is consistent, in that the result does not depend of the application order. There are three possible kinds of results: convergence, arity mismatches, and non-termination of application.

Remark 1. If $P \equiv (\nu\tilde{x})\prod_{i=1}^{n} M_i$ and $\neg P \xrightarrow[nd]{err} \bot$ then $P \equiv P' \Leftrightarrow P \Downarrow P'$.

Proof. Suppose that $P \equiv (\nu\tilde{x})\prod_{i=1}^{n} M_i$ and $\neg P \xrightarrow[nd]{err} \bot$. If $P \equiv P'$ then $P \xrightarrow[nd]{app}{}^{*} P'$ by definition of reflexivity so that $P \Downarrow P'$. Conversely, suppose that $P \Downarrow P'$. By definition of convergence, this implies $P \xrightarrow[nd]{app}{}^{*} P'$, which yields $P \equiv P'$ since $[P]_{\equiv}$ is irreducible with respect to $\xrightarrow[nd]{app}$ by Lemma 1.

2.5 Stochastic Operational Semantics

We present a stochastic operational semantics for the π-calculus with priorities, under the assumption that stochastic rates in $\mathbb{R}_+^\infty$ are used as priorities with two levels, the lower level for numbers in $\mathbb{R}_+$ and the higher for ∞.

In contrast to most previous approaches, the syntax of processes remains without change. This means in particular, that stochastic rates are annotated to

Labeled communication steps ($r \in \mathbb{R}_+^\infty$ and $\ell \in \mathbb{N}^4$)

$$(\text{COM}_\ell)\quad \frac{\ell = (i_1, j_1, i_2, j_2) \quad i_1 \neq i_2 \qquad \pi_{i_1}^{j_1} = x?\tilde{y} \quad \pi_{i_2}^{j_2} = x{:}r!\tilde{z} \quad |\tilde{y}| = |\tilde{z}|}{(\nu\tilde{x}) \prod_{i=1}^{n} \sum_{j=1}^{m_i} \pi_i^j.P_i^j \xrightarrow[\ell]{r} (\nu\tilde{x})(\prod_{i=1, i\neq i_1, i_2}^{n} \sum_{j=1}^{m_i} \pi_i^j.P_i^j \mid P_{i_1}^{j_1}[\tilde{v}/\tilde{y}] \mid P_{i_2}^{j_2})}$$

Fig. 7. Axioms of operational semantics of stochastic π-calculus

Markov chain ($r, r' \in \mathbb{R}^+$)

$$(\text{SUM})\quad \frac{P \Downarrow P_1 \quad \sum_{\{(r',\ell) \mid P_1 \xrightarrow[\ell]{r'} P_2 \equiv P'\}} r' = r \neq 0 \quad \neg\exists\ell\exists P''.P_1 \xrightarrow[\ell]{\infty} P''}{P \xrightarrow{r} P'}$$

$$(\text{COUNT})\quad \frac{P \Downarrow P_1 \quad n = \sharp\{\ell \mid P_1 \xrightarrow[\ell]{\infty} P_2 \equiv P'\} \neq 0}{P \xrightarrow{\infty(n)} P'}$$

Fig. 8. Rules of the stochastic semantics of the π-calculus with priorities

output prefixes rather than to channel names as in [12,21,20], or to both input and output prefixes [19].

The stochastic semantics of a process P in the stochastic π-calculus is a continuous time Markov chain (CTMC). The states of such CTMCs are classes of processes $[P]_\equiv$. A priori, the state space may be infinite, even though only finitely many states may be reachable in many cases. The purpose of the transitions of a CTMC is to define the probability of a reduction step $P \rightarrow P'$. We will use transitions $P \xrightarrow{r} P'$ that are labeled by propensities $r \in \mathbb{R}_+^\infty$. If r is finite, then the probability of a reduction step from P to P' is r/s, where s is the sum of all propensities r' of transitions starting in P. If s is infinite, then the transition is impossible, since a transition exists with infinite rate which has priority.

The probability of a reduction step follows the *Chemical Law of Mass Action* according to which the propensity of a chemical reaction in a solution is proportional to the number of possible interactions of its reactants in the solution (when assuming a fixed volume). Given a source process P and a target process P', the propensity of $P \rightarrow P'$ depends on the number of ways in which P may reduce to P'. For instance, consider $P_1 = x?().\mathbf{0}$ and $P_2 = x{:}r!().\mathbf{0}$ for some rate $r \in \mathbb{R}_+$. If we fix $P = P_1 \mid P_1 \mid P_2$ and $P' = P_1$ then we have two possible interactions of rate r, so we have $P \xrightarrow{2r} P'$ where $2r$ is the propensity of this reaction.

In order to discriminate interactions leading to the same state, rule (COM$_\ell$) in Fig. 7 defines communication steps labeled by positions $\ell \in \mathbb{N}^4$, where the interaction occurs. Given a prenex normal form $P = (\nu\tilde{x}) \prod_{i=1}^{n} \sum \pi_i^j.P_i^j$, a tuple $\ell = (i_i, j_1, i_2, j_2)$ defines the pair of communication prefixes $\pi_{i_1}^{j_1}.P_{i_1}^{j_1}$ and $\pi_{i_2}^{j_2}.P_{i_2}^{j_2}$.

As before, a communication step can only be applied to senders and receivers on the same channel. We write $P \xrightarrow[\ell]{r} P'$ if there exists a potential interaction at position ℓ, where r is the rate annotated to the sender.

The transitions of the CTMC are defined in Fig. 8. Transitions $[P]_\equiv \xrightarrow[nd]{r} [P']_\equiv$ with finite propensities $r \in \mathbb{R}_+$ are obtained by rule (SUM). First, convergence of P with respect to application is tested. If this test fails then no transition is possible. Otherwise, the unique equivalence class $[P_1]_\equiv$ is computed such that $P \Downarrow P_1$. Second, an arbitrary representative in prenex normal form P_1 of this congruence class is fixed. Third, all pairs (r', ℓ) of P are computed such that there exists $P_2 \equiv P'$ and a communication step $P_1 \xrightarrow[\ell]{r'} P_2$. Finally, all such rates r' are summed up into propensity r. Going back to our previous example, we have $P_1 \mid P_1 \mid P_2 \xrightarrow[(1,1,3,1)]{r} P_1$ and $P_1 \mid P_1 \mid P_2 \xrightarrow[(2,1,3,1)]{r} P_1$, so that $P_1 \mid P_1 \mid P_2 \xrightarrow{2r} P_1$ as expected.

Communication steps with infinite propensities are treated by rule (COUNT). These are given highest priority as stated already in rule (SUM). The probability of a reduction $P \xrightarrow{\infty(n)} P'$ is n/m where n is the number of interactions with rate ∞ leading from P to a process congruent to P', and m the overall number of interactions with rate ∞ starting from P. Given these probabilities, and provided that no infinite sequence of immediate transitions is reachable, one can build a reduction, without immediate transitions, that defines a proper CTMC and preserves the *probabilities of transitions* and *sojourn times* (see e.g. [10] for details).

For illustration, consider a system of two chemical reactions, $x : A, B \xrightarrow{0.5} A, C$ and the inverse $y : A, C \xrightarrow{5} A, B$ whose rate is 10-fold higher. In Fig. 9, we define species A, B, C as processes in the stochastic π-calculus that act according to these chemical reactions. A chemical solution with species A, B, C is a multiset of molecules, i.e. a multiset with these species. In the π-calculus, it can be expressed by a parallel compositions of defined processes, such as for instance $A^2 \mid B^2 \mid C^1$, where we write P^n instead of $\prod_{i=1}^n P$. The reachable part of the CTMC of this chemical system is shown in Fig. 9.

The stochastic semantics of the π-calculus with priorities does indeed properly refine the non-deterministic operational semantics.

Proposition 2. *If the set of priorities $(R, <)$ is equal to $(\mathbb{R}_+^\infty, <_2)$ where $<_2$ defines the usual two levels of priorities (i.e. $r <_2 \infty$ for all $r \in \mathbb{R}_+$), then for all processes P, Q:*

$$P \to Q \textit{ iff } (\exists r \in \mathbb{R}_+ : P \xrightarrow{r} Q \vee \exists n \in \mathbb{N} : P \xrightarrow{\infty(n)} Q)$$

Proof. The implication from the right to the left is quite obvious, since $P \xrightarrow[\ell]{r} Q$ implies $P \to Q$. For the direction from the left to the right, we start with a claim that relates communication steps to labeled communication steps in this direction:

Chemical reactions:

$$x : A, B \xrightarrow{0.5} A, C$$
$$y : A, C \xrightarrow{5} A, B$$

π-calculus definitions:

$$A \triangleq x{:}0.5!().A + y{:}5!().A$$
$$B \triangleq x?().C$$
$$C \triangleq y?().B$$

Transitions of CTMC fragment reachable from $A^2 \mid B^2 \mid C^1$:

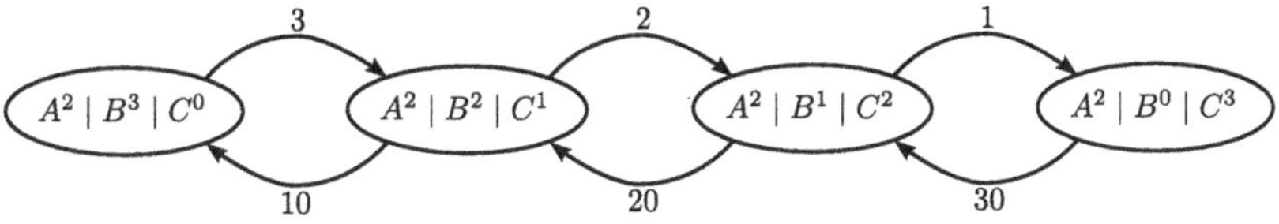

Fig. 9. Example of a CTMC generated by the stochastic π-calculus

Claim. If $P_1 \xrightarrow[nd]{r} Q$ and $P_1 = (\nu x) \prod_{j=1}^{n} \sum_{i=1}^{m_j} \pi_i^j.P_i^j$ then there exists a label $\ell = (i_1, j_1, i_2, j_2)$ and a process Q' such that $Q' \equiv Q$ and $P_1 \xrightarrow[\ell]{r} Q'$.

This follows from a standard analysis of the structural congruence. Suppose now, that $P \rightarrow Q$ holds. In this case, the following rule must be applicable:

$$(\text{PRIOR})\ \frac{P \Downarrow P_1 \quad P_1 \xrightarrow[nd]{r} Q \quad \neg\exists r_1 \in R.\exists Q_1.\ r < r_1 \wedge P_1 \xrightarrow[nd]{r_1} Q_1}{P \rightarrow Q}$$

Without loss of generality, we can assume that P_1 is in prenex normal form, since relation $\xrightarrow[nd]{r}$ is closed under structural congruence by rule (STRUCT). The second hypothesis and the above claim show that $P_1 \xrightarrow[\ell]{r} Q'$ for some process Q' with $Q' \equiv Q$. The third hypothesis holds if and only if either $r = \infty$ or else $r \in \mathbb{R}_+$ and $\neg\exists Q_1.\ P_1 \xrightarrow[nd]{\infty} Q_1$.

- In the case $r = \infty$, we can create a transition with infinite propensity:

$$(\text{COUNT})\ \frac{P \Downarrow P_1 \quad n = \sharp\{\ell \mid P_1 \xrightarrow[\ell]{\infty} Q' \equiv Q\} \neq 0}{P \xrightarrow{\infty(n)} Q}$$

- In the case $r \in \mathbb{R}_+$, property $P_1 \xrightarrow[\ell]{r} Q'$ shows that $\sum_{\{(r',\ell) \mid P_1 \xrightarrow[\ell]{r'} Q' \equiv Q\}} r' \neq 0$. We can thus create a transition of the Markov chain with finite propensity:

$$(\text{SUM})\ \frac{P \Downarrow P_1 \quad \sum_{\{\ell \mid P_1 \xrightarrow[\ell]{r'} Q' \equiv Q\}} r' = r \neq 0 \quad \neg\exists\ell\exists Q_1.P_1 \xrightarrow[\ell]{\infty} Q_1}{P \xrightarrow{r} Q}$$

□

2.6 Type System

We present a type system for the π-calculus with priorities, that prevents from arity mismatches in communication attempts as defined by rule (E.COM). Non-immediate errors, like nonterminating applications of unguarded process definitions are not captured though. These can be detected by a simple cycle check.

Channels are the only values over our calculus. In order to exclude arity mismatches on channels, we introduce channel types, that fix the types of all arguments that can be communicated:

$$\text{types} \qquad \tau ::= ch(\tilde{\tau})$$

A channel of type $ch(\tilde{\tau})$ may only be used to receive and send tuples of values of type $\tilde{\tau}$. A defined process $A(\tilde{x})$ of process type $\tilde{\tau}$ must receive arguments of type $\tilde{\tau}$ for $\tilde{x}$. In our typed setting, we assume that channel creation is typed, by adding types into the syntax of new operators:

$$\text{typed processes} \qquad P ::= (\nu x{:}\tau)P \ \mid \ \ldots$$

A *type environment* Γ is a set of type assignments from channel names to types $x{:}\tau$ and from process names to tuples of types $A{:}\tilde{\tau}$. Thereby we fix the types of the arguments of parametrized process definitions.

Example 2. Consider the process $P = x{:}r!(z).z?(y).P_1 \mid x?(y).y{:}r!().P_2$. The arities of the sender and receiver for x coincide in that they both have 1 argument. After communication, however, P becomes $z?(y).P_1 \mid z{:}r!().P_2$ which has an arity mismatch on z. These kinds of situations are excluded for well-typed processes, so P cannot be well-typed. Indeed, the first subprocess of P is well-typed for type environments containing $x{:}ch(ch(\tau)), z{:}ch(\tau)$ for some type τ. The second subprocess of P, requires type environments containing $x{:}ch(ch())$. Both conditions together are unsatisfiable.

Example 3. Processes $List_2$ and $Ring_3$ are well-typed in environments, where channel z is given an arbitrary type, say $\tau = ch()$, while process Fwd must be assigned type $(ch(\tau), ch(\tau))$. Furthermore the three channels x_1, x_2, x_3 that connect the forwarders must be of type $ch(\tau)$ too. Valid type environments Γ for $Ring_3$ thus must contain the following assumptions:

$$z{:}\tau, Fwd{:}(ch(\tau), ch(\tau)), List_2{:}(), Ring_3{:}(), x_1{:}ch(\tau), x_2{:}ch(\tau), x_3{:}ch(\tau)$$

Type checking rules for processes are given in Fig. 10. They infer judgments $\Gamma \vdash P$ that state the consistency of a process P with a type environment Γ as usual. Inconsistencies may arise from arity mismatches of receivers (T.REC), sender (T.SEND), process applications (T.APP), and process definitions (T.DEF). Furthermore, there are three structural rules, and a rule for the null process $\mathbf{0}$.

Proposition 3 (Type safety). *If $\Gamma \vdash P$ and $P \to Q$ then $\Gamma \vdash Q$.*

The proof works as usual. See the proof of Theorem 1 for a more general instance.

$$(\text{T.REC})\ \frac{\Gamma \vdash x{:}ch(\tilde{\tau}) \quad \Gamma, \tilde{y}{:}\tilde{\tau} \vdash P}{\Gamma \vdash x?\tilde{y}.P} \qquad (\text{T.SEND})\ \frac{\Gamma \vdash x{:}ch(\tilde{\tau}) \quad \Gamma \vdash \tilde{z}{:}\tilde{\tau} \quad \Gamma \vdash P}{\Gamma \vdash x{:}r!\tilde{z}.P}$$

$$(\text{T.PAR})\ \frac{\Gamma \vdash P \quad \Gamma \vdash Q}{\Gamma \vdash P \mid Q} \qquad (\text{T.SUM})\ \frac{\Gamma \vdash M \quad \Gamma \vdash M'}{\Gamma \vdash M + M'} \qquad (\text{T.NEW})\ \frac{\Gamma, x{:}\tau \vdash P}{\Gamma \vdash (\nu x{:}\tau)P}$$

$$(\text{T.APP})\ \frac{\Gamma \vdash A{:}\tilde{\tau} \quad \Gamma \vdash \tilde{x}{:}\tilde{\tau}}{\Gamma \vdash A(\tilde{x})} \qquad (\text{T.DEF})\ \frac{\Gamma \vdash A{:}\tilde{\tau} \quad \Gamma, \tilde{x}{:}\tilde{\tau} \vdash P}{\Gamma \vdash A(\tilde{x}) \triangleq P} \qquad (\text{T.NIL})\ \frac{}{\Gamma \vdash \mathbf{0}}$$

Fig. 10. Type system for π-calculus with priorities

Corollary 1 (Error freeness). *If $\Gamma \vdash P$ and $P \to^* Q$ then $\neg Q \xrightarrow[nd]{err} \bot$.*

Proof. Assuming $\Gamma \vdash P$ and $P \to^n Q$, the proof is by induction on n. The inductive step follows from Proposition 3; it thus remains to prove the initial case that is $Q = P$. We proceed by contradiction assuming that there exists some process P_0 such that $\Gamma \vdash P_0$ and $P_0 \xrightarrow[nd]{err} \bot$.

As for the proof of Lemma 1, a standard analysis of the structural congruence shows the following claim: $P_0 \equiv (\nu\tilde{x}{:}\tilde{\tau}) \prod_{i=1}^{n} P_i$ be a prenex normal form in which all bound variables are renamed apart, and such that all P_i are sums or defined processes, and $\exists j,k.1 \leq j < k \leq n$, $P_j = x_0?\tilde{y}.Q_1 + M_1$, $P_k = x_0{:}r!\tilde{z}.Q_2 + M_2$, and $|\tilde{y}| \neq |\tilde{z}|$.

From $\Gamma \vdash P_0$, we have $\Gamma \vdash (\nu\tilde{x}{:}\tilde{\tau}) \prod_{i=1}^{n} P_i$. This statement follows from a series of applications of rules (T.NEW) and (T.PAR) and from statements $\Gamma, \tilde{x}{:}\tilde{\tau} \vdash P_i$ for all $i \in \{1, \ldots, n\}$. In particular, $\Gamma, \tilde{x}{:}\tilde{\tau} \vdash x_0?\tilde{y}.Q_1 + M_1$, and $\Gamma, \tilde{x}{:}\tilde{\tau} \vdash x_0{:}r!\tilde{z}.Q_2 + M_2$. Therefore, by rules (T.REC) and (T.SEND), $\Gamma, \tilde{x}{:}\tilde{\tau} \vdash x_0{:}ch(\tilde{\tau})$ and, $\Gamma, \tilde{x}{:}\tilde{\tau} \vdash \tilde{z}{:}\tilde{\tau}$ and, $\Gamma, \tilde{x}{:}\tilde{\tau}, \tilde{y}{:}\tilde{\tau} \vdash Q_1$, thus $|\tilde{z}| = |\tilde{\tau}|$ and $|\tilde{y}| = |\tilde{\tau}|$ which contradicts $|\tilde{y}| \neq |\tilde{z}|$. □

3 Attributed π-Calculus with Priorities

We introduce the attributed π-calculus with priorities $\pi(\mathcal{L})$, by extending the π-calculus with priorities with richer sets of values and expressions in some call-by-value λ-calculus $\mathcal{L}$ that we call the attribute language $\mathcal{L}$. We permit λ-expressions in generalized senders and receivers, in order to impose constraints on communication steps, subsuming priorities and stochastic rates. As before, we will present both a non-deterministic and a stochastic operational semantics (except that the set of successful values must be $\mathbb{R}_+^\infty$ with 2 levels of priorities in the stochastic case).

3.1 Idea of Communication Constraints

For illustration of communication constraints, we reconsider the example from the introduction. Proteins *Prot*(x) can bind to operators *Op*(y) only if they have equal sorts $x{=}y$. Expressing such constraints in object-centered languages such as the π-calculus is difficult, since it concerns the attribute values of two independent processes. Our solution proposed for the attributed π-calculus is to use functions such as $\lambda x.x{=}y$ on receiver side, and to apply them to the value of x on the sender side:

$$\begin{aligned} \mathit{Prot}(x) &\triangleq \mathit{bind}[x]!().0 \\ \mathit{Op}(y) &\triangleq \mathit{bind}[\lambda x.x{=}y]?().\mathit{OpBound}(y) \end{aligned}$$

These definitions allow reduction steps $\mathit{Prot}(\mathtt{b}) \mid \mathit{Op}(\mathtt{b}) \to \mathit{OpBound}(\mathtt{b})$ for arbitrary values $\mathtt{b}$, since the application $(\lambda x.x{=}\mathtt{b})\mathtt{b}$ evaluates to $\mathtt{true}$. In order to permit richer sets of values and constraints, such as for instance arithmetic values and constraints, we define use call-by-value λ-calculi as attribute languages $\mathcal{L}$. We keep the choice of constants parametric, in order to avoid reinventing independent calculi for the many useful choices in practice. We will define our semantics such that they are independent of the concrete choice of the attribute language.

3.2 Attribute Languages

An attribute language is a functional programming language that provides expressions by which to compute values. Expressions are built from constants – for numbers, functions, relations, or biological entities (such as 0, 1, $+$, $*$, $\geq$, `fst`, `snd`, `repressor`, ...) – and from variables $x, y \in \mathit{Vars}$, which may play the name channels in particular. Whenever ambiguities might arise, we typeset constants in courier font (such as `fst` or `repressor`) and variables in italics such as *bind*.

As an example, consider the expression $(\mathtt{snd}\ x) + (\mathtt{fst}\ y)$ in which `fst`, `snd`, and $+$ are constants, while x and y are variables. In the following process definition, this expression defines a stochastic rate or priority: $A(x, y, z) \triangleq z[(\mathtt{snd}\ x) + (\mathtt{fst}\ y)]!().P$.

An attribute language with variables in *Vars* is a tuple $\mathcal{L} = (\mathit{Consts}, \Downarrow, R, <)$, that contains a set of constants $c \in \mathit{Consts}$, a big-step evaluator $\Downarrow$ for λ-expressions with constants, pairs, and conditionals, and a partially ordered set $(R, <)$ of successful values, that enables communication steps. More precisely, the first component *Consts* is a finite set that fixes the constants of a call-by-value λ-calculus. The set of expressions $e \in \mathit{Exprs}$ of $\mathcal{L}$ is defined as the set of all λ expressions with constants in $c \in \mathit{Consts}$ and variables in $x \in \mathit{Vars}$. The set of values $v \in \mathit{Vals}$ of $\mathcal{L}$ is the subset of all values of this λ-calculus.

$$\begin{aligned} c \in \mathit{Consts} &::= \mathtt{false} \mid \mathtt{true} \mid \mathtt{fst} \mid \mathtt{snd} \mid \ldots \\ v \in \mathit{Vals} &::= x \mid c \mid \lambda x.e \mid \langle v_1, v_2 \rangle \\ e \in \mathit{Exprs} &::= v \mid e_1 e_2 \mid \langle e_1, e_2 \rangle \mid \mathtt{if}\ e\ \mathtt{then}\ e_1\ \mathtt{else}\ e_2 \end{aligned}$$

As usual, λ-expressions provide abstractions $\lambda x.e$ and applications $e_1 e_2$ for specifying function definitions and function application. We assume that the set of constants *Consts* contains the Booleans `true` and `false`, and pair projections `fst` and `snd`. We also assume that there are expressions for pairs $\langle e_1, e_2 \rangle$ and

$$(\text{VAL})\ \frac{v \in \mathit{Vals}}{v \Downarrow v} \qquad (\text{FUN})\ \frac{e_1 \Downarrow \lambda x.e_1' \quad e_2 \Downarrow v' \quad e_1'[v'/x] \Downarrow v}{e_1 e_2 \Downarrow v}$$

$$(\text{PAIR})\ \frac{e_1 \Downarrow v_1 \quad e_2 \Downarrow v_2}{\langle e_1, e_2\rangle \Downarrow \langle v_1, v_2\rangle} \qquad (\text{SELECT})\ \frac{e \Downarrow \langle v_1, v_2\rangle}{\texttt{fst}\ e \Downarrow v_1 \qquad \texttt{snd}\ e \Downarrow v_2}$$

$$(\text{COND}_1)\ \frac{e \Downarrow \texttt{true} \quad e_1 \Downarrow v_1}{\texttt{if}\ e\ \texttt{then}\ e_1\ \texttt{else}\ e_2 \Downarrow v_1} \qquad (\text{COND}_2)\ \frac{e \Downarrow \texttt{false} \quad e_2 \Downarrow v_2}{\texttt{if}\ e\ \texttt{then}\ e_1\ \texttt{else}\ e_2 \Downarrow v_2}$$

Fig. 11. Big-step evaluators of call-by-value λ-calculus with pairs and conditionals

$$(\text{EQ}_1)\ \frac{e_1 \Downarrow v \quad e_2 \Downarrow v \quad v \in \mathit{Vars} \cup \mathit{Consts}}{e_1{=}e_2 \Downarrow \texttt{true}} \qquad (+_{\mathbb{N}})\ \frac{e_1 \Downarrow n_1 \quad e_2 \Downarrow n_2 \quad n_1 +_{\mathbb{N}} n_2 = n}{e_1 + e_2 \Downarrow n}$$

$$(\text{EQ}_2)\ \frac{e_1 \Downarrow v_1 \quad e_2 \Downarrow v_2 \quad v_1 \neq v_2 \in \mathit{Vars} \cup \mathit{Consts}}{e_1{=}e_2 \Downarrow \texttt{false}}$$

Fig. 12. Additional rules of big-step evaluator of the attribute language $\lambda(\mathbb{N}_0, +, =)_{<_1}$

Boolean conditionals `if` e `then` e_1 `else` e_2. In examples, we will freely write `if` e `then` e_1 instead of `if` e `then` e_1 `else false`.

The third component is a set $R \subseteq \mathit{Vals}$ of successful values enabling communication steps, such as priorities or stochastic rates. The fourth component $<$ is a partial order on successful values R that defines priorities. The last component of $\mathcal{L}$ is a big-step evaluator λ expressions, i.e. a partial function $\Downarrow : \mathit{dom}(\Downarrow) \to \mathit{Vals}$ from λ-expressions in a domain $\mathit{dom}(\Downarrow) \subseteq \mathit{Exprs}$ to value. It can be understood as a black box algorithm that evaluates all expressions to values or failure, in case of program errors or nontermination. Instead of $\Downarrow(e) = v$ we will write $e \Downarrow v$ and call v the value of e.

We assume that the big-step evaluator satisfies the usual rules of the call-by-value λ-calculus with conditionals in Fig. 11. Rule (VAL) states that all values evaluate to themselves. Rule (FUN) defines the usual meaning of call-by-value function application. It says that the value of an application $e_1 e_2$ is obtained by by evaluating e_1 to some function $\lambda x.e_1'$ and e_2 to some value v', and then returning the value of $e_1'[v'/x]$. Rule (PAIR) states that the value of a pair is the pair of values of its components (as usual in call-by-value languages). Rule (SELECT) states the usual meaning of pair selectors. Rule (COND_1) and (COND_2) defines the semantics of conditionals such that only the needed branch is evaluated. For richer attribute language with further constants (such as $+$, $*$, or call-by-value fixed point operator `fix`), we need to add further rules. This is possible, since we keep the big-step evaluator abstract.

As a first example, we consider the attribute language $\lambda(R)_<$ for some partially ordered set $(R, <)$ of priorities. These are the only successful values and ordered according to $<$. No new rules are needed for the big-step evaluator, since no function constants are added.

As a second example, we introduce the attribute language $\lambda(\mathbb{N}_0, +, =)_{<_1}$ where the call-by-value λ-calculus is extended by constants for natural numbers with 0 and addition $+$. The successful values are nonzero natural numbers and there is a single level of priorities, fixed by the empty partial order that we denote as $<_1$. Constant $=$ defines equality on all constants and variables, i.e. on Booleans, natural numbers, channel names, and the function constants. The required extensions of the big-step evaluator are given in Fig. 12. Note that addition and equality are treated as curried binary functions, since introduced by constant. We freely write $e + e'$ instead of $(+\ e)\ e'$ and respectively, $e{=}e'$ instead of $(=\ e)\ e'$.

As third example, we consider the attribute language $\lambda(\mathbb{R}_+^\infty)_{<_2}$, whose successful values are the stochastic rates in $\mathbb{R}_+^\infty$. There are two levels of priorities, lower priority for all positive real numbers in $\mathbb{R}_+$ and higher priority for ∞. We write $<_2$ for the obvious order that introduces these two levels of priorities.

Further extensions of the attribute language might be useful in various applications, such as n-tuples (beyond pairs), lists, or case statements, but cannot be obtained by adding new constants, since they require new forms of expressions and values. Even though we do not expect particular difficulties, we refrain from generalizing or extending the attribute language any further for the sake of simplicity.

Values of expressions may be undefined, i.e. for some expressions e there might not exist any value v with $e \Downarrow v$. This may have two possible reasons. The first are program errors, like division by 0, or type errors, like sending or receiving on non-channels. The second reason is non-termination, which may arise in an untyped setting or in rich attribute languages with fixed point operators.

In Section 3.7 we will present a type system for the attributed π-calculus which prevents us from type errors. If not adding any constants to the attribute language, it even excludes non-termination. For more general attribute languages, however, the type systems may neither exclude all program errors (e.g. division by 0) nor ensure termination (e.g. fixed point operators).

3.3 Syntax of Attributed Processes

Let $\mathcal{L}$ be an attribute language over some infinite set of variables $x \in \mathit{Vars}$, with expressions $e \in \mathit{Exprs}$ and values $v \in \mathit{Vals}$.

The syntax of the attributed π-calculus $\pi(\mathcal{L})$ is defined in Fig. 13. Compared to before, we use variables x of various types instead of channel names (which correspond to variables of channel type), permit expressions e in all non-binding positions where previously only channel names were allowed, extend priorities "$:r$" in senders to expressions "$[e]$", and introduce such expressions in the symmetric position in receivers. Receiver prefixes thus have the form $e_1[e_1']?\tilde{x}$ and sender prefixes the form $e_2[e_2']!\tilde{e}$. Prefixes in which e_1 resp. e_2 do not evaluate to channels are erroneous. The application $e_1'e_2'$ imposes a constraint on the ability to communicate, in addition to that e_1 and e_2 must evaluate to the same channel. Communication is permitted only if $e_1'e_2' \Downarrow v$ for some successful value $v \in R$. This value then fixes the priority or stochastic rates of the communication step.

Prefixes	$\pi ::= e_1[e_2]?\tilde{x}$	receiver
	$\mid\ e_1[e_2]!\tilde{e}$	sender
Sums	$M ::= \pi.P$	guarded process
	$\mid\ M_1 + M_2$	choice
Processes	$P ::= M$	sums
	$\mid\ A(\tilde{e})$	defined process
	$\mid\ P_1 \mid P_2$	parallel composition
	$\mid\ (\nu x)P$	channel creation
	$\mid\ \mathbf{0}$	empty solution
Definitions	$D ::= A(\tilde{x}) \triangleq P$	parametric process definition

Fig. 13. Syntax of $\pi(\mathcal{L})$ where $x, \tilde{x} \in \mathit{Vars}$, and $e_1, e_2, \tilde{e} \in \mathit{Exprs}$

For illustration, we consider 3 instances of the attributed π-calculus with 3 different attribute languages. As a first example, we consider the calculus $\pi(\lambda(R)_<)$ whose attribute language is a λ-calculus with priorities. The second example is $\pi(\lambda(\mathbb{N}_0, +, =)_{<_1})$ which provides for natural numbers with addition and equality. The third examples is $\pi(\lambda(\mathbb{R}_+^\infty)_{<_2})$, where λ-expressions can be used in order to compute priorities on 2 levels.

For illustration, we consider process definitions in $\pi(\lambda(\mathbb{N}_0, +, =)_{<_1})$, which express schemes for chemical reactions *react*: $A(x), B(y) \xrightarrow{x+y} A(x+1), B(y)$ in which x and y can be instantiated by all possible natural numbers. Similar reaction schemes are used in [38], in order to model chemical reactions in biochemical systems[1].

$$A(x) \triangleq \mathit{react}[x]!().A(x{+}1) \qquad B(y) \triangleq \mathit{react}[\lambda x.x + y]?().B(y)$$

The process $A(2)|B(5)$ may communicate on channel z and become $A(3) \mid B(5)$, since the sum of the x-attribute of $A(2)$ and the value of the y-attribute of $B(5)$ is the successful value 7. More formally, we can compute this number by evaluating the interaction constraint $(\lambda x.x + 5)2$ via (FUN), (VAL), and $(+_\mathbb{N})$:

$$\frac{\dfrac{\lambda x.x + 5 \in \mathit{Vals}}{\lambda x.x + 5 \Downarrow \lambda x.x + 5} \quad \dfrac{2 \in \mathit{Vals}}{2 \Downarrow 2} \quad \dfrac{\dots}{2 + 5 \Downarrow 7}}{(\lambda x.x + 5)2 \Downarrow 7}$$

Free variables $\mathit{fv}(P)$ are defined as before, except that we now need to account for free variables $\mathit{fv}(e)$ in λ-expressions e too, i.e., those occurring out of the scope of all λ binders in e.

$$\mathit{fv}(e_1[e_2]?\tilde{y}.P) = \mathit{fv}(e_1) \cup \mathit{fv}(e_2) \cup (\mathit{fv}(P) \setminus \{\tilde{y}\})$$
$$\mathit{fv}(e_1[e_2]!\tilde{e}.P) = \mathit{fv}(e_1) \cup \mathit{fv}(e_2) \cup \mathit{fv}(\tilde{e}) \cup \mathit{fv}(P) \qquad \mathit{fv}(A(\tilde{e})) = \mathit{fv}(\tilde{e})$$

[1] In contrast to here, variables in [38] quantify over finite sets of values, while rules there permit more then 2 reactants in contrast to the π-calculus.

Communication and application steps

$$(\text{SEND})\ \frac{e_1 \Downarrow x \quad e_2 \Downarrow v_2 \quad \tilde{e} \Downarrow \tilde{v}}{e_1[e_2]!\tilde{e} \Downarrow x[v_2]!\tilde{v}} \qquad (\text{REC})\ \frac{e_1 \Downarrow x \quad e_2 \Downarrow v_2}{e_1[e_2]?\tilde{y} \Downarrow x[v_2]?\tilde{y}} \qquad (\text{TUP})\ \frac{\bigwedge_{i=1}^n e_i \Downarrow v_i}{(e_i)_{i=1}^n \Downarrow (v_i)_{i=1}^n}$$

$$(\text{COM})\ \frac{\pi_1 \Downarrow x[v_1]?\tilde{y} \quad \pi_2 \Downarrow x[v_2]!\tilde{v} \quad v_1 v_2 \Downarrow r \in R \quad |\tilde{v}| = |\tilde{y}|}{\pi_1.P_1 + M_1 \mid \pi_2.P_2 + M_2 \xrightarrow[nd]{r} P_1[\tilde{v}/\tilde{y}] \mid P_2} \qquad (\text{APP})\ \frac{\tilde{e} \Downarrow \tilde{v} \quad A(\tilde{x}) \triangleq P}{A(\tilde{e}) \xrightarrow[nd]{app} P[\tilde{v}/\tilde{x}]}$$

Program errors

$$(\text{E.COM})\ \frac{\pi_1 \Downarrow x[v_1]?\tilde{y} \quad \pi_2 \Downarrow x[v_2]!\tilde{v} \quad |\tilde{y}| \neq |\tilde{v}|}{\pi_1.P_1 + M_1 \mid \pi_2.P_2 + M_2 \xrightarrow[nd]{err} \bot}$$

$$(\text{E.PREF})\ \frac{\neg\exists\pi'.\pi \Downarrow \pi'}{\pi.P + M \xrightarrow[nd]{err} \bot} \qquad (\text{E.CONSTR})\ \frac{\pi_1 \Downarrow x[v_1]?\tilde{y} \quad \pi_2 \Downarrow x[v_2]!\tilde{v} \quad \neg\exists v.v_1 v_2 \Downarrow v}{\pi_1.P_1 + M_1 \mid \pi_2.P_2 + M_2 \xrightarrow[nd]{err} \bot}$$

Fig. 14. Non-deterministic operational semantics of $\pi(\mathcal{L})$ with priorities: all rules of the π-calculus with priorities in Fig. 6 remain valid, too

Bound variables $bv(P)$ are defined as before, except that λ-binders in expressions $e \in \mathit{Exprs}$ are included too. The structural congruence on processes $\equiv$ remains unchanged, except that α-conversion becomes applicable to bound variables in λ-expressions.

3.4 Non-deterministic Operational Semantics

The non-deterministic operational semantics of the attributed π-calculus with priorities is given by the rules in Fig. 14 and the previous rules in Fig. 6. The new rules always evaluate expressions to values before applying communication or application steps (COM) and (APP). This is done by using the big-step evaluator of the attribute language according to axioms (SEND), (REC), and (TUP). Note that evaluation of expressions may get stuck – in contrast to the pure π-calculus with priorities. For instance, an application $A(\tilde{e})$ gets stuck if the evaluation of one of the expressions in $\tilde{e}$ does not succeed. In this case, application does not converge, so that communication gets blocked.

The communication rule (COM) permits receivers $x[v_1]?\tilde{y}.P_1$ and senders $x[v_2]!\tilde{v}.P_2$ to interact only if expression $v_1 v_2$ evaluates to a successful value $v_1 v_2 \Downarrow r \in R$. This value defines the priority level of the communication step. Communication steps perform substitutions $[\tilde{v}/\tilde{y}]$ replacing variables by values. The application of substitutions is well-defined for all processes, since our syntax permits values in all positions, where free variables may be used. Note however, that substitution may raise program errors as specified by rule (E.PREF), where non-channel values arise in sender or receiver position. As before, we write $\bot$ for an arbitrary erroneous expression. Rule (E.CONSTR) specifies constraint errors, where the evaluation of communication constraints $v_1 v_2$ fails.

The closure rules in Fig. 6 remain unchanged. As before, all relations are closed under the structural rules, while (CONV) applies definitions exhaustively and continues to require error-freeness. The overall reduction relation $P \to P'$ is defined in rule (PRIOR) without change. All changes are imported from the changes in communication, application, and error steps.

Example 4. Let us consider a client server system in the attributed π-calculus with integers and strings and two levels of priorities $\pi(\lambda(\mathtt{Int}, \mathtt{String})_{<_2})$, such that there are two successful values $R = \{1, 2\}$ ordered by least ordering $<_2$ that satisfies $1 <_2 2$. We fix a value $\mathit{never} =_{df} 0$ that is not successful and name the two successful values as follows: $\mathit{low} =_{df} 1$ and $\mathit{high} =_{df} 2$. Furthermore, let function $\mathit{price} : \mathtt{String} \to \mathtt{Int}$ be defined by the following expression:

$$\mathit{price} =_{df} \lambda x.\mathtt{if}\ x{=}\mathtt{chicken}\ \mathtt{then}\ 10\ \mathtt{else}\ \mathtt{if}\ x{=}\mathtt{fish}\ \mathtt{then}\ 14\ \mathtt{else}\ 0$$

Servers are accessible on a public channel *connect* to all clients that know a password *key* of type String. The server applies function *price* to a string value received from the client and returns the value on a private channel, that was also provided by the client. We define servers and clients as follows:

$$\begin{aligned}
\mathit{Server}() &\triangleq \mathit{connect}[\lambda k.\mathtt{if}\ k{=}\mathit{key}\ \mathtt{then}\ \mathit{low}\ \mathtt{else}\ \mathit{never}]?(x, \mathit{ret}).\\
&\qquad\qquad\qquad (\mathit{ret}[\mathit{high}]!(\mathit{price}\ x).\mathbf{0}) \mid \mathit{Server}())\\
\mathit{Client}(s) &\triangleq (\nu \mathit{ret})\mathit{connect}[\mathit{key}]!(s, \mathit{ret}).\mathit{ret}[\lambda z.z]?(y).P
\end{aligned}$$

We can then reduce a process with two clients and one server as follows:

$$\begin{aligned}
&\mathit{Server}() \mid \mathit{Client}(\mathtt{chicken}) \mid \mathit{Client}(\mathtt{fish})\\
\to\ &(\nu \mathit{ret})(\mathit{ret}[\mathit{high}]!(\mathit{price}\ \mathtt{chicken}).\mathbf{0} \mid \mathit{Server}() \mid \mathit{ret}[\lambda x.x]?(y).P) \mid \mathit{Client}(\mathtt{fish})\\
\equiv\ &\mathit{Server}() \mid \mathit{Client}(\mathtt{fish}) \mid (\nu \mathit{ret})(\mathit{ret}[\mathit{high}]!(\mathit{price}\ \mathtt{chicken}).\mathbf{0} \mid \mathit{ret}[\lambda x.x]?(y).P)\\
\to\ &\mathit{Server}() \mid \mathit{Client}(\mathtt{fish}) \mid (\nu \mathit{ret})\ P[10/y]
\end{aligned}$$

No unrelated client can access a client-server dialog, since private channels are used for communication. Note however, that the second communication action gets highest priority, so that client *Client*(fish) can not act before *Client*(chicken) obtained the price for the chicken.

3.5 Convergence

The next lemma extends on Lemma 1. It states that application of definitions is confluent, so that exhaustive application must lead to a unique outcome (including non-termination).

Lemma 2. *The rewrite relation* $\xrightarrow[nd]{app}$ *is confluent modulo structural congruence. Irreducible processes are congruent to processes of the form* $(\nu\tilde{x})\prod_{i=1}^{n} P_i$ *such that all* P_i *are sums or match some defined process* $A_i(\tilde{e}_i)$ *with* $\neg\exists\tilde{v}.\tilde{e}_i \Downarrow \tilde{v}$.

Proof. A standard analysis of the structural congruence yields the following:

Labeled communication steps ($\ell \in \mathbb{N}^4$, $r \in \mathbb{R}_+^\infty$)

$$(\text{COM}_\ell)\quad \frac{\ell = (i_1, j_1, i_2, j_2) \quad i_1 \neq i_2 \qquad \pi_{i_1}^{j_1} \Downarrow x[v_1]?\tilde{y} \quad \pi_{i_2}^{j_2} \Downarrow x[v_2]!\tilde{v} \quad v_1 v_2 \Downarrow r \in \mathbb{R}_+^\infty \quad |\tilde{y}| = |\tilde{v}|}{(\nu\tilde{x}) \prod_{i=1}^n \sum_{j=1}^{m_i} \pi_i^j.P_i^j \xrightarrow[\ell]{r} (\nu\tilde{x})(\prod_{i=1, i\neq i_1, i_2}^n \sum_{j=1}^{m_i} \pi_i^j.P_i^j \mid P_{i_1}^{j_1}[\tilde{v}/\tilde{y}] \mid P_{i_2}^{j_2})}$$

Fig. 15. Axioms of stochastic semantics of $\pi(\mathcal{L})$. The rules of the stochastic semantics for defining CTMCs are the same as those for the stochastic π-calculus in Fig. 8.

Claim. Let $P = (\nu\tilde{x}) \prod_{i=1}^n P_i$ be a process in prenex normal form in which all bound variables are renamed apart, and such that all P_i are sums or defined processes. In this case, $P \xrightarrow[nd]{app} P'$ if and only if the following rule applies:

$$\frac{1 \leq j \leq n \quad A_j(\tilde{y}_j) \triangleq Q_j \qquad \tilde{e}_j \Downarrow \tilde{v}_j \quad P_j = A_j(\tilde{e}_j) \quad P' \equiv (\nu\tilde{x})(\prod_{i=1, i\neq j}^n P_i \mid Q_j[\tilde{v}_j/\tilde{y}_j])}{P \xrightarrow[nd]{app} P'}$$

We consider the rewrite system on congruence classes of processes defined by $[P]_\equiv \xrightarrow[nd]{app} [P']_\equiv$ if $P \xrightarrow[nd]{app} P'$. The above claim shows that this rewrite system terminates on equivalence classes of processes of the form $(\nu\tilde{x}) \prod_{i=1}^n P_i$ where all P_i are either sums or irreducible defined processes $A(\tilde{e})$. We can prove the uniform confluence of this rewrite system by minor adaptation of the proof in Lemma 1. □

The following two properties of $\pi(\mathcal{L})$ are precisely the same that we obtained for the π-calculus with priorities (but now from Lemma 2 instead of Lemma 1).

Proposition 4 (Error-free convergence). *For every P there exists at most one class $[P']_\equiv$ such that $P \Downarrow P'$.*

Proof. This follows immediately from the confluence result in Lemma 2.

Remark 2. If $P \equiv (\nu\tilde{x}) \prod_{i=1}^n M_i$ and $\neg P \xrightarrow[nd]{err} \bot$ then $P \equiv P' \Leftrightarrow P \Downarrow P'$.

3.6 Stochastic Operational Semantics

We present a stochastic semantics for the attributed $\pi(\mathcal{L})$, under the condition that the set of successful values of the attribute language are the stochastic rates $R \subseteq \mathbb{R}_+^\infty$. As in the stochastic π-calculus, we assign highest priority to communication steps with infinite rates, and lowest priority to all others.

The axioms of the stochastic semantics of $\pi(\mathcal{L})$ is given in Fig. 15. The whole presentation of the article is done such that only very few changes are needed with respect to the stochastic π-calculus. In particular, both calculi have the same closure rules, that were already presented in Fig. 8. This is a major improvement compared to the conference version of the present article [25].

The main difference concerns the new communication rule (COM_ℓ), where we have to evaluate all expressions, in order to compute the stochastic rate. All other difference are hidden in the convergence predicate, as defined in the non-deterministic operational semantics.

The stochastic version remains a proper refinement of the non-deterministic version of the attributed π-calculus with priorities.

Proposition 5. *If the successful values of $\mathcal{L}$ are $\in \mathbb{R}_+^\infty$ with the usual two levels of priority, then for all processes P, P':*

$$P \rightarrow P' \text{ iff } (\exists r \in \mathbb{R}_+.\ P \xrightarrow{r} P' \vee \exists n \in \mathbb{N}.\ P \xrightarrow{\infty(n)} P')$$

The proof is mostly the same as for Proposition 2, which relates the two operational semantics of the stochastic π-calculus. The only minor difference is in the treatment of basic interaction steps.

3.7 Type System

Higher-order attribute languages add much expressive power to the π-calculus, but at the price of introducing many new error situations, that systems biology users will be faced with during modeling and simulation. The most frequent errors are type errors. In this section, we present a type system by which to exclude type errors in attributed processes. Note that well-typedness does not exclude all kinds of errors, such as division by 0 for instance.

We present a type system for the $\pi(\mathcal{L})$, which integrates the simple type system for the λ-calculus $\mathcal{L}$ into the type system of the π-calculus from Section 2.6. We will show that the type system of $\pi(\mathcal{L})$ is safe if the type system of $\mathcal{L}$ is. Whether this holds depends on the precise definition of the big-step evaluator of $\mathcal{L}$ that we left open. The attribute languages $\lambda(R)_<$ from Fig. 11 and $\lambda(\mathbb{N}_0, +, =)_{<_1}$ in Fig. 12 are type safe. These two attribute languages are strongly normalizing (i.e., always terminating), as usual for the simply typed lambda calculus. General termination may fail, however, once we add new rules to the big-step evaluators for new constants with functional type, as for instance, in order to define the semantics of fixed-point combinators. In this case, the type safety of the attribute language must be checked again.

We assume a set of type constants such as `Int`, `Bool`, and `String` and define *Types* with these constants by the following grammar:

type constants	$\iota ::= \texttt{Int} \mid \texttt{Bool} \mid \ldots$	
types	$\tau, \sigma ::= \iota$	constants
	$\mid\ \tau \rightarrow \sigma$	function type
	$\mid\ [\tau] \Rightarrow \tilde{\sigma}$	channel type
	$\mid\ \tau \times \sigma$	pair type

Channel types $[\tau] \Rightarrow \tilde{\sigma}$ now type channel constraints by τ and channel arguments by $\tilde{\sigma}$. More precisely, a channel x of type $[\tau_1 \rightarrow \tau_2] \Rightarrow \tilde{\sigma}$ can be used as follows:

- in input prefixes $x[e]?y$, the type of expressions e must be $\tau_1 \to \tau_2$ and the types of $\tilde{y}$ must be $\tilde{\sigma}$.
- in output prefixes $x[e]!\tilde{e}'$, the type of expression e must by τ_1 and the types of $\tilde{e}'$ must be $\tilde{\sigma}$.

Type constants and the types for functions and pairs are standard. As before, we assume that *type environments* Γ and Δ are sets of type assignments for variables $x{:}\tau$ and process names $A{:}\tilde{\tau}$.

In the typed version of the attributed π-calculus, we need to assume type annotation in the syntax. In order to do so, we add types to all occurrences of constants in processes and to all channel creators.

$$\text{typed processes} \qquad P ::= c_\tau \mid (\nu x{:}\tau)P \mid \ldots$$

In examples, we will often ignore type annotations, if they are clear from the context. It is particularly useful to annotate functional types to constants, for instance, in order to type pair selectors $\texttt{fst}_{\tau\times\sigma\to\tau}$ and $\texttt{snd}_{\tau\times\sigma\to\sigma}$ or fixed point operators. Note also, that we can use pairs of different types, since we may use different annotations for the same constant.

Example 5. In the client server example, we used a λ-expressions *price* of type $\texttt{Int} \to \texttt{Int}$. In a typed version of this example, we have to annotate all constants by their types, i.e., $never =_{df} 0_{\texttt{Int}}$, $low =_{df} 1_{\texttt{Int}}$ and $high =_{df} 2_{\texttt{Int}}$. Furthermore, we need to annotate the new binder in the definition of client by its type:

$$\begin{aligned} &Server() \triangleq connect[\lambda k.\texttt{if}\ k{=}key_{\texttt{Int}}\ \texttt{then}\ low\ \texttt{else}\ never]?(x, ret). \\ &\qquad\qquad\qquad\qquad\qquad\qquad (ret[high]!(price\ x).\mathbf{0}) \mid Server()) \\ &Client(s) \triangleq (\nu ret{:}[\texttt{Int} \to \texttt{Int}] \Rightarrow (\texttt{Int}))connect[key_{\texttt{Int}}]!(s, ret).ret[\lambda z.z]?(y).P \end{aligned}$$

The definitions are well typed if in the following type environment:

$$\begin{aligned} &connect : [\texttt{Int} \to \texttt{Int}] \Rightarrow (\texttt{String}, [\texttt{Int} \to \texttt{Int}] \Rightarrow (\texttt{Int})), \\ &Client{:}(\texttt{String}), Server{:}() \end{aligned}$$

Rules for typing expressions and processes are given in Fig. 16. Typing rules for expressions are standard from simply typed λ-calculus. Rule (T.AXIOMS) reveals for instance, that we treat equality and addition as curried binary functions. They receive their arguments in two steps, rather than at once. Typing communication prefixes (T.REC) and (T.SEND) derive directly from the above explanations of channel types. Rules for process application (T.APP) and definition (T.DEF) are similar to those for π-calculus with priorities in Fig. 10. Typing rule (T.NEW) now checks explicitly that a new channel name is created and nothing else; previously all values were channel names. Finally, typing rules (T.PAR) and (T.SUM) remain as in Fig. 10 and are not repeated here.

Proposition 6 (Type safety for expressions). *The attribute language* $\lambda(\mathbb{N}_0, +, =)$ *in Fig. 11 is type safe, i.e., if* $\Gamma \vdash e{:}\tau$ *and* $e \Downarrow v$ *then* $\Gamma \vdash v{:}\tau$.

Typing rules for expressions

$$(\text{T.CONST})\ \frac{c \in \mathit{Consts} \quad c{:}\tau}{\Gamma \vdash c_\tau{:}\tau} \qquad (\text{T.AXIOMS})\ \frac{\tau, \sigma \text{ types}}{\begin{array}{ll} \texttt{fst} : \tau \times \sigma \to \tau & \texttt{false}{:}\texttt{Bool} \\ \texttt{snd} : \tau \times \sigma \to \sigma & \texttt{true}{:}\texttt{Bool} \\ = : \tau \to \sigma \to \texttt{Bool} & + : \mathbb{N} \to \mathbb{N} \to \mathbb{N} \end{array}}$$

$$(\text{T.VAR})\ \frac{x{:}\tau \in \Gamma}{\Gamma \vdash x{:}\tau} \qquad (\text{T.PAIR})\ \frac{\Gamma \vdash e{:}\tau \qquad \Gamma \vdash e'{:}\sigma}{\Gamma \vdash \langle e, e' \rangle : \tau \times \sigma}$$

$$(\text{T.COND})\ \frac{\Gamma \vdash e{:}\texttt{Bool} \qquad \Gamma \vdash e_1{:}\tau \qquad \Gamma \vdash e_2{:}\tau}{\Gamma \vdash \texttt{if}\ e\ \texttt{then}\ e_1\ \texttt{else}\ e_2 : \tau}$$

$$(\text{T.FUNDEF})\ \frac{\Gamma, x{:}\tau \vdash e{:}\sigma}{\Gamma \vdash \lambda x.e : \tau \to \sigma} \qquad (\text{T.FUNAPP})\ \frac{\Gamma \vdash e : \tau \to \sigma \qquad \Gamma \vdash e'{:}\tau}{\Gamma \vdash e\ e' : \sigma}$$

Typing rules for processes

$$(\text{T.REC})\ \frac{\Gamma \vdash e_1 : [\tau] \Rightarrow \tilde{\sigma} \quad \Gamma \vdash e_2{:}\tau \quad \Gamma, \tilde{x}{:}\tilde{\sigma} \vdash P}{\Gamma \vdash e_1[e_2]?\tilde{x}.P} \qquad (\text{T.NEW})\ \frac{\Gamma, x{:}[\tau] \Rightarrow \tilde{\sigma} \vdash P}{\Gamma \vdash (\nu x{:}[\tau] \Rightarrow \tilde{\sigma})P}$$

$$(\text{T.SEND})\ \frac{\Gamma \vdash e_1 : [\tau_1 \to \tau_2] \Rightarrow \tilde{\sigma} \qquad \Gamma \vdash e_2{:}\tau_1 \qquad \Gamma \vdash \tilde{e}_3{:}\tilde{\sigma} \qquad \Gamma \vdash P}{\Gamma \vdash e_1[e_2]!\tilde{e}_3.P}$$

$$(\text{T.APP})\ \frac{\Gamma \vdash A{:}\tilde{\tau} \qquad \Gamma \vdash \tilde{e}{:}\tilde{\tau}}{\Gamma \vdash A(\tilde{e})} \qquad (\text{T.DEF})\ \frac{\Gamma \vdash A{:}\tilde{\tau} \qquad \Gamma, \tilde{x}{:}\tilde{\tau} \vdash P}{\Gamma \vdash A(\tilde{x}) \triangleq P}$$

Fig. 16. Type system

The proof is standard and proceeds by induction on the proof of $\Gamma \vdash e{:}\tau$ and follows from a substitution lemma stating that: if $\Gamma, x{:}\tau \vdash e{:}\sigma$ and $\Gamma \vdash v{:}\tau$ then $\Gamma \vdash e[v/x] : \sigma$.

Proposition 7 (Normalization). *In the attribute language $\lambda(\mathbb{N}_0, +, =)$ every typable expression evaluates to some value, i.e., if $\Gamma \vdash e{:}\tau$, then there exists v such that $e \Downarrow v$.*

Typings of terms in $\lambda(\mathbb{N}_0, +, =)$ that make use of rule (T.CONST) always do so with a constant type (`Int`) for τ. Therefore, $\lambda(\mathbb{N}_0, +, =)$ is a simply-typed λ-calculus (e.g. fixed point operators are not typable) that is known to have the normalization property. A proof of this result by Tait's methods [39] can be found in many text books (e.g. [40]).

Lemma 3. *The following properties holds for the typing rules of processes:*

1. *(strengthening) if $\Gamma, x{:}\tau \vdash P$ and $x \notin fv(P)$ then $\Gamma \vdash P$,*
2. *(weakening) if $\Gamma \vdash P$ and $x \notin fv(P)$ then $\Gamma, x{:}\tau \vdash P$,*
3. *(substitution) if $\Gamma, x{:}\tau \vdash P$ and $\Gamma \vdash v{:}\tau$ then $\Gamma \vdash P[v/x]$,*
4. *if $\Gamma \vdash P$ and $P \equiv Q$ then $\Gamma \vdash Q$.*

Strengthening and weakening also hold for the typing of definitions.

Proof. The proofs of the three first properties are straightforward inductions on the derivation of $\Gamma, x{:}\tau \vdash P$ (strengthening and substitution) and of $\Gamma \vdash P$ (weakening). They easily extend to (and depend on) the same properties for expressions. The proof of the last property is by induction of the definition of the structural congruence. The only interesting case is for scope extrusion, that is, assuming $x \notin fv(Q)$, $\Gamma \vdash (\nu x{:}\tau)(P \mid Q) \Leftrightarrow \Gamma \vdash (\nu x{:}\tau)P \mid Q$.
($\Rightarrow$) By rules (T.NEW) and (T.PAR), we have $\Gamma, x{:}\tau \vdash P$ and $\Gamma, x{:}\tau \vdash Q$. Since $x \notin fv(Q)$, by strengthening, $\Gamma \vdash Q$ and, by rule (T.NEW), $\Gamma \vdash (\nu x{:}\tau)P$. Finally, by rule (T.PAR) $\Gamma \vdash (\nu x{:}\tau)P \mid Q$.
($\Leftarrow$) By rules (T.PAR) and then (T.NEW), we have $\Gamma, x{:}\tau \vdash P$ and $\Gamma \vdash Q$. By weakening, we have $\Gamma, x{:}\tau \vdash Q$ and, by (T.PAR) and (T.NEW) we conclude that $\Gamma \vdash (\nu x{:}\tau)(P \mid Q)$. □

Lemma 4. *Let P be a process with definitions $\mathcal{D}$ in the attributed π-calculus with a type safe attribute language, and Δ a type environment such that $\Delta \vdash D$ for all $D \in \mathcal{D}$. If $\Gamma \vdash P$ with $\Delta \subseteq \Gamma$ and $P \Downarrow Q$ then $\Gamma \vdash Q$.*

Proof. By reduction rule (CONV), there exists $n \geq 0$ such that $P(\xrightarrow[nd]{app})^n Q$. Thus, we reduce to the proof, by induction on n, that if $\Gamma \vdash P$ and $P(\xrightarrow[nd]{app})^n Q$ then $\Gamma \vdash Q$. The case $n = 0$ is straightforward, so we only need to prove the case $n = 1$ by induction on the derivation of $P \xrightarrow[nd]{app} Q$. The induction cases (PAR) and (NEW) are straightforward and (STRUCT) follows from Lemma 3(4). In the (APP) case, we have $A(\tilde{e}) \xrightarrow[nd]{app} P[\tilde{v}/\tilde{e}]$ with $\tilde{e} \Downarrow \tilde{v}$ and $A(\tilde{x}) \triangleq P$. Since $\Delta \vdash A(\tilde{x}) \triangleq P$, by weakening Lemma 3(2), $\Gamma \vdash A(\tilde{x}) \triangleq P$ and, by rule (T.DEF), $\Gamma, \tilde{x}{:}\tilde{\sigma} \vdash P$ (†) and $\Gamma \vdash A{:}\tilde{\sigma}$. Moreover, by hypothesis, $\Gamma \vdash A(\tilde{e})$, thus $\Gamma \vdash \tilde{e}{:}\tilde{\sigma}$. Since $\tilde{e} \Downarrow \tilde{v}$, the type safety of attribute language yields $\Gamma \vdash \tilde{v}{:}\tilde{\sigma}$. Property (†) and substitution Lemma 3(3) yield $\Gamma \vdash P[\tilde{v}/\tilde{x}]$. □

Theorem 1 (Type safety for processes). *If $\mathcal{L}$ is a type safe attribute language then $\pi(\mathcal{L})$ is type safe in that if $\Gamma \vdash P$ and $P \rightarrow Q$ then $\Gamma \vdash Q$.*

Proof. More precisely, we assume that P is a process with definitions in $\mathcal{D}$ and that Γ is a type environment with $\Gamma \vdash P$ and $\Gamma \vdash D$ for any $D \in \mathcal{D}$. By reduction rule (PRIOR), there exists P' such that $P \Downarrow P'$ and $P' \xrightarrow[nd]{r} Q$ where $r \in R$. By Lemma 4, it is thus sufficient to prove the theorem for reduction $\xrightarrow[nd]{r}$ by induction on the derivation. The inductive cases (PAR), (STRUCT) and (NEW) are straightforward. In the (COM) case, we have $P = e_1[e_2]?\tilde{y}.P_1 + M_1 \mid e_1'[e_2']?\tilde{e}.P_2 + M_2$, $Q = P_1[\tilde{v}/\tilde{y}] \mid P_2$, such that $e_1 \Downarrow x$, $e_1' \Downarrow x$, $e_2 \Downarrow v_2$, $e_2' \Downarrow v_2'$, $v_2 v_2' \Downarrow r$ and $\tilde{e} \Downarrow \tilde{v}$. By rules (T.PAR), (T.SUM), (T.REC) and (T.SEND), we have $\Gamma \vdash e_1 : [\tau] \Rightarrow \tilde{\sigma}$ and $\Gamma \vdash e_1 : [\tau_1 \rightarrow \tau_2] \Rightarrow \tilde{\sigma'}$. Since $e_1 \Downarrow x$ and $e_1' \Downarrow x$, type safety ensures that x has the same type as e_1 and e_1', thus $\tau = \tau_1 \rightarrow \tau_2$, $\sigma' = \tau$ and $\Gamma \vdash x : [\tau_1 \rightarrow \tau_2] \Rightarrow \tilde{\sigma}$. Moreover, since $\Gamma \vdash \tilde{e}{:}\tilde{\sigma}$, type safety yields

$\Gamma \vdash \tilde{v}{:}\tilde{\sigma}$. In addition, we have $\Gamma, \tilde{y}{:}\tilde{\sigma} \vdash P_1$ and, by the substitution Lemma 3(3), $\Gamma \vdash P_1[\tilde{v}/\tilde{y}]$. Finally, from $\Gamma \vdash P_2$ and rule (T.PAR), $\Gamma \vdash P_1[\tilde{v}/\tilde{y}] \mid P_2$. □

Corollary 2 (Error freeness). *If $\mathcal{L}$ is an attribute language that is both type safe and normalizing (see Propositions 6 and 7) then $\pi(\mathcal{L})$ is error free in that if $\Gamma \vdash P$ and $P \to^* Q$ then $\neg Q \xrightarrow[nd]{err} \bot$.*

The proof is elaborated in Appendix B.

4 Modeling Techniques and Biological Examples

We illustrate the usefulness of the attributed π-calculus for modeling biological systems, and extract useful modeling techniques. As examples we consider spatial aspect of Euglena's phototaxis and cooperative enhancement for gene regulation at the lambda switch. Furthermore, we discuss population-based modeling in the attributed π-calculus, in contrast to modeling population aspects in individual-based modeling.

4.1 Spatial Aspects: Euglena's Phototaxis

We start with simple spatial aspects by considering location dependent motion at the example of Euglena's phototaxis [41]. The general relevance of spatial aspects in molecular biology is discussed e.g. in [42]. For the modeling of systems with dynamic compartment structures, we refer to Section 5.2.

Euglena is a single cell organism that lives in inland water and performs photosynthesis. Depending on the brightness, it swims up and down in order to reach a zone with just the right amount of light, [43]. In our model, the probability that an Euglena moves upwards is constant, since it always tries to reach regions with more light. However, in order to avoid too intense light, Euglena moves downwards whenever it receives light for photosynthesis, i.e. with a probability proportional to the light intensity of its current position. We assume that light photons travel top-down and that light intensity degrades exponentially with respect to the depth (repeated filtering).

Given a light source with initial intensity $I \in \mathbb{R}^+$ at depth 0, and a transparency factor for filtering $\sigma \in]0,1]$, this means that the light intensity at depth $d \in \mathbb{R}^+$ equals $\sigma^d * I$. Our model comprises two light sources with initial intensities $\mathtt{I}_1$ and $\mathtt{I}_2$, such that the overall amount of light yields $\mathtt{I} = \mathtt{I}_1 + \mathtt{I}_2$. Furthermore, we assume a constant rate $\mathtt{u} \in \mathbb{R}$ for upward motion.

We consider discrete depth levels $\{0, \ldots, \mathtt{m}\}$ where level 0 denotes the surface and level $\mathtt{m} \in \mathbb{N}_0$ the ground. Euglenas may move up and down in steps of exactly 1 level. Continuous depth levels and movement steps could be modeled similarly, but would increase simulation costs. For the initial system we assume, that $\mathtt{n} \in \mathbb{N}_0$ Euglenas are on every level, summing up to totally $\mathtt{n}*(\mathtt{m}+1)$ Euglenas in the water. Our model comprises two light sources, which are located at the water surface. Their respective intensities are given by real numbers $\mathtt{I}_1, \mathtt{I}_2 \in \mathbb{R}^+$. The

Parameters

```
n ∈ ℕ0       // initial number of Euglenas per water level
m ∈ ℕ0       // deepest water level I1,I2 ∈ ℝ+ //
intensity rates of light sources 1 and 2
σ ∈ [0,1] // transparency of water
u ∈ ℝ+       // Euglena's upwards speed
```

Process definitions

$$\mathsf{Euglena}(d) \triangleq up[\lambda\_.\ \textbf{if}\ d \geq 1\ \textbf{then}\ \mathtt{u}]?().\mathsf{Euglena}(d-1)$$
$$+\ down[\lambda i.\ \textbf{if}\ d \leq \mathtt{m}-1\ \textbf{then}\ \sigma^{d}{*}i]?().\mathsf{Euglena}(d+1)$$
$$\mathsf{Light}(i) \triangleq down[i]!().\mathsf{Light}(i)$$
$$\mathsf{Dummy}() \triangleq up[\_]!().\mathsf{Dummy}()$$

Example solution

$$\prod_{d=0}^{\mathtt{m}}\prod_{i=1}^{\mathtt{n}}\mathsf{Euglena}(d)\ \mid\ \mathsf{Light}(\mathtt{I}_1)\ \mid\ \mathsf{Light}(\mathtt{I}_2)\ \mid\ \mathsf{Dummy}()$$

Fig. 17. A discrete model of Euglena's light-dependent motion with two light sources

probability of an interaction with a light source is proportional to its intensity. The values $\mathtt{u}, \sigma, \mathtt{m}, \mathtt{n}, \mathtt{I}_1, \mathtt{I}_2$ are model parameters.

We define our model in $\pi(\lambda(\mathbb{R}, +, -, *, /, pow, \leq)_{<_1})$, the attributed π-calculus with constants for real numbers and the usual arithmetic operations. For convenience, we write x^y instead of $(pow\ x)\ y$. The successful values are the positive real numbers, that all have the same level of priority. The big-step evaluator for these operators can be defined as usual (in analogy to natural numbers, see Section 3.2). We consider, non-zero positive real numbers to be the only successful values.

Our model is given in Fig. 17. An Euglena at depth level d may interact with a light source of intensity i and go down by one level:

$$down : \mathsf{Euglena}(d), \mathsf{Light}(i) \xrightarrow{\sigma^d * i} \mathsf{Euglena}(d+1), \mathsf{Light}(i)$$

Such interactions happen on channel *down* with rate $\sigma^d * i$ under the condition that $d \leq m-1$. An Euglena can also move up with constant rate $\mathtt{u}$ by interacting with a dummy interaction partner on channel *up*:

$$up : \mathsf{Euglena}(d) \xrightarrow{u} \mathsf{Euglena}(d-1)$$

If Euglena is at the surface, i.e. the constraint $d \geq 1$ is not satisfied, it cannot move upwards any further.

Based on the master equation, the amounts of Euglenas on each depth level in equilibrium can be computed. For illustration, we fix $m = 4$. The equilibrium happens if the system is free of change, such that its derivation is 0. Let $l_0, \ldots, l_4$ be the amounts of Euglena per level. The master equation yields:

$$\begin{pmatrix} -\sigma^0 * \mathtt{I} & u & 0 & 0 & 0 \\ \sigma^0 * \mathtt{I} & -(\sigma^1 * \mathtt{I} + u) & u & 0 & 0 \\ 0 & \sigma^1 * \mathtt{I} & -(\sigma^2 * \mathtt{I} + u) & u & 0 \\ 0 & 0 & \sigma^2 * \mathtt{I} & -(\sigma^3 * \mathtt{I} + u) & u \\ 0 & 0 & 0 & \sigma^3 * \mathtt{I} & -u \\ 1 & 1 & 1 & 1 & 1 \end{pmatrix} \cdot \begin{pmatrix} l_0 \\ l_1 \\ l_2 \\ l_3 \\ l_4 \end{pmatrix} = \begin{pmatrix} 0 \\ 0 \\ 0 \\ 0 \\ 0 \\ 5n \end{pmatrix}$$

The first equation $-\sigma^0 * \mathtt{I} * l_0 + u * l_1 = 0 = l_0'$ states that the change in the amount of Euglena at level 0 is obtained by summing up a loss of $\sigma^0 * \mathtt{I} * l_0$ due to Euglena's downward motion to level 1, and a gain of $u * l_1$ due to Euglenas upwards motion from level 1. The last equation $\sum_{i=0}^{4} l_i = 5n$ denotes that the overall amount of Euglena is constant and equals the initial amount.

In order to verify the behavior of our model with respect to predictions obtained from the Master Equation, we performed two simulation experiments, named A and B. There are constantly five depth levels ($\mathtt{m} = 4$), 100 Euglenas on each depth level ($\mathtt{n} = 100$), a rate of upward motion $\mathtt{u} = 0.4$, intensity rates $\mathtt{I}_1 = 5.0, \mathtt{I}_2 = 15.0$ ($\mathtt{I} = 20.0$), and transparency factors $\sigma = 0.1$ in experiment A and $\sigma = 0.2$ in experiment B. Each experiment consists in a single simulation run, all of them performed until simulation time $t = 10.0$.

The simulation results are presented in Fig. 18. Heat maps and line charts show the amounts of Euglena on each depth level over time. Below them, the solutions of the Master Equation with the respective model parameters are given. The simulation results confirm the predictions, with slight derivations due to stochasticity. The comparison of both experiments shows, that with a higher transparency Euglena accumulates on a deeper level, as a consequence of more light being available.

We can translate our Euglena model with attributed processes into the stochastic π-calculus without attributes, since all parameters are finitely valued. The idea is to duplicate the *down*-channels for all depth levels, so that its rates can be made dependent for the depth. This leads to processes $\mathsf{Euglena}_d()$, $\mathsf{Light}_{1,d}()$, and $\mathsf{Light}_{2,d}()$ for all possible depth levels, see Fig. 19.

4.2 Cooperative Enhancement: Gene Regulation at Lambda Switch

Cooperative binding is a frequent and often decisive aspect in gene regulatory networks, where proteins stabilize each other's binding to neighboring DNA sites by adhesive contacts. In quantitative terms, the unbinding rate of one DNA-protein complex decreases by the existence of another. This is an instance of cooperative enhancement of reaction rates by third partners. As shown in [27,26], cooperative enhancement can be modeled in the stochastic π-calculus. It however requires nontrivial encodings, that can be alleviated within the attributed π-calculus.

A well understood instance of cooperative binding occurs during transcription initiation control at the lambda switch. The lambda switch is a segment of the DNA of bacteriophage lambda. It contains two binding sites OR_1 and OR_2, where rep and cro proteins can bind. An unstable binding of a rep molecule to

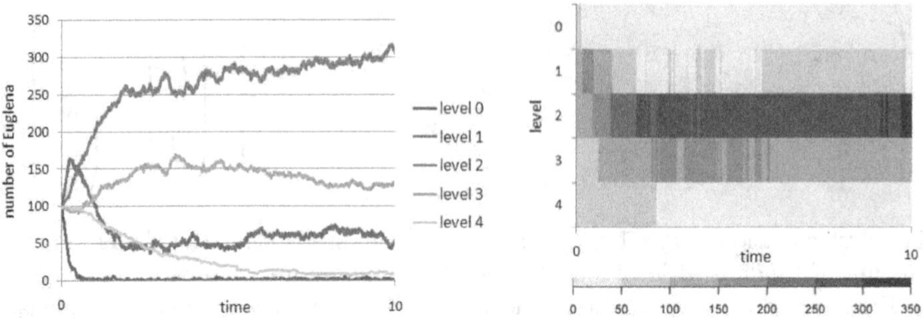

Experiment A, Predictions: $l_0 = 1.16, l_1 = 57.84, l_2 = 289.20, l_3 = 144.65, l_4 = 7.15$

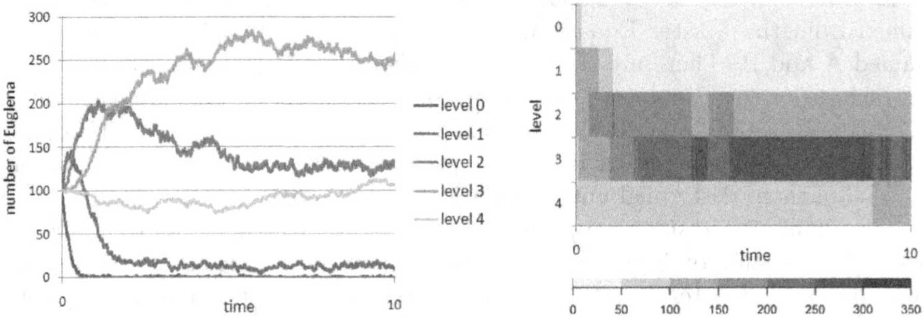

Experiment B, Predictions: $l_0 = 0.26, l_1 = 12.81, l_2 = 128.14, l_3 = 256.28, l_4 = 102.51$

Fig. 18. Experiments with Euglena: two experiments (A,B), with $\mathtt{m} = 4$, $\mathtt{n} = 100$, $\mathtt{u} = 0.4$, $\mathtt{I}_1 = 5.0, \mathtt{I}_2 = 15.0$ ($\mathtt{I} = 20.0$), and $\sigma = 0.1$ in experiment A and $\sigma = 0.2$ in experiment B. For each experiment a single simulation run until simulation time $t = 10.0$. The line charts and heat maps show the development of the number of Euglenas on each depth level over time. Predictions for the amounts on the different depth levels in equilibrium as obtained by solving the Master Equation are shown below each experiment.

OR_2 is stabilized by the simultaneous presence of another rep at the neighboring site OR_1. As illustrated in Fig. 20, the two proteins actually touch each other.

A model of cooperative binding at OR_2 in $\pi(\lambda(\mathbb{R}_+^\infty, =)_{<_2})$ is presented in Fig. 21. It contains the parametric definition Prot(*type*), which emulates the behavior of the proteins. The parameter *type* can be instantiated by *rep* or *cro*, modeling either protein sort. Proteins can bind to both sites OR_1 and OR_2. Free sites are defined by processes $\mathrm{OR}_1()$ and $\mathrm{OR}_2()$, where proteins can attach via channel *bind*. As this occurs the channel *release* is created, and henceforth connects the protein to the site (complexation). Later communication on *release* breaks the complex. The reaction rate of complexation is fixed to `0.098`. For decomplexation the rate is determined by the sender, i.e. the binding site, the receiving protein accepts it by applying the identity function.

Now consider the models for the protein bound DNA sites. $\mathrm{OR_iB}$(*type*,*release*) describes the unbinding from the occupied site $\mathrm{OR_i}$, where *type* indicates the

```
// Euglenas on different depth levels
Euglena_0() ≜ down_0?().Euglena_1()
Euglena_1() ≜ up?().Euglena_0() + down_1?().Euglena_2()
...
Euglena_m() ≜ up?().Euglena_m-1()
// light from first source on different levels
Light_1,0() ≜ down_0:σ^0*I_1!().Light_1,0()
...
Light_1,m() ≜ down_m :σ^m*I_1!().Light_1,m()
// light from second source on different levels
Light_2,0() ≜ down_0:σ^0*I_2!().Light_2,0()
...
Light_2,m() ≜ down_m :σ^m*I_2!().Light_2,m() Dummy() ≜ up:u!().Dummy()
```

Example solution

$\prod_{d=0}^{m} (\prod_{i=0}^{n}$ Euglena$_d$() | Light$_{1,d}$() | Light$_{2,d}$())

Fig. 19. An equivalent model of Euglena in the stochastic π-calculus

type of the bound protein. For $i = 1$ the rate of the unbinding reaction merely depends on the protein type.

For the second site ($i = 2$) decomplexation is influenced by cooperative binding. To model this, OR_1 and OR_2 are linked via the channel *or2Delay*, illustrated in Fig. 20. Additionally, the release operation is decomposed into an interaction on channel *or2Delay*, with a reaction rate defining the actual unbinding delay, and an immediate communication on *release*. As stated in the definition of the global channel *or2Delay* the unbinding delay depends not only on the type of the bound protein, but also on the state of OR_1, which can be either *free*, bound to *rep* or bound to *cro*.

A previous model [26] in the stochastic π-calculus requires to keep OR_2 constantly informed about state changes of OR_1, which is implemented by immediate communication steps. Keeping state information consistent in this manner

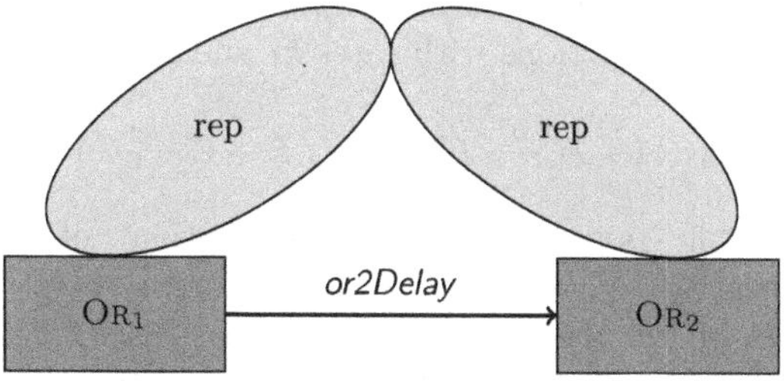

Fig. 20. The decay of the rep-OR_2 complex: in order to make the decay rate of the rep-OR_2 complex dependent on OR_1's state, the two sites communicate on *or2Delay* before OR_2 unbinds

Process definitions

```
Prot(type) ≜ (νrelease) bind [ _ ] ! ( type , release ) . release [λr . r] ? ( ) . Prot(type)
OR1() ≜ bind [λ_ .  0.098] ? ( type , release ) . OR1B(type,release)
      + or2Delay [ free ] ! . OR1 ( )
OR1B(type,release) ≜ release [
                          if type = rep then 0.155 else
                          if type = cro then 2.45
                          ] ! ( ) . OR1()
                       + or2Delay [ type ] ! ( ) . OR1B(type,release)
OR2() ≜ bind [λ_ .  0.098] ? ( type , release ) . OR2B(type,release)
OR2B(type,release) ≜
   or2delay [λt .
     if t = rep then
       if type = rep then 0.155 else // big delay (cooperative)
       if type = cro then 3.99 // small delay
     else 2.45 // cro or free
     ] ? ( ) . release [∞] ! ( ) . OR2()
```

Example process

$\mathrm{OR_1}() \mid \mathrm{OR_2}() \mid \prod_{i=1}^{28} \mathrm{Prot}(rep) \mid \prod_{i=1}^{67} \mathrm{Prot}(cro)$

Fig. 21. A model of cooperative binding between $\mathrm{OR_1}$ and $\mathrm{OR_2}$ at the λ switch

is error-prone, it may easily lead to deadlocks. A subsequent model [27] in SPiCO requires significantly fewer updates. In $\pi(\lambda(\mathbb{R}_+^\infty, =)_{<_2})$, reaction rates directly depend on the attribute values of the interaction partners. State changes are propagated without additional communication steps.

4.3 Population-Based Modeling

The stochastic π-calculus supports individual-based modeling where molecules are mapped to objects. The attributed π-calculus enables in addition a population-based modeling style, where reactions are mapped to objects.

For illustration, we consider a chemical system with three species A, B, and C and the following two reactions with rates k_1 and k_2:

$$r_1 : A + B \xrightarrow{k_1} C \qquad r_2 : B + C \xrightarrow{k_2} A$$

Figure 22 shows a population-based description of this system in $\pi(\lambda(\mathbb{R}, +, -, *, /, pow, \leq)_{<_1})$. The model parameters $\mathsf{a}_0, \mathsf{b}_0, \mathsf{c}_0 \in \mathbb{N}_0$ represent the initial amounts of the three species. A process $\mathsf{Reac}(f, d_a, d_b, d_c)$ defines a reaction with kinetic function f, while the other parameters d_a, d_b, and d_c reflect

Parameters

$a_0, b_0, c_0 \in \mathbb{N}_0$ // initial amounts of A, B, and C
$k_1, k_2 \in \mathbb{R}$ // reaction rates

Process definitions

$\mathsf{Reac}(f, d_a, d_b, d_c) \triangleq r[f]!(d_a, d_b, d_c).\mathsf{Reac}(f, d_a, d_b, d_c)$
$\mathsf{Pop}(a, b, c) \triangleq r[\lambda f.(f\ a\ b\ c)]?(d_a, d_b, d_c).\mathsf{Pop}(a+d_a, b+d_b, c+d_c)$

Example solution

$\mathsf{Reac}(\lambda a.\lambda b.\lambda c.\, a*b*k_1, -1, -1, 1) \mid \mathsf{Reac}(\lambda a.\lambda b.\lambda c.\, b*c*k_2, 1, -1, -1) \mid$
$\mathsf{Pop}(a_0, b_0, c_0)$

Fig. 22. A population-based model of three species and two reactions in $\pi(\lambda(\mathbb{R}, +, -, *, /, pow, \leq)_{<_1})$. Process $\mathsf{Reac}(f, d_a, d_b, d_c)$ defines reactions, where parameter f is a function reflecting the reaction kinetics and parameters d_a, d_b, d_c account for the way species amounts are changed when the reaction occurs. Process $\mathsf{Pop}(a, b, c)$ reflects species amounts and interacts with process $\mathsf{Reac}(f, d_a, d_b, d_c)$ for reaction execution.

how the reaction affects the population, i.e. the differences in the amounts of the species. The latter parameters could also be regarded as stoichiometric factors, except that reactants are always associated with negative numbers. The kinetics of all reactions in this example follow the law of Mass action. For instance, the kinetics of r_1 yields the product of the amounts of species A and B and rate k_1. It changes the population by decreasing the amount of A and B by one each and increasing the amount of C by one. Thus, to represent reaction r_1, our initial solution comprises a process $\mathsf{Reac}(\lambda a.\lambda b.\lambda c.a * b * k_1, -1, -1, 1)$, where the function parameters a, b, c represent the amounts of A, B, C, respectively. Consequently, we also start with one process $\mathsf{Reac}(\lambda a.\lambda b.\lambda c.b * c * k_2, 1, -1, -1)$ to model reaction r_2. Notice, that it is also possible to account for different kinetic laws and different stoichiometric factors.

In order to represent populations, we define processes $\mathsf{Pop}(a, b, c)$, whose parameters stand for the amounts of the three species. In addition to the two reactions, the initial process thus also comprises $\mathsf{Pop}(a_0, b_0, c_0)$. Interactions on channel r indicate the occurrence of a reaction. For the computation of reaction kinetics $\mathsf{Reac}(f, d_a, d_b, d_c)$ provides its kinetics function f as the constraint argument. $\mathsf{Pop}(a, b, c)$ defines a constraint function, which applies f to the current amounts of the species. When an interaction occurs, $\mathsf{Reac}(f, d_a, d_b, d_c)$ sends its population changes and $\mathsf{Pop}(a, b, c)$ applies them to the current amounts, which is implemented by recursively calling $\mathsf{Pop}(a + d_a, b + d_b, c + d_c)$. Afterward, the next reaction can happen. Function f will evaluate to 0 whenever one of the populations becomes 0.

The model can be generalized to systems in which new species can be created dynamically, by using property lists as parameters, where each element contains a pair of a species and its amount. Even new reactions could be dynamically introduced. Such a model is close to the way biochemistry is expressed in sCCP, pointing to the possibility of generally encoding sCCP in attributed processes, which is, however, still subject to future work.

4.4 Global Information in Individual-Based Modeling

Priorities are necessary in order to track global information in individual-based models in a consistent manner, for instance in order to model changes in compartment structures [18]. The general idea is to propagate changes globally by a sequence of prioritized local interactions, before enabling the next possible reactions, since these are given lower priority.

In the example Fig. 23, we trace global information on population sizes, i.e. numbers of individuals. It rephrases the population-based model in Section 4.3 in an individual-based way. We model molecules of the three species by processes A(),B(), and C() and define a process Pop(a,b,c) such that it tracks molecule numbers as obtained by interactions between molecules. Chemical reactions are mimicked by interactions on channel r, where A() or C() send to B() with rates k_1 and k_2, respectively. The changes in the population are updated by prioritized interaction over channel u once a reaction occurred. Process Pop(a,b,c) receives such changes with infinite rate, i.e. with priority, such that no reaction can occur before the population information is updated. By this, it is ensured, that the effect of each reaction is correctly reflected in the species amounts, i.e. that the population information is consistent.

This example indicates, that priorities might permit to implement individual-based models with global stores, as proposed in [44].

Parameters

$a_0, b_0, c_0 \in \mathbb{N}_0$ // initial amounts of A, B, and C
$k_1, k_2 \in \mathbb{R}$ // reaction rates

Process definitions

$\mathsf{A}() \triangleq$
$r[k_1]!().u[\_]!(-1,-1,1).\mathsf{C}()$
$\mathsf{B}() \triangleq r[\lambda k.k]?().0$
$\mathsf{C}() \triangleq r[k_2]!().u[\_]!(1,-1,-1).\mathsf{A}()$
$\mathsf{Pop}(a,b,c) \triangleq u[\lambda\_.\infty]?(d_a,d_b,d_c).\mathsf{Pop}(a+d_a,b+d_b,c+d_c)$

Example solution

$\prod_{i=1}^{a_0} \mathsf{A}() \mid \prod_{i=1}^{b_0} \mathsf{B}() \mid \prod_{i=1}^{c_0} \mathsf{C}() \mid \mathsf{Pop}(a_0,b_0,c_0)$

Fig. 23. An individual-based variant of the population-based model in Figure 22. Processes A(),B(), and C() represent species, process Pop(a,b,c) accounts for species amounts, which are updated by prioritized interactions on channel u.

4.5 Species-Based Modeling

As a last example, we show how one can rephrase the model given in Section 4.3 in a *species-based* style, see Figure 24. We make use of priorities (i.e. immediate communications) and the fact that reactions have at most two reactants. A process $\mathsf{A}(a)$ represents the species A which is attributed by the number a of molecules of A. Species B and C are implemented analogously. In contrast to the individual-based model, the solution always contains a single process for each species.

Reaction r_1 is modeled as an interaction between processes $\mathsf{A}(a)$ and $\mathsf{B}(b)$ on channel r_1. The corresponding communication constraint yields a rate, which follows Mass action kinetics. After an interaction, $\mathsf{A}(a\text{-}1)$ and $\mathsf{B}(b\text{-}1)$ are called recursively, thus decreasing the number of molecules of species A and B. In parallel, a request is sent with priority on channel u_c in order to increase the number of molecules of species C. Reaction r_2 is implemented analogously.

5 Expressiveness of the Attributed Pi-Calculus

We show that the attributed π-calculus provides a unifying framework that generalizes on various dialects of the π-calculus in the literature.

5.1 Encoding of the π-Calculus with Priorities

We start with an encoding of the π-calculus with priorities, and prove its correctness with respect to both semantics, non-deterministic and stochastic. The encoding can be refined such that it preserves well-typing.

The translation of the π-calculus with priorities in $(R, <)$ into the attributed π-calculus $\pi(\lambda(R)_<)$ is given in Fig. 25. Senders $x{:}r!\tilde{y}.P$ are mapped to $x[r]!\tilde{y}.P$ and receivers $x?\tilde{y}.P$ to $x[\lambda z.z]?\tilde{y}.P$.

Parameters

$\mathtt{a}_0, \mathtt{b}_0, \mathtt{c}_0 \in \mathbb{N}_0$ // initial amounts of A, B, and C
$\mathtt{k}_1, \mathtt{k}_2 \in \mathbb{R}$ // reaction rates

Process definitions

$$\mathsf{A}(a) \triangleq r_1[a]!().(\mathsf{A}(a-1) \mid u_c[\_]!(1)) + u_a[\lambda\_.\infty]?(d).\mathsf{A}(a+d)$$
$$\mathsf{B}(b) \triangleq r_1[\lambda a.\mathtt{k}_1 * a * b]?().\mathsf{B}(b-1) + r_2[\lambda c.\mathtt{k}_2 * c * b]?().\mathsf{B}(b-1)$$
$$\mathsf{C}(c) \triangleq r_2[c]!().(\mathsf{C}(c-1) \mid u_a[\_]!(1)) + u_c[\lambda\_.\infty]?(d).\mathsf{C}(c+d)$$

Example solution

$$\mathsf{A}(\mathtt{a}_0) \mid \mathsf{B}(\mathtt{b}_0) \mid \mathsf{C}(\mathtt{c}_0)$$

Fig. 24. An species-based variant of the population-based model in Figure 22, also in $\pi(\lambda(\mathbb{R}, +, -, *, /, pow, \leq)_{<_1})$. Processes $\mathsf{A}(a)$, $\mathsf{B}(b)$, and $\mathsf{C}(c)$ represent species parametrized by their multiplicities possibly updated through channel u_a and u_c.

$$[\![x?\tilde{y}.P]\!] = x[\lambda z.z]?\tilde{y}.[\![P]\!] \qquad [\![P_1 \mid P_2]\!] = [\![P_1]\!] \mid [\![P_2]\!]$$

$$[\![x{:}r!\tilde{y}.P]\!] = x[r]!\tilde{y}.[\![P]\!] \qquad [\![M_1 + M_2]\!] = [\![M_1]\!] + [\![M_2]\!] \qquad [\![\mathbf{0}]\!] = \mathbf{0}$$

$$[\![(\nu x)P]\!] = (\nu x)[\![P]\!] \qquad [\![A(\tilde{x}) \triangleq P]\!] = A(\tilde{x}) \triangleq [\![P]\!]$$

Fig. 25. Encoding the π-calculus with priorities $(R,<)$ into $\pi(\lambda(R)_<)$, and the stochastic π-calculus into the attributed π-calculus with stochastic semantics

Theorem 2. *The encoding of the π-calculus with priorities $(R,<)$ into the attributed π-calculus with priorities $\pi(\lambda(R)_<)$ is correct in that for all processes P, P' and attributed processes Q:*

1. *if $P \to P'$ then $[\![P]\!] \to [\![P']\!]$.*
2. *if $[\![P]\!] \to Q$ then there exists a process $\hat{Q}$ of the π-calculus with priorities such that $[\![\hat{Q}]\!] \equiv Q$ and $P \to \hat{Q}$.*

The proof is elaborated in Appendix B. It is mostly straightforward, but covers several pages since all rules of both calculi must be inspected in detail.

The same encoding is also correct with respect to the stochastic operational semantics, under the assumption that we choose the set of stochastic rates $(\mathbb{R}_+^\infty, <_2)$ as priorities.

Theorem 3. *The encoding of the π-calculus with priorities in $(\mathbb{R}_+^\infty, <_2)$ into the attributed π-calculus $\pi(\lambda(\mathbb{R}_+^\infty)_{<_2})$ is correct with respect to the stochastic operational semantics. For all processes $P, P', \hat{P}$, attributed processes Q and labels $\alpha \in \{r, \infty(n) \mid r \in \mathbb{R}_+, n \in \mathbb{N}\}$:*

1. *if $P \xrightarrow{\alpha} P'$ then $[\![P]\!] \xrightarrow{\alpha} [\![P']\!]$*
2. *if $[\![\hat{P}]\!] \xrightarrow{\alpha} Q$ then there exists a process $\hat{Q}$ such that $\hat{P} \xrightarrow{\alpha} \hat{Q}$ and $[\![\hat{Q}]\!] \equiv Q$.*

Proof. The stochastic semantics of both calculi are build on top of their on their non-deterministic semantics. We prove in the appendix (see the proof of Theorem 2) that the translation is invariant under substitution and that it reflects and preserves the structural congruence and errors. Furthermore, we proved there that if $P \; \rho \; Q$ then $[\![P]\!] \; \rho \; [\![Q]\!]$, for $\rho \in \{\Downarrow, \to\} \cup \{\xrightarrow[nd]{\alpha} \mid \alpha \in \{app\} \cup R\}$.

Claim. Relation $\xrightarrow[\ell]{r}$ is preserved and reflected by translation (i.e., positions ℓ of redexes remain unchanged):

1. if $P \xrightarrow[\ell]{r} Q$ then $[\![P]\!] \xrightarrow[\ell]{r} [\![Q]\!]$,
2. if $[\![\hat{P}]\!] \xrightarrow[\ell]{r} Q$ then exists $\hat{Q}$ such that $\hat{P} \xrightarrow[\ell]{r} \hat{Q}$ and $[\![\hat{Q}]\!] \equiv Q$.

Proof. 1. If $P \xrightarrow[\ell]{r} Q$ then rule ($\textsc{com}_\ell$) can be applied as follows:

$$\frac{\ell = (i_1, j_1, i_2, j_2) \quad \pi_{i_1}^{j_1} = x?\tilde{y} \quad \pi_{i_2}^{j_2} = x{:}r!\tilde{z} \quad |\tilde{y}| = |\tilde{v}|}{P \xrightarrow[\ell]{r} Q}$$

where

$$P = (\nu\tilde{x})\prod_{i=1}^{n}\sum_{j=1}^{m_i}\pi_i^j.P_i^j \text{ and } Q = (\nu\tilde{x})(\prod_{i=1,i\neq i_1,i_2}^{n}\sum_{j=1}^{m_i}\pi_i^j.P_i^j \mid P_{i_1}^{j_1}[\tilde{v}/\tilde{y}] \mid P_{i_2}^{j_2})$$

Thus, $[\![P]\!] = (\nu\tilde{x})\prod_{i=1}^{n}\sum_{j=1}^{m_i}[\![\pi_i^j]\!].[\![P_i^j]\!]$, with $[\![\pi_{i_1}^{j_1}]\!] = x[\lambda y.y]?\tilde{y}$ and $[\![\pi_{i_2}^{j_2}]\!] = x[r]!\tilde{z}$. Now, rule (COM$_\ell$) of $\pi(\mathcal{L})$ applies to the translations, while using (VAL) and (FUN):

$$\frac{\ell = (i_1, j_1, i_2, j_2) \quad \begin{array}{l}[\![\pi_{i_1}^{j_1}]\!] \Downarrow x[\lambda y.y]?\tilde{y} \\ [\![\pi_{i_2}^{j_2}]\!] \Downarrow x[r]!\tilde{z}\end{array} \quad (\lambda y.y)r \Downarrow r \in \mathbb{R}_+^\infty \quad |\tilde{y}| = |\tilde{z}|}{[\![P]\!] \xrightarrow[\ell]{r} [\![Q]\!]}$$

where

$$[\![P]\!] = (\nu\tilde{x})\prod_{i=1}^{n}\sum_{j=1}^{m_i}[\![\pi_i^j]\!].[\![P_i^j]\!] \text{ and}$$

$$[\![Q]\!] = (\nu\tilde{x})\Big(\prod_{i=1,i\neq i_1,i_2}^{n}\sum_{j=1}^{m_i}[\![\pi_i^j]\!].[\![P_i^j]\!] \mid [\![P_{i_1}^{j_1}]\!][\tilde{v}/\tilde{y}] \mid [\![P_{i_2}^{j_2}]\!]\Big)$$

The last equality follows from the substitution claim $[\![P_{i_1}^{j_1}[\tilde{v}/\tilde{y}]]\!] = [\![P_{i_1}^{j_1}]\!][\tilde{v}/\tilde{y}]$ and the compositionality of the translation.

2. If $[\![\hat{P}]\!] \xrightarrow[\ell]{r} Q$ then rule (COM$_\ell$) must be applicable as follows:

$$\frac{\ell = (i_1, j_1, i_2, j_2) \quad \pi_{i_1}^{j_1} \Downarrow x[v_1]?\tilde{y} \quad \pi_{i_2}^{j_2} \Downarrow x[v_2]!\tilde{v} \quad v_1 v_2 \Downarrow r \in \mathbb{R}_+^\infty \quad |\tilde{y}| = |\tilde{v}|}{[\![\hat{P}]\!] \xrightarrow[\ell]{r} Q}$$

where

$$[\![\hat{P}]\!] = (\nu\tilde{x})\prod_{i=1}^{n}\sum_{j=1}^{m_i}\pi_i^j.P_i^j \text{ and}$$

$$Q = (\nu\tilde{x})\Big(\prod_{i=1,i\neq i_1,i_2}^{n}\sum_{j=1}^{m_i}\pi_i^j.P_i^j \mid P_{i_1}^{j_1}[\tilde{v}/\tilde{y}] \mid P_{i_2}^{j_2}\Big)$$

Since the translation is compositional, process $\hat{P}$ must have the form $\hat{P} = (\nu\tilde{x})\prod_{i=1}^{n}\sum_{j=1}^{m_i}\hat{\pi}_i^j.\hat{P}_i^j$, with $[\![\hat{\pi}_i^j]\!] = \pi_i^j$ and $[\![\hat{P}_i^j]\!] = P_i^j$. Furthermore, we have that $v_1 = \lambda y.y$, $v_2 = r$, such that $\hat{\pi}_{i_1}^{j_1} = x?\tilde{y}$ and $\hat{\pi}_{i_2}^{j_2} = x{:}r!\tilde{z}$, with $\tilde{v} = \tilde{z}$. We define $\hat{Q} = (\nu\tilde{x})(\prod_{i=1,\ i\neq i_1,i_2}^{n}\sum_{j=1}^{m_i}\hat{\pi}_i^j.\hat{P}_i^j \mid \hat{P}_{i_1}^{j_1}[\tilde{v}/\tilde{y}] \mid \hat{P}_{i_2}^{j_2})$. Since the translation is substitution invariant, we obtain $[\![\hat{Q}]\!] = Q$. Rule (COM$_\ell$) applies as follows:

$$\frac{\ell = (i_1, j_1, i_2, j_2) \quad \hat{\pi}_{i_1}^{j_1} = x?\tilde{y} \quad \hat{\pi}_{i_2}^{j_2} = x{:}r!\tilde{z} \quad |\tilde{y}| = |\tilde{z}|}{\hat{P} \xrightarrow[\ell]{r} \hat{Q}}$$

where

$$\hat{P} = (\nu\tilde{x})\prod_{i=1}^{n}\sum_{j=1}^{m_i}\hat{\pi}_i^j.\hat{P}_i^j \text{ and } \hat{Q} = (\nu\tilde{x})\Big(\prod_{\substack{i=1\\ i\neq i_1,i_2}}^{n}\sum_{j=1}^{m_i}\hat{\pi}_i^j.\hat{P}_i^j \mid \hat{P}_{i_1}^{j_1}[\tilde{v}/\tilde{y}] \mid \hat{P}_{i_2}^{j_2}\Big)$$

Given two processes P, Q we define a set $I(P,Q) \subseteq \mathbb{R}_+^\infty \times \mathbb{N}^4$ and a number $S(P,Q) \in \mathbb{R}_+^\infty$ as used in rule (SUM) as follows:

$$I(P,Q) = \{(r,\ell) \mid \exists Q'.\ P \xrightarrow[\ell]{r} Q' \equiv Q\} \qquad \text{and} \qquad S(P,Q) = \sum_{(r,\ell)\in I(P,Q)} r$$

Claim. $S(P,Q) = S([\![P]\!],[\![Q]\!])$

Proof. It is sufficient to show that $I(P,Q) = I([\![P]\!],[\![Q]\!])$. There are two inclusions to be shown:

"$\subseteq$" If $(r,\ell) \in I(P,Q)$ then there exists Q' such that $P \xrightarrow[\ell]{r} Q' \equiv Q$. The first part of the previous claim shows that $[\![P]\!] \xrightarrow[\ell]{r} [\![Q']\!]$, and since translation preserves structural congruence also $[\![Q']\!] \equiv [\![Q]\!]$. Hence $(r,\ell) \in I([\![P]\!],[\![Q]\!])$.

"$\supseteq$" If $(r,\ell) \in I([\![P]\!],[\![Q]\!])$ then there exists Q'' such that $[\![P]\!] \xrightarrow[\ell]{r} Q'' \equiv [\![Q]\!]$. The second part of the previous claim shows that there exists Q' such that $P \xrightarrow[\ell]{r} Q'$ with $[\![Q']\!] \equiv Q'' \equiv [\![Q]\!]$. This implies $Q' \equiv Q$ since translation reflects structural congruence, so that $(r,\ell) \in I(P,Q)$.

Claim. Let Q be a process and $P_1 \equiv P_2$ processes. If P_1 and P_2 are prenex normal forms in which all bound variables are renamed apart, then $S(P_1,Q) = S(P_2,Q)$.

Proof. Suppose that $P_1 = (\nu x_1)\dots(\nu x_k)\prod_{i=1}^{m}\sum_{j=1}^{n_i} M_i^j$ for guarded processes M_i^j. An analysis of the structural congruence shows that there exists a sequence of variables $(y_1,\dots,y_k)$ and permutations $\sigma : \{1,\dots,k\} \to \{1,\dots,k\}$, $\theta : \{1,\dots,m\} \to \{1,\dots,m\}$, and $\theta_i : \{1,\dots,n_i\} \to \{1,\dots,n_i\}$ such that:

$$P_2 = (\nu y_{\sigma(1)})\dots(\nu y_{\sigma(k)})\prod_{i=1}^{m}\sum_{j=1}^{n_i} {M'}_{\theta(i)}^{\theta_i(j)} \text{ and } M_i^j \equiv {M'}_i^j[y_{\sigma(1)}/x_1,\dots,y_{\sigma(k)}/x_k]$$

Given this representation of P_2, and since all bound variables are renamed apart, it is easy to check that $(r,(\theta(i_1),\theta_{i_1}(j_1),\theta(i_2),\theta_{i_2}(j_2))) \in I(P_1,Q)$ iff $(r,(i_1,j_1,i_2,j_2)) \in I(P_2,Q)$.

We next prove the theorem for reductions with finite rates.

Claim. Translation preserves and reflects relations $\xrightarrow{r}$ for all $r \in \mathbb{R}_+$.

1. if $P \xrightarrow{r} P'$ then $[\![P]\!] \xrightarrow{r} [\![P']\!]$
2. if $[\![\hat{P}]\!] \xrightarrow{r} Q$ then there exists $\hat{Q}$ such that $\hat{P} \xrightarrow{r} \hat{Q}$ and $[\![\hat{Q}]\!] \equiv Q$.

Proof. 1. Assumption $P \xrightarrow{r} Q$ must be inferred by rule (SUM) as follows:

$$\frac{P \Downarrow P_1 \quad S(P_1, Q) = r \neq 0 \quad \neg\exists\ell\exists Q'.P_1 \xrightarrow[\ell]{\infty} Q'}{P \xrightarrow{r} Q}$$

We have shown in the proof of Theorem 2 that $P \Downarrow P_1$ implies $[\![P]\!] \Downarrow [\![P_1]\!]$. The second claim above shows that $S(P_1, Q) = S([\![P_1]\!], [\![Q]\!])$. The second part of the first claim above ensures that $\neg\exists\ell\exists Q'.[\![P_1]\!] \xrightarrow[\ell]{\infty} Q'$. Thus, the following rule is applicable:

$$\frac{[\![P]\!] \Downarrow [\![P_1]\!] \quad S([\![P_1]\!], [\![Q]\!]) = r \neq 0 \quad \neg\exists\ell\exists Q'.[\![P_1]\!] \xrightarrow[\ell]{\infty} Q'}{[\![P]\!] \xrightarrow{r} [\![Q]\!]}$$

2. We assume $[\![\hat{P}]\!] \xrightarrow{r} Q$ for $r \in \mathbb{R}_+$. Since the stochastic semantics refines the nondeterministic semantics by Proposition 5 we know that $[\![\hat{P}]\!] \to Q$. Theorem 2 on the preservation of the non-deterministic semantics shows that there exists a process $\hat{Q}$ such that $[\![\hat{Q}]\!] \equiv Q$ and $\hat{P} \to \hat{Q}$. In the following we only make use of $[\![\hat{Q}]\!] \equiv Q$. Assumption $[\![\hat{P}]\!] \xrightarrow{r} Q$ must be inferred by rule (SUM):

$$\frac{[\![\hat{P}]\!] \Downarrow P_1 \quad S(P_1, Q) = r \neq 0 \quad \neg\exists\ell\exists Q'.P_1 \xrightarrow[\ell]{\infty} Q'}{[\![\hat{P}]\!] \xrightarrow{r} Q}$$

In particular, P_1 must be in prenex normal form, and w.l.o.g. all its variables are renamed apart. Since $[\![\hat{P}]\!] \Downarrow P_1$ there exists $\hat{P}_1$ such that $\hat{P} \Downarrow \hat{P}_1$ and $[\![\hat{P}_1]\!] \equiv P_1$, as we showed in the proof of Theorem 2. Process $\hat{P}_1$ is a prenex normal form, and we can assume w.l.o.g. that all its bound variables are renamed apart. The above claims show that:

$$S(P_1, Q) = S([\![\hat{P}_1]\!], [\![\hat{Q}]\!]) = S(\hat{P}_1, \hat{Q})$$

Since the translation reflects $\xrightarrow[\ell]{\infty}$ steps, we can apply rule (SUM) as follows:

$$\frac{\hat{P} \Downarrow \hat{P}_1 \quad S(\hat{P}_1, \hat{Q}) = r \neq 0 \quad \neg\exists\ell\exists Q'.\hat{P}_1 \xrightarrow[\ell]{\infty} Q'}{\hat{P} \xrightarrow{r} \hat{Q}}$$

Claim. The translation preserves and reflects immediate reactions:

1. if $P \xrightarrow{\infty(n)} P'$ then $[\![P]\!] \xrightarrow{\infty(n)} [\![P']\!]$
2. if $[\![\hat{P}]\!] \xrightarrow{\infty(n)} Q$ then exists $\hat{Q}$ such that $\hat{P} \xrightarrow{\infty(n)} \hat{Q}$ and $[\![\hat{Q}]\!] \equiv Q$.

We omit the proof of this claim. It concerns rule (COUNT), which can be treated quite similarly to rule (SUM) above. □

We can refine our translation such that types are preserved. This is necessary, since we need to translate type annotation on constants in λ-expressions and on newly created channels.

In order to do so, we assume that there exists a type constant R by which to type priorities $r \in R$ during translation. We refine the translation for restriction and output prefixes as follows:

$$\begin{aligned}[\![(\nu x{:}\tau)P]\!] &= (\nu x{:}[\![\tau]\!])[\![P]\!] \\ [\![x{:}r!\tilde{y}.P]\!] &= x[r_R]!\tilde{y}.[\![P]\!]\end{aligned}$$

Types of the π-calculus with priorities are translated to types of $\pi(\lambda(R)_<)$:

$$[\![ch(\tau_1,\ldots,\tau_n)]\!] = [R \to R] \Rightarrow ([\![\tau_1]\!],\ldots,[\![\tau_n]\!])$$

The translation can be lifted homomorphically to type environments $[\![\Gamma,\Delta]\!] = [\![\Gamma]\!],[\![\Delta]\!]$.

Proposition 8 (Type preservation). *Let P be a process of the π-calculus with priorities and Γ a type environment such that $\Gamma \vdash P$ then $[\![\Gamma]\!] \vdash [\![P]\!]$.*

The proof is straightforward by structural induction over type derivations.

5.2 Encoding π@ for Dynamic Compartments

We present an encoding of the π-calculus with polyadic synchronization and priorities π@ [18] into the attributed π-calculus, such that it inherits encodings of BioAmbients [13] and Brane calculus [14], in which different systems with dynamic compartment organizations can be defined. The correctness of the encoding of π@ can be shown only wrt. the non-deterministic semantics, since π@ lacks a stochastic semantics in terms of CTMCs (there only exists a stochastic simulation algorithm for sπ@ that can deal with dynamic compartments of variable volumes [22]).

We show how to encode π@ with priorities in an ordered set $(R,<)$ into the attributed π-calculus with priorities $\pi(\lambda(R,=)_<)$. This result shows that the attributed π-calculus with 3 levels of priorities inherits correct encodings of BioAmbients and Brane from π@.

The syntax of π@ is the same as for the π-calculus with priorities, except that communication now acts on nonempty tuples of channels and that priorities are assigned to both senders and receivers. This means that prefixes now have the following form, where $|\tilde{x}| \geq 1$:

$$\text{polyadic prefixes} \qquad \pi ::= \tilde{x}{:}r?\tilde{y} \mid \tilde{x}{:}r!\tilde{z}$$

The communication rule (COM) is adapted in such a way that tuples of channels and priorities are tested for equality before communication. Otherwise, the non-deterministic semantics of the π-calculus with priorities remains unchanged:

$$(\text{COM}_@)\ \frac{|\tilde{y}| = |\tilde{z}|}{\tilde{x}{:}r?\tilde{y}.P_1 + M_1 \mid \tilde{x}{:}r!\tilde{z}.P_2 + M_2 \xrightarrow[nd]{r} P_1[\tilde{z}/\tilde{y}] \mid P_2}$$

We decompose the encoding of $\pi@$ in two parts. The first part is a preprocessing step that rewrites all tuples in sending or receiving positions, such that they obtain the same arity. Given a process P of $\pi@$, let n be the maximal arity of tuples in subject position of polyadic prefixes and x a fresh channel name not occurring in P (which exists since *Vars* is infinite). Sending and receiving tuples in P are completed by x'es until they are of arity n:

$$(x_1, \ldots, x_m) \Rightarrow (x_1, \ldots, x_m, \underbrace{x, \ldots, x}_{n-m})$$

We recursively define functions eq_n that check equality of n-tuples of constants or variables (and thus channel names):

$$\begin{array}{l} eq_0() =_{df} \texttt{true} \\ eq_n(x_1, \ldots, x_n, y_1, \ldots, y_n) =_{df} \\ \qquad \texttt{if } x_1{=}y_1 \texttt{ then } eq_{n-1}(x_2 \ldots, x_n, y_2, \ldots, y_n) \texttt{ else false} \end{array}$$

Lemma 5. *For all constants or variables $v_1, \ldots, v_n, v'_1, \ldots, v'_n$ it is true that:*

1. $eq_n(v_1, \ldots, v_n, v_1, \ldots, v_n) \Downarrow \texttt{true}$.
2. $eq_n(v_1, \ldots, v_n, v'_1, \ldots, v'_n) \Downarrow \texttt{false}$ *if* $v_i \neq v'_i$ *for some* $1 \leq i \leq n$.

The proof is straightforward by induction on n. It relies on the definition of conditionals and equality in the big-step evaluator in Figs. 11 and 12. See Appendix B for details.

The main translation $[\![\_]\!] : \pi@ \to \pi(\lambda(R, =)_<)$ maps to an attributed π-calculus with additional constants for priorities and equality. Only priorities are successful values. We define encoding $[\![\_]\!]$ to be compositional, so that we have only to specify the encoding of communication prefixes. Here, we assume that all subject tuples have the same arity n.

$$\begin{array}{rl} [\![(x_1, \ldots, x_n){:}r!\tilde{z}.P]\!] = & x_1[\lambda y_2 \ldots \lambda y_n \lambda r'. \\ & \texttt{if } eq_n(x_2, \ldots, x_n, r, y_2, \ldots, y_n, r') \texttt{ then } r]!\tilde{z}.[\![P]\!] \\ [\![(y_1, \ldots, y_n){:}r'?\tilde{z}.P]\!] = & y_1[\lambda u.u\ y_2 \ldots y_n r']?\tilde{z}.[\![P]\!] \end{array}$$

A sender on channel tuple $(x_1, \ldots, x_n)$ with priority r is translated to a sender on a single channel x_1 and a constraint that checks equality with $(x_2 \ldots x_n, r)$ and if successful returns r. Symmetrically, we translate a receiver on channel tuple $(y_1, \ldots, y_n)$ with priority r' to a receiver on a single channel y_1 that receives the constraint and applies it to $(y_2, \ldots, y_n, r')$. Of course, we also need to translate all process definitions:

$$[\![A(\tilde{x}) \triangleq P]\!] = A(\tilde{x}) \triangleq [\![P]\!]$$

Theorem 4. *The encoding of $\pi@$ with priorities in $(R, <)$ to the attributed π-calculus with priorities $\pi(\lambda(R, =)_<)$ is correct, in that all reprocessed processes P and P' of $\pi@$ and attributed processes Q satisfy:*

(a) if $P \to Q$ *then* $[\![P]\!] \to [\![Q]\!]$

(b) if $[\![P]\!] \rightarrow Q$ *then exists* $\hat{Q}$ *such that* $Q \equiv [\![\hat{Q}]\!]$ *and* $P \rightarrow \hat{Q}$.

The proof is elaborated in Appendix B. It checks that communication steps correspond in both calculi, i.e., that polyadic synchronization of $\pi@$ is translated properly to equality testing in $\pi(\lambda(R,=)_{<})$. This mostly follows from Lemma 5 on the correctness of encoding equality of n-tuples.

Finally, notice that the encoding of $\pi@$ does not preserve types in any obvious sense. Finding a convincing type system for $\pi@$ is nontrivial, since there the capabilities of tuples and channels are overloaded while usual type system separate tuples and channel types properly.

5.3 Variants of the Stochastic Pi-Calculus

It remains to discuss the relationship to variants of the stochastic π-calculus where rates are annotated to channels.

BioSpi and spim. The syntax of BioSpi [12] and SPiM [21] differs from ours in that stochastic rates are annotated to channels at creation time, rather than to communication prefixes. The rates of the prefixes can then be deduced from the rate of the communicating channel.

The idea of encoding this variant of the stochastic π-calculus into $\pi(\lambda(\mathbb{R}_{+}^{\infty})_{<_2})$ is to replace channels x with rate r by pairs $\langle x, r\rangle$, that are decomposed at communication time. Here it is relevant that the attributed π-calculus permits pairs, and that it allows for expressions in sender and receiver positions.

We obtain the encoding below that we claim to be correct with respect to both semantics, non-deterministic and stochastic (without proof).

$$
\begin{aligned}
[\![(\nu x{:}r)P]\!] &= (\nu x)[\![P]\!][\langle x,r\rangle/x] \\
[\![x?\tilde{y}.P]\!] &= (\texttt{fst}\ x)[\lambda z.z]?\tilde{y}.[\![P]\!] \\
[\![x!\tilde{y}.P]\!] &= (\texttt{fst}\ x)[\texttt{snd}\ x]?\tilde{y}.[\![P]\!]
\end{aligned}
$$

The first line states, how to remove the annotation r of a channel x, and to substitute all occurrences of x by the $\langle x, r\rangle$. The next two lines state that the channel is extracted from the pair before communication. In the third line, the rate is extracted in the communication constraint.

Stochastic Pi-Calculus with Concurrent Objects. The next more expressive language is SPiCO, the stochastic π-calculus for concurrent objects [20]. It supports a static form of polyadic synchronization, called pattern guarded inputs.

Patterns are tuples $a(\tilde{y})$ that are built from a finite set of function symbols a in some set Σ and a sequence of channels. Senders send tuples $b(\tilde{z})$ to receivers, which match it against a pattern $a(\tilde{y})$. A communication step is allowed only if the function symbol b of the tuple sent matches the function symbol a of the receiving pattern:

$$x?a(\tilde{y}).P \mid x!b(\tilde{z}).P' \rightarrow P[\tilde{z}/\tilde{y}] \mid P' \text{ if } a = b$$

The communication constraint is thus equality $a{=}b$. This is a weak form of polyadic synchronization, since the sending and receiving channels must be checked for equality too. As before, stochastic rates are annotated to channels at creation time. We can encode SPiCO into $\pi(\lambda(\mathbb{R}_+^\infty, \Sigma, =))_{<_2}$ similarly as before, where $a, b \in \Sigma$:

$$\begin{aligned}
[\![(\nu x{:}r)P]\!] &= (\nu x)[\![P]\!][\langle x, r\rangle/x] \\
[\![x?a(\tilde{y}).P]\!] &= (\texttt{fst}\ x)[\lambda z.\texttt{if}\ z{=}a\ \texttt{then}\ (\texttt{snd}\ x)]?\tilde{y}.[\![P]\!] \\
[\![x!b(\tilde{y}).P]\!] &= (\texttt{fst}\ x)[b]?\tilde{y}.[\![P]\!]
\end{aligned}$$

The only new aspect here is that we have to check the communication constraint $a{=}b$ in addition.

Original Attributed Pi-Calculus. Finally, in the conference version of the attributed π-calculus at CMSB'08 [25], we annotated stochastic rates to channels, and used a fixed function `val` mapping channels to their rates. This version of $\pi(\mathcal{L})$ can be encoded into the version of $\pi(\mathcal{L})$ presented here:

$$\begin{array}{lll}
[\![(\nu x{:}v)P]\!] = (\nu x)[\![P[\langle x, v\rangle/x]]\!] & [\![\texttt{val}]\!] = \texttt{snd} & [\![x]\!] = x \\
[\![v[e]?\tilde{v}.P]\!] = (\texttt{fst}\ [\![v]\!])[[\![e]\!]]?[\![\tilde{v}]\!].[\![P]\!] & [\![\lambda x.e]\!] = \lambda x.[\![e]\!] & \\
[\![v[e]!\tilde{y}.P]\!] = (\texttt{fst}\ [\![v]\!])[[\![e]\!]]!\tilde{y}.[\![P]\!] & [\![e\ e']\!] = [\![e]\!][\![e']\!] &
\end{array}$$

6 Stochastic Simulator

We develop a stochastic simulation algorithm that closely follows the stochastic semantics of the attributed π-calculus in terms of CTMCs. Thereby, we show that a simulator for $\pi(\mathcal{L})$ can be obtained independently of the choice of $\mathcal{L}$ by extending previous simulators for the stochastic π-calculus or SPiCO [11,10,12].

The stochastic semantics for $\pi(\mathcal{L})$ induces the naive stochastic simulator given in Fig. 26. A simulator's input comprises a process P and a time point $t \in \mathbb{R}$. The next reduction step for process P is chosen in a memoryless stochastic manner. The sojourn time $\Delta \in \mathbb{R}_+$ of P is inferred, and the simulator proceeds with the resulting solution at time point $t{+}\Delta$. This loop continues until no next reduction step can be found, in fact it may run for ever, if not interrupted externally or equipped with some additional termination condition.

The first step of the simulation algorithm is to apply definitions of P exhaustively. This computation may run into an infinite loop or raise errors, in case of non well-founded definitions or if the evaluation of some expressions diverges ($\neg\exists v.e \Downarrow v$). If application raises an immediate error $P_1 \xrightarrow[nd]{err} \bot$ by rules (E.COM), (E.PREF), or (E.CONSTR), then the simulator throws an exception (which kills its continuation). Note that error checking may run into infinite loops or raise an error too. If P does converge to an error-free process P_1 then P_1 is uniquely determined up to structural congruence (Proposition 4) and must be congruent to some prenex form $(\nu\tilde{x}) \prod_{i=0}^{n} M_i$. The remainder of the algorithm is independent of the concrete representative of congruence class $[P_1]_\equiv$, so that we can chose

```
Simulate-naive(P,t)
  // process P, time point t ∈ ℝ
  let P1 such that P ⇓ P1
      // P1 is obtained from P by exhaustively applying definitions.
      // This computation may diverge.
  if P1 --err-->_nd ⊥ then raise error
     // Apply all rules (E.COM), (E.PREF), (E.CONSTR).
     // This computation may diverge since expressions are to be evaluated.
  let Reacts = {(ℓ,r) ∈ ℕ^4 × ℝ^∞_+} | ∃P2. P1 --r-->_ℓ P2}  // (COM_ℓ)
  if Reacts ∩ (ℕ4 × {∞}) = ∅
  then
     let ((ℓ,r),Δ) = Gillespie(Reacts)  // (SUM)
     let P2 such that P1 --r-->_ℓ P2
     Simulate-naive(P2,t + Δ)
  else
     select (ℓ,∞) ∈ Reacts with equal probability //(COUNT)
     let P2 such that P1 --r-->_ℓ P2
     Simulate-naive(P2,t)
```

Fig. 26. Naive simulator interpreting the stochastic semantics

this representative arbitrarily. The next step is to compute the set of all labeled reactions of P_1:

$$Reacts = \{(\ell, r) \in \mathbb{N}^4 \times \mathbb{R}^{\infty}_{+} \mid \exists P_2.\ P_1 \xrightarrow[\ell]{r} P_2\}$$

Labeled reactions with rate $r = \infty$ are executed with priority and without time consumption. If no reaction with rate $r = \infty$ exists, we apply Gillespie's algorithm [45] to select a reaction $(\ell, r) \in Reacts$ with probability r/s where $s = \sum_{(\ell,r') \in Reacts} r'$. The sojourn time in P is $\Delta = -ln(1/U)/s$ for some uniformly distributed random number $0 < U \leq 1$.

In order to compute *Reacts*, we have to enumerate all possible instances of the communication rule (COM$_\ell$). This requires to evaluate all evaluation constraints, by applying the evaluation algorithm of the attribute language $\mathcal{L}$.

Most fortunately, the CTMC itself does not need to be computed by the simulation algorithm. This would be largely unfeasible, since the number of possible outcomes of non-deterministic interactions may grow exponentially. Furthermore, it would require to decide structural congruence (rules (SUM) and (COUNT)), which is a graph isomorphism complete problem [46].

In order to increase efficiency of the naive simulation algorithm, we apply an idea exploited already in the BioSpi implementation [12]. The objective is to avoid the enumeration of all pairs of alternatives (and thus redexes), since there may be quadratically many in the size of P_1. The strategy is to *group* all reactions on the same channel with the same rate. We apply Gillespie's algorithm to such *grouped reactions* and then choose a specific interaction with equal distribution.

In order to identify grouped reactions, we introduce group labels. A *group label* of a process P_1 is a triple in $fv(P_1) \times Vals(P_1)^2$. The group of reactions for $P_1 = \prod_{i=1}^{n} \sum_{j=1}^{m} \pi_i^j.P_i^j$ with label $L = (x, v, r)$ is defined as follows:

$$\begin{aligned} Reacts(L) = \{((i_1, j_1, i_2, j_2), r) \in Reacts \mid \\ \exists v' \exists \tilde{y} \exists \tilde{v}. \pi_{i_1}^{j_1} \Downarrow x[v]!\tilde{y},\ \pi_{i_2}^{j_2} \Downarrow x[v']?\tilde{v},\ v'v \Downarrow r\} \end{aligned}$$

A triple L identifies reaction groups by a communication channel x, a constraint value of senders v, and a rate r yielding the application of the receivers λ-abstraction v' to v.

The stochastic rate for a grouping label L is usually called propensity $prop(L) \in \mathbb{R}^+ \uplus \{\infty(n) \mid n \in \mathbb{N}\}$. It sums up all rates of the labeled reactions that are grouped together, or counts the number of labels of infinite rate reactions if there are any:

$$prop(L) = \begin{cases} \infty(n) & \text{if } n = \#\{\ell \mid (\ell, \infty) \in Reacts(L)\} \geq 1 \\ \sum_{(\ell,r) \in Reacts(L)} r & \text{otherwise} \end{cases}$$

We define the set of grouped reactions with their propensities as follows. These will be used as input of the Gillespie's algorithm:

$$GReacts = \{(L, prop(L)) \mid L \in Vars(P_1) \times Vals(P_1)^2\}$$

The cardinality of *GReacts* is linear in the size of P_1. In many practically relevant cases, only a fixed number of values will ever be used. This is e.g. the case in our example models of Euglena's movement and cooperative enhancement, where none of the processes succeeding the initial solution introduces new channels or new constraint values, see Section 4. By contrast, the cardinality of set *Reacts* becomes quadratic in the size of P_1, e.g., if all senders and receivers may interact.

Fig. 27 gives a simulation algorithm based on grouped reactions. In contrast to the naive simulator, it first selects a grouped reaction based on Gillespie's algorithm, and then a label of a reaction within this group with equal distribution.

What remains is to compute the propensities of all labels of grouped reactions in a process P_1. These can be derived from the values below if $P_1 = \prod_{i=1}^{n} \sum_{j=1}^{m} \pi_i^j.P_i^j$:

$$\begin{aligned} out(x, v) &= \#\{(i, j) \mid \exists \tilde{v} : \ \pi_i^j \Downarrow x[v]!\tilde{v}\} \\ in(x, v, r) &= \#\{(i, j) \mid \exists v' \exists \tilde{y} : \pi_i^j \Downarrow x[v']?\tilde{y},\ v'v \Downarrow r\} \\ mixin(x, v, r) &= \#\{(i, j_1, j_2) \mid \exists v' \exists \tilde{v} \exists \tilde{y} : \ \pi_i^j \Downarrow x[v]!\tilde{v},\ \pi_i^j \Downarrow x[v']?\tilde{y},\ v'v \Downarrow r\} \end{aligned}$$

Lemma 6. *$prop(x, v, r) = (out(x, v) * in(x, v, r) - mixin(x, v, r)) * r$, if the solution does not contain infinite rates.*

```
Simulate(P,t) //solution P, time point t ∈ ℝ
  let P₁ be such that P ⇓ P₁
      // P₁ is obtained from P by exhaustively applying definitions.
      // This computation may diverge.
  if P₁ --err/nd--> ⊥ then raise error
      // Apply all rules (E.COM), (E.PREF), (E.CONSTR).
      // This computation may diverge since expressions are to be evaluated.
  let GReacts = {(L, prop(L)) | L ∈ Vars(P₁) × Vals(P₁)²}
  if {(L,r) ∈ GReacts | r = ∞(n)} = ∅
  then
     let ((L,r),Δ) = Gillespie(GReacts)
     select (ℓ,r) ∈ Reacts(L) with equal probability
     let P₂ such that P₁ --r/ℓ--> P₂
     Simulate(P₂,t+Δ)
  else
     select (L,∞(n)) ∈ GReacts
        with probability n/m where m = Σ_{(L',∞(n'))∈GReacts} n'
     select (ℓ,∞) ∈ Reacts(L) with equal probability
     let P₂ such that P₁ --∞/ℓ--> P₂
     Simulate(P₂,t)
```

Fig. 27. Stochastic simulator for $\pi(\mathcal{L})$ (to be implemented incrementally)

Proof. Let $L = (x, v, r)$. It is enough to show that $out(x, v) * in(x, v, r) - mixin(x, v, r) = \#Reacts(L)$. This follows, since all pair of indices counted by $out(x, v) * in(x, v, r)$ form a redex according to rule (COM) except for those that are counted by $mixin(x, v, r)$.

The computation of mixins can still produce an output of quadratic size and thus need quadratic time. The square factor, however, is in the maximal number of alternatives in sums defining molecules of P_1, which will be small in practice. All other needed values can be computed in linear time in the size of P_1, when ignoring the time for evaluating expressions, which is justified in many practical cases.

The final step toward an efficient simulator consists in computing the propensities $prop(x, v, r)$ incrementally, so that they do not need to be recomputed from scratch in every reduction step. This can be based on Lemma 6, since the values of $out(x, v)$, $in(x, v, r)$, $mixin(x, v, r)$ can be updated incrementally, when adding new solutions or canceling alternative choices by communication.

7 Implementation and Performance Evaluation

We discuss our implementation of the stochastic simulator for $\pi(\mathcal{L})$, and present some experimental results in order to give an impression of its performance.

We implemented the $\pi(\mathcal{L})$ simulator on top of an existing simulator for the stochastic π-calculus in the modeling and simulation framework JAMES II [29]. The implementation is freely available[2]. We deployed a two layer approach: the base layer is the simulator of the stochastic π-calculus along the lines of [21], i.e. for each communication channel the propensity is calculated under consideration of the corresponding senders and receivers. The obtained propensities are passed to a Stochastic Simulation Algorithm (SSA) that determines the next communication to perform and the sojourn time. There are three alternative SSAs that can be freely chosen, the First Reaction Method (the original version), the Direct Reaction Method [47], and the Next Reaction Method [48]. The top layer implements the grouping as explained in Section 6, i.e. it groups the communication pairs in a solution by the combinations of channels and rate constants resulting from the application of receiver abstractions to sender arguments. For each group it creates a communication channel and assigns the rate constant and the corresponding senders and receivers to it. The set of thus obtained communication channels is passed to the base layer in order to determine the following solution.

In our performance experiments, we compare the simulators for $\pi(\mathcal{L})$ and the stochastic π-calculus based on JAMES II, with the stochastic Pi Machine (SPiM) by Phillips and Cardelli [11]. We only consider the Direct Reaction Method, since this is the SSA chosen by SPiM. Our experiments are performed on a WindowsXP machine, with an Intel Core 2 Duo 2.00 GHz processor and 2 GB RAM providing a SciMark 2.0 Java benchmark score [49] of 383.9 Mflops. Notice, that there exists a faster version of SPiM for Linux based on native code compilation, an aspect that is irrelevant for our comparison.

As test example, we use the Euglena model from Section 4.1, since it allows us to gradually raise the number of grouped reactions and process definitions by increasing the number of depth levels. Furthermore, it can be implemented in both the stochastic π-calculus and $\pi(\mathcal{L})$. Our Euglena benchmark model comprises two light source with intensity rates $\mathtt{I}_1 = 5.0$ and $\mathtt{I}_2 = 15.0$ and 100 Euglenas on each depth level ($\mathtt{n} = 100$). The rate of Euglena's upwards motion equals $\mathtt{u} = 2.0$ and the water opacity is set to $\sigma = 0.2$. Among the experiments, we gradually increased the number of depth levels from 10 to 100 by steps of 10, i.e. $\mathtt{m} \in \{9, 19, \ldots, 99\}$. Implementations in the stochastic π-calculus are obtained by enumerating the Euglena processes for different depth levels in the same way as shown in Section 4.1. To ensure comparability, we used two $\pi(\mathcal{L})$ implementations for each experiment, one enumerating the depth levels as in the stochastic π-calculus (enum) and one in the more compact form with the depth level as a parameter of Euglena (comp). We measured the time needed to simulate until time point 100.0, see Appendix A. For each experiment, we averaged over three simulation runs with small deviations resulting from both the stochastic nature of the simulation and the work load of the machine. The results of the experiment sets are shown in Figs. 28. The implementations are

[2] See the James-Imp-Pi web page at `http://biopi-lille-ros.gforge.inria.fr`.

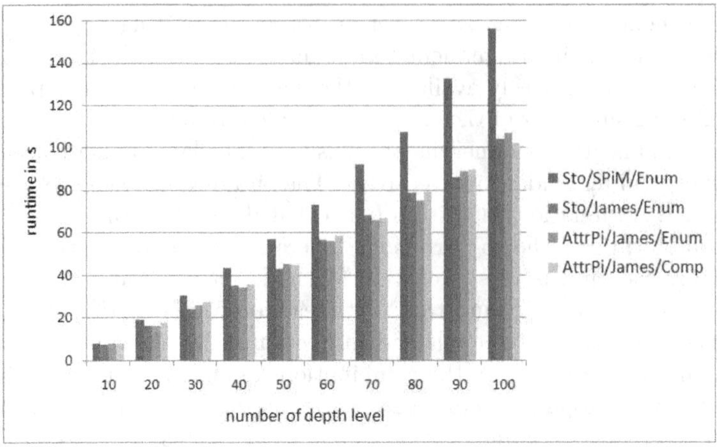

Fig. 28. Runtime of different simulators in s for the Euglena model with two light sources with intensity rates $\mathtt{I}_1 = 5.0, \mathtt{I}_2 = 15.0$, 100 Euglenas on each depth level ($\mathtt{n} = 100$), an opacity $\sigma = 0.2$, a rate for upward motion $\mathtt{u} = 2.0$, and different numbers of depth levels ranging from 10 to 100, i.e. $\mathtt{m} \in \{9, 19, \ldots, 99\}$. Simulation runs were performed until simulation time $t = 100.0$: Sto = stochastic π-calculus, Attr = attributed π-calculus, SPiM = SPiM, James = JamesII, Enum = model with enumerated depth levels, Comp = model with depth level as species parameter.

labeled according to the used formalism, *Sto* for the stochastic π-calculus or *attr* for $\pi(\mathcal{L})$, the tool, SPiM or JAMES II, and the implementation, *Enum* or *Comp*.

The results in Fig. 28 show a general increase of simulation time with a rising number of depth levels. Presumably, due to our choice of operating system, SPiM performs slower. All other implementations need similar amounts of time. The maximal simulation time required is around $160s$. Our results indicate that the computational complexity of the $\pi(\mathcal{L})$ simulator is moderate.

8 Conclusion and Outlook

We presented the attributed π-calculus in order to define attribute processes with interaction constraints depending on attribute values. We used the call-by-value λ-calculus as a sequential language in which to define data values and constraints for concurrent interactions.

The attributed π-calculus forms a unifying framework, which extends on the one hand the π-calculus with priorities and its extension by polyadic synchronization π@, and, on the other hand, the stochastic π-calculus, and its extension by concurrent objects SPiCO. This allows us to express existing π-calculus models of biological systems in the attributed π-calculus. Furthermore, the attributed π-calculus permits to model spatial aspects in systems biology depending on numeric attributes, and dynamic compartments with various nesting structures such as in BioAmbients and Brane.

Due to this combination of a concurrent process language, i.e. the π-calculus, and a sequential core language, i.e. the λ-calculus, the modeling of biological systems becomes possible on multiple levels. In addition to the traditional individual-based modeling of the stochastic π-calculus which maps molecules to objects, the attributed π-calculus enables a population-based modeling style. In combining both levels within one model priorities play a key role. Thereby, n-ary chemical reactions with $n > 2$ reactions and diverse kinetics can be realized. E.g. in a recent model of the Wnt-pathway, which is aimed at analyzing the cytoplasmic-nuclear shuttling of β-catenin, Michaelis-Menten dynamics that are defined at population level interact with other processes described at individual-based level. Thereby the objective of the simulation study and data availability are taken into account [51].

We have presented and implemented a stochastic simulator for the attributed π-calculus, which shows that this general approach is feasible in practice with moderate performance.

The imperative π-calculus is a recent extension of the attributed π-calculus with a global imperative store [44]. The original motivation was to enable a simpler encoding of dynamic compartments, that does not rely on priorities. Note that a global imperative store is equally supported by sCCP [16], but in sCCP all interactions are forced to pass through the global store.

Quite some questions remain for future research. The first question refers to the precise relationship of individual-based modeling in the attributed π-calculus to rule-based modeling in the kappa calculus [52,7] or Bigraphs [9]. Independently of population-based modeling, it would be interesting to know whether sCCP can be expressed by the attributed π-calculus, and similarly, independently of species-based modeling, it would be of interest to know whether Bio-PEPA can be encoded into the attributed π-calculus. There is also the question, whether there exists a fragment of $\pi(\lambda(R,=)_<)$ that corresponds precisely to π@ with priorities in $(R,<)$. Finally, there is an open issue to develop and implement object-oriented abstractions with inheritance for attributed π-calculus, in the spirit of SPiCO [10].

Acknowledgements

We thank Francois Lemaire for helping us to compute the equilibrium level for our Euglena models, Stefan Rybacki for his support in implementing the attributed π-calculus, and Céline Kuttler for having initiated the present cooperation between Lille and Rostock. The anonymous reviewers send us valuable comments, that helped us improving the article.

This research has been supported by the ANR jeune project BioSpace (2009-2012) of the national French research founding agency and as part of the Research Training School "dIEM oSiRiS" (2006-2011) of the Deutsche Forschungsgemeinschaft (DFG).

References

1. Hillston, J.: Process algebras for quantitative analysis. In: Proceedings of 20th IEEE Symposium on Logic in Computer Science (LICS 2005), Chicago, IL, USA, June 26-29, pp. 239–248. IEEE Comp. Soc. Press, Los Alamitos (2005)
2. Cardelli, L.: On process rate semantics. Theoretical Computer Science 391, 190–215 (2008)
3. Chabrier-Rivier, N., Fages, F., Soliman, S.: The Biochemical Abstract Machine BIOCHAM. In: Computational Methods in Systems Biology, pp. 172–191 (2004)
4. Regev, A.: Computational Systems Biology: A Calculus for Biomolecular Knowledge. Tel Aviv University, PhD thesis (2003)
5. Regev, A., Shapiro, E.: Cells as Computation. Nature 419, 343 (2002)
6. Gilbert, D., Heiner, M., Lehrack, S.: A unifying framework for modelling and analysing biochemical pathways using petri nets. In: Calder, M., Gilmore, S. (eds.) CMSB 2007. LNCS (LNBI), vol. 4695, pp. 200–216. Springer, Heidelberg (2007)
7. Danos, V., Feret, J., Fontana, W., Harmer, R., Krivine, J.: Rule-based modelling of cellular signalling. In: Caires, L., Vasconcelos, V.T. (eds.) CONCUR 2007. LNCS, vol. 4703, pp. 17–41. Springer, Heidelberg (2007)
8. Faeder, J.R., Blinov, M.L., Goldstein, B., Hlavacek, W.S.: Rule-Based Modeling of Biochemical Networks. Complexity 10, 22–41 (2005)
9. Krivine, J., Milner, R., Troina, A.: Stochastic bigraphs. In: 24th Conference on the Mathematical Foundations of Programming Semantics. Electronical notes in theoretical computer science, vol. 218, pp. 73–96. Elsevier, Amsterdam (2008)
10. Kuttler, C., Lhoussaine, C., Niehren, J.: A stochastic pi calculus for concurrent objects. In: Anai, H., Horimoto, K., Kutsia, T. (eds.) AB 2007. LNCS, vol. 4545, pp. 232–246. Springer, Heidelberg (2007)
11. Phillips, A., Cardelli, L.: Efficient, correct simulation of biological processes in the stochastic pi-calculus. In: Calder, M., Gilmore, S. (eds.) CMSB 2007. LNCS (LNBI), vol. 4695, pp. 184–199. Springer, Heidelberg (2007)
12. Priami, C., Regev, A., Shapiro, E., Silverman, W.: Application of a Stochastic Name-Passing Calculus to Representation and Simulation of Molecular Processes. Information Processing Letters 80, 25–31 (2001)
13. Regev, A., Panina, E.M., Silverman, W., Cardelli, L., Shapiro, E.: BioAmbients: An Abstraction for Biological Compartments. TCS 325, 141–167 (2004)
14. Cardelli, L.: Brane calculi. In: Danos, V., Schachter, V. (eds.) CMSB 2004. LNCS (LNBI), vol. 3082, pp. 257–278. Springer, Heidelberg (2005)
15. Ciocchetta, F., Hillston, J.: Bio-PEPA: An Extension of the Process Algebra PEPA for Biochemical Networks. ENTCS 194, 103–117 (2008)
16. Bortolussi, L., Policriti, A.: Modeling biological systems in stochastic concurrent constraint programming. Constraints, an International Journal 13, 66–90 (2008)
17. Carbone, M., Maffeis, S.: On the expressive power of polyadic synchronisation in pi-calculus. Nordic Journal of Computing 10, 70–98 (2003)
18. Versari, C.: A Core Calculus for a Comparative Analysis of Bio-inspired Calculi. In: Programming Languages and Systems, pp. 411–425 (2007)
19. Priami, C.: Stochastic π-calculus. Computer Journal 6, 578–589 (1995)
20. Kuttler, C., Lhoussaine, C., Niehren, J.: A stochastic pi calculus for concurrent objects. In: 1st International Workshop on Probabilistic Automata and Logics (2006)
21. Phillips, A., Cardelli, L.: A correct abstract machine for the stochastic pi-calculus. In: Proceedings of BioConcur 2004 (2004)

22. Versari, C., Busi, N.: Stochastic simulation of biological systems with dynamical compartment structure. In: Calder, M., Gilmore, S. (eds.) CMSB 2007. LNCS (LNBI), vol. 4695, pp. 80–95. Springer, Heidelberg (2007)
23. Jaffar, J., Lassez, J.L.: Constraint Logic Programming. In: POPL 1987: Proceedings of the 14th ACM SIGACT-SIGPLAN Symposium on Principles of Programming Languages, pp. 111–119. ACM, New York (1987)
24. Saraswat, V.A., Rinard, M.C.: Concurrent constraint programming. In: ACM SICPLAN-SIGACT Symposium on Principles of Programming Languages, pp. 232–245. ACM Press, New York (1990)
25. John, M., Lhoussaine, C., Niehren, J., Uhrmacher, A.: The attributed pi calculus. In: Heiner, M., Uhrmacher, A.M. (eds.) CMSB 2008. LNCS (LNBI), vol. 5307, pp. 83–102. Springer, Heidelberg (2008)
26. Kuttler, C., Niehren, J.: Gene regulation in the pi calculus: Simulating cooperativity at the lambda switch. Transactions on Computational Systems Biology, 24–55 (2006)
27. Kuttler, C.: Modeling Bacterial Gene Expression in a Stochastic Pi Calculus with Concurrent Objects. PhD thesis, Université des Sciences et Technologies de Lille - Lille 1 (2007)
28. Versari, C.: A Core Calculus for the Analysis and Implementation of Biologically Inspired Languages. PhD thesis, University of Bologna (2009)
29. Himmelspach, J., Uhrmacher, A.M.: Plug'n Simulate. In: ANSS 2007: Proceedings of the 40th Annual Simulation Symposium, Washington, DC, USA, pp. 137–143. IEEE Computer Society, Los Alamitos (2007)
30. Baldamus, M., Parrow, J., Victor, B.: A fully abstract encoding of the pi-calculus with data terms. In: Caires, L., Italiano, G.F., Monteiro, L., Palamidessi, C., Yung, M. (eds.) ICALP 2005. LNCS, vol. 3580, pp. 1202–1213. Springer, Heidelberg (2005)
31. Johansson, M., Parrow, J., Victor, B., Bengtson, J.: Extended pi-calculi. In: Aceto, L., Damgård, I., Goldberg, L.A., Halldórsson, M.M., Ingólfsdóttir, A., Walukiewicz, I. (eds.) ICALP 2008, Part II. LNCS, vol. 5126, pp. 87–98. Springer, Heidelberg (2008)
32. Guerriero, M.L., Priami, C., Romanel, A.: Modeling static biological compartments with beta-binders. In: Anai, H., Horimoto, K., Kutsia, T. (eds.) AB 2007. LNCS, vol. 4545, pp. 247–261. Springer, Heidelberg (2007)
33. Priami, C., Quaglia, P., Romanel, A.: Blenx static and dynamic semantics. In: Bravetti, M., Zavattaro, G. (eds.) CONCUR 2009. LNCS, vol. 5710, pp. 37–52. Springer, Heidelberg (2009)
34. Maurin, M., Magnin, M., Roux, O.H.: Modeling of genetic regulatory network in stochastic pi-calculus. In: Rajasekaran, S. (ed.) BICoB 2009. LNCS (LNBI), vol. 5462, pp. 282–294. Springer, Heidelberg (2009)
35. Lecca, P.: Stochastic pi-calculus models of the molecular bases of parkinson's disease. In: International Conference on Bioinformatics and Computational Biology, pp. 298–304 (2008)
36. Niehren, J.: Uniform confluence in concurrent computation. Journal of Functional Programming 10, 453–499 (2000)
37. Huet, G.P.: Confluent reductions: Abstract properties and applications to term rewriting systems. Journal of the ACM 27, 797–821 (1980)
38. Kuttler, C., Lhoussaine, C., Nebut, M.: Rule-based modeling of transcriptional attenuation at the tryptophan operon. In: Transactions on Computational Systems Biology (2009)

39. Tait, W.W.: Intensional interpretations of functionals of finite type i. Journal of Symbolic Logic 32, 198–212 (1967)
40. Mitchell, J.C.: Foundations for Programming Languages. MIT Press, Cambridge (1996)
41. John, M., Ewald, R., Uhrmacher, A.M.: A Spatial Extension to the Pi Calculus. ENTCS 194, 133–148 (2008)
42. Kholodenko, B.N.: Cell-Signalling Dynamics in Time and Space. Nature Reviews Molecular Cell Biology 7, 165–176 (2006)
43. Grell, K.G.: Protozoologie. Springer, Heidelberg (1968)
44. John, M., Lhoussaine, C., Niehren, J.: Dynamic compartments in the imperative pi calculus. In: Degano, P., Gorrieri, R. (eds.) CMSB 2009. LNCS (LNBI), vol. 5688, pp. 235–250. Springer, Heidelberg (2009)
45. Gillespie, D.T.: A General Method for Numerically Simulating the Stochastic Time Evolution of Coupled Chemical Reactions. Journal of Computational Physics 22, 403–434 (1976)
46. Khomenko, V., Meyer, R.: Checking pi-calculus structural congruence is graph isomorphism complete. Technical Report CS-TR: 1100, School of Computing Science, Newcastle University, 20 pages (2008)
47. Gillespie, D.T.: Exact stochastic simulation of coupled chemical reactions. Journal of Physical Chemistry 81, 2340–2361 (1977)
48. Gibson, M.A., Bruck, J.: Efficient exact stochastic simulation of chemical systems with many species and many channels. J. Phys. Chem. 104, 1876–1889 (2000)
49. Pozo, R., Miller, B.: SciMark 2.0 (2009), `http://math.nist.gov/scimark2/`
50. Degenring, D., Roehl, M., Uhrmacher, A.: Discrete event, multi-level simulation of metabolite channeling. BioSystems 1-3, 29–41 (2004)
51. Mazemondet, O., John, M., Maus, C., Uhrmacher, A.M., Rolfs, A.: Integrating diverse reaction types into stochastic models - a signaling pathway case study in the imperative pi-calculus. In: Rossetti, M.D., Hill, R.R., Johansson, B., Dunkin, A., Ingalls, R.G. (eds.) Proceedings of the Winter Simulation Conference (to appear)
52. Danos, V., Laneve, C.: Formal molecular biology. Theoretical Computer Science 325, 69–110 (2004)

A Experiment Results

Table 1. Runtime of different simulators in s for the Euglena model with two light sources with intensity rates $\mathtt{I}_1 = 5.0, \mathtt{I}_2 = 15.0$, 100 Euglenas on each depth level ($\mathtt{n} = 100$), an opacity $\sigma = 0.2$, a rate for upward motion $\mathtt{u} = 1.0$, and different numbers of depth levels ranging from 10 to 100, i.e. $\mathtt{m} \in \{9, 19, \ldots, 99\}$. Simulation runs were performed until simulation time $t = 100.0$: Sto = stochastic π-calculus, Attr = attributed π-calculus, SPiM = SPiM, James = JAMES II, Enum = model with enumerated depth levels, Comp = model with depth level as species parameter. Standard deviation in parentheses.

Levels	Sto/SPiM/Enum	Sto/James/Enum	AttrPi/James/Enum	AttrPi/James/Comp
10	8.10 (0.09)	7.33 (0.58)	8.00 (0.00)	8.00 (0.00)
20	19.12 (0.12)	16.00 (0.00)	16.33 (0.58)	17.33 (0.58)
30	30.42 (0.12)	24.00 (1.00)	25.67 (0.58)	27.00 (0.00)
40	43.37 (0.13)	35.00 (1.00)	34.33 (0.58)	35.33 (0.58)
50	56.99 (0.18)	43.00 (1.73)	45.00 (2.00)	44.67 (2.00)
60	73.40 (0.20)	56.67 (1.53)	56.33 (0.58)	58.33 (1.16)
70	92.32 (0.34)	68.00 (1.73)	66.00 (0.00)	67.00 (3.00)
80	107.41 (0.16)	78.67 (2.31)	75.00 (2.65)	79.33 (1.53)
90	132.77 (0.26)	86.33 (2.52)	89.00 (1.73)	89.67 (1.53)
100	156.13 (2.16)	104.00 (2.65)	106.67 (1.15)	102.00 (1.00)

B Remaining Proofs

B.1 Subsection 3.7 (Type System)

Corollary 2 (Error freeness). *If $\mathcal{L}$ is an attribute language that is both type safe and normalizing (see $\mathcal{L}$ Propositions* 6 *and* 7*) then $\pi(\mathcal{L})$ is error free in that if $\Gamma \vdash P$ and $P \rightarrow^* Q$ then $\neg Q \xrightarrow[nd]{err} \bot$.*

Proof. Assuming that $\Gamma \vdash P$ and $P \rightarrow^n Q$ the proof proceeds by induction on n. The inductive step follows from Theorem 1; it thus remains to prove the initial case that is $P \equiv Q$. We proceed by contradiction, assuming that there exists some process P_0 such that $\Gamma \vdash P_0$ and $P_0 \xrightarrow[nd]{err} \bot$.

Again, a standard analysis of the structural congruence shows the following claim: $P_0 \equiv (\nu\tilde{x}{:}\tilde{\tau}) \prod_{i=1}^{n} P_i$ be a prenex normal form in which all bound variables are renamed apart, and such that all P_i are sums or defined processes. A derivation of $P_0 \xrightarrow[nd]{err} \bot$ necessarily involves one the error axioms given in Figure 14. We now show that neither of them is applicable by case analysis.

(E.COM) In that case, $\exists j, k. 1 \leq j < k \leq n$, $P_j = \pi_1.P_1 + M_1$, $P_k = \pi_2.Q_2 + M_2$, $\pi_1 \Downarrow x[v_1]?\tilde{y}$, $\pi_2 \Downarrow x[v_2]!\tilde{v}$ and $|\tilde{y}| \neq |\tilde{v}|$. From $\Gamma \vdash P_0$, by the lemma 3(4), we have $\Gamma \vdash (\nu\tilde{x}{:}\tilde{\tau}) \prod_{i=1}^{n} P_i$ that derives from a series of applications of rules (T.NEW) and (T.PAR) and from statements $\Gamma, \tilde{x}{:}\tilde{\tau} \vdash P_i$ for all $i \in \{1, \ldots, n\}$. In particular,

$\Gamma, \tilde{x}{:}\tilde{\tau} \vdash \pi_1.Q_1 + M_1$, and $\Gamma, \tilde{x}{:}\tilde{\tau} \vdash \pi_2.Q_2 + M_2$. By (T.REC), $\pi_1 = e_1[e_1']?\tilde{y}$ and $\Gamma, \tilde{x}{:}\tilde{\tau} \vdash e_1{:}[\tau_1] \Rightarrow \tilde{\sigma_1}$, $\Gamma, \tilde{x}{:}\tilde{\tau} \vdash e_1'{:}\tau_1$ and $|\tilde{y}| = |\tilde{\sigma_1}|$. Similarly, $\pi_2 = e_2[e_2']!\tilde{e_2''}$ and $\Gamma, \tilde{x}{:}\tilde{\tau} \vdash e_2{:}[\tau_2 \rightarrow \tau_2'] \Rightarrow \tilde{\sigma_2}$ and $\Gamma, \tilde{x}{:}\tilde{\tau} \vdash e_2'{:}\tau_2$ and $\Gamma, \tilde{x}{:}\tilde{\tau} \vdash \tilde{e_2''}{:}\tilde{\sigma_2}$. Because $e_1[e_1']?\tilde{y} \Downarrow x[v_1]?\tilde{y}$ and $e_2[e_2']!\tilde{e_2''} \Downarrow x[v_2]!\tilde{v}$ and, by Proposition 6, e_1 and e_2 have the same type, that is $\tau_1 = \tau_2 \rightarrow \tau_2'$ and $\tilde{\sigma_1} = \tilde{\sigma_2}$. Because, $\tilde{e_2''} \Downarrow \tilde{v}$, by Proposition 6, $\Gamma, \tilde{x}{:}\tilde{\tau} \vdash \tilde{v} : \tilde{\sigma_2}$, therefore $|\tilde{v}| = |\tilde{\sigma_2}| = |\tilde{\sigma_1}| = |\tilde{y}|$ which contradicts $|\tilde{y}| \neq |\tilde{v}|$.

(E.PREF) In this case, $\exists j.1 \leq j \leq n$, $P_j = \pi_1.P_1 + M_1$ and $\neg\exists\pi_1'.\pi_1 \Downarrow \pi_1'$. Similarly to the previous case one can show that $\Gamma, \tilde{x}{:}\tilde{\tau} \vdash \pi_1.P_1 + M_1$ which is either derived from rule (T.REC) or (T.SEND). Suppose the later (the case (T.REC) is similar), then $\pi_1 = e_1[e_2]!\tilde{e_3}$ and $\Gamma, \tilde{x}{:}\tilde{\tau} \vdash e_1 : [\tau_1 \rightarrow \tau_2] \Rightarrow \tilde{\sigma}$ and $\Gamma, \tilde{x}{:}\tilde{\tau} \vdash e_2 : \tau_1$ and $\Gamma, \tilde{x}{:}\tilde{\tau} \vdash \tilde{e_3} : \tilde{\sigma}$. By Propositions 6 and 7, each of e_1, e_2 and $\tilde{e_3}$ evaluate to some typable value and rule (SEND) is applicable which contradicts $\neg\exists\pi_1'.\pi_1 \Downarrow \pi_1'$.

(E.CONSTR) In this case, $\exists j,k.1 \leq j < k \leq n$, $P_j = \pi_1.P_1 + M_1$, $P_k = \pi_2.Q_2 + M_2$, $\pi_1 \Downarrow x[v_1]?\tilde{y}$, $\pi_2 \Downarrow x[v_2]!\tilde{v}$ and $\neg\exists v.v_1v_2 \Downarrow v$. Similarly to the case (E.COM) one can show that v_1 has some type $\tau_1 \rightarrow \tau_2$ and v_2 has type τ_1. Thus, by rule (T.FUNAPP), v_1v_2 is typable with type τ_2 and, by Proposition 7, evaluates to some value v which contradicts $\neg\exists v.v_1v_2 \Downarrow v$.

We conclude that none of the error axioms is applicable and thus $P_0 \xrightarrow[nd]{err} \bot$ does not hold. □

B.2 Subsection 5.1 (Encoding the Pi-Calculus with Priorities)

Theorem 2. *The encoding of the π-calculus with priorities $(R, <)$ into the attributed π-calculus $\pi(\lambda(R)_<)$ is correct in that for all processes P, P' with priorities and attributed processes Q:*

1. *$P \rightarrow P'$ then $[\![P]\!] \rightarrow [\![P']\!]$.*
2. *$[\![P]\!] \rightarrow Q$ then there exists a process $\hat{Q}$ of the π-calculus with priorities such that $[\![\hat{Q}]\!] \equiv Q$ and $P \rightarrow \hat{Q}$.*

Proof. First, we need to show that the encoding is invariant under substitutions.

Claim. $[\![P[\tilde{v}/\tilde{y}]]\!] = [\![P]\!][\tilde{v}/\tilde{y}]$ where $\tilde{v}$ is a tuple of values.

The proof is by induction on the structure of P. Second, we claim that the translation preserves and reflects structural congruence.

Claim. $P \equiv Q \Leftrightarrow [\![P]\!] \equiv [\![Q]\!]$.

The proof is by structural induction on derivations respectively of $P \equiv Q$ and $[\![P]\!] \equiv [\![Q]\!]$. We have to inspect all axioms of the structural congruence and all structural rules. We omit the details. Third we have to prove that errors are preserved and reflected by the translation.

Claim. $P \xrightarrow[nd]{err} \bot \Leftrightarrow [\![P]\!] \xrightarrow[nd]{err} \bot$.

(E.COM) This case is obvious, since this both calculi provide analogous rules. We omit the details.

(E.PREF) This rule exists only in the attributed π-calculus. Suppose that $[\![P]\!] \xrightarrow[nd]{err} \bot$ is inferred by (E.PREF):

$$\frac{\neg\exists\pi'.\pi \Downarrow \pi'}{[\![P]\!] = \pi.Q + M \xrightarrow[nd]{err} \bot}$$

Since sums can only be obtained by translating sums, P must match $\hat{\pi}.\hat{Q}+\hat{M}$ for some $\hat{Q}$ and $\hat{M}$. Here, $\hat{\pi}$ must be a prefix of the π-calculus with priorities, it follows that $\pi = [\![\hat{\pi}]\!]$ converges to itself, in contradiction to the hypothesis of the rule.

(E.CONSTR) This rules exists only in the attributed π-calculus. So suppose that (E.CONSTR) proves $[\![P]\!] \xrightarrow[nd]{err} \bot$, so it is applied as follows:

$$\frac{\pi_1 \Downarrow x[v_1]?\tilde{y} \quad \pi_2 \Downarrow x[v_2]!\tilde{v} \quad \neg\exists v.v_1 v_2 \Downarrow v}{[\![P]\!] = \pi_1.P_1 + M_1 \mid \pi_2.P_2 + M_2 \xrightarrow[nd]{err} \bot}$$

By inspection of the translation and the first two premises of the rule, we see that P must have the form $x?\tilde{y}.\hat{P}_1 + \hat{M}_1 \mid x{:}r!\tilde{z}.\hat{P}_2 + \hat{M}_2$. Thus, $[\![P]\!] = x[\lambda z.z]?\tilde{y}.P_1 + M_1 \mid x[r]!\tilde{z}.P_2 + M_2$. This contradicts the third premise, however, since $v_1 v_2 = (\lambda z.z)\ r \Downarrow r$ by rule (FUN) and since $r \in R$.

The treatment of structural rules is as before.

Fourth, we generalize the theorem.

Claim. For any relations $\rho \in \{\Downarrow, \xrightarrow[nd]{app}, \xrightarrow[nd]{r}, \rightarrow \mid r \in R\}$, and all processes $P', \hat{P}$ of the π-calculus with priorities and attributed processes P, Q:

1. if $P \,\rho\, P'$ then $[\![P]\!] \,\rho\, [\![P']\!]$.
2. if $[\![\hat{P}]\!] \equiv P$ and $P \,\rho\, Q$ then there exists an attributed process $\hat{Q}$ such that $[\![\hat{Q}]\!] \equiv Q$ and $\hat{P} \,\rho\, \hat{Q}$.

We proof the claim for all above relations ρ in the order in which they are given. The proof of point 1. is by structural induction on derivations of $P \,\rho\, P'$. We have to consider all rules of the operational semantics of the π-calculus with priorities.

(COM) This rule yields $P \xrightarrow[nd]{r} P'$ as follows:

$$\frac{|\tilde{y}| = |\tilde{z}|}{P = x?\tilde{y}.P_1 + M_1 \mid x{:}r!\tilde{z}.P_2 + M_2 \xrightarrow[nd]{r} P_1[\tilde{z}/\tilde{y}] \mid P_2 = P'}$$

Thus $[\![P]\!] = x[\lambda y.y]?\tilde{y}.[\![P_1]\!] + [\![M_1]\!] \mid x[r]!\tilde{z}.[\![P_2]\!] + [\![M_2]\!]$, so that the (COM) rule of $\pi(\lambda(R)_<)$ applies while using (VAL) and (FUN):

$$\frac{x[\lambda y.y]?\tilde{y} \Downarrow x[\lambda y.y]?\tilde{y} \quad x[r]!\tilde{z} \Downarrow x[r]!\tilde{z} \quad (\lambda y.y)r \Downarrow r \in R \quad |\tilde{y}| = |\tilde{z}|}{[\![P]\!] \xrightarrow[nd]{r} [\![P_1]\!][\tilde{z}/\tilde{y}] \mid [\![P_2]\!]}$$

We can now apply our first claim on substitutions above to show that the reduction result is $[\![P_1[\tilde{z}/\tilde{y}]]\!] \mid [\![P_2]\!] = [\![P']\!]$ as required.

(APP) Suppose the following rule is applicable.

$$\frac{A(\tilde{x}) \triangleq P}{A(\tilde{v}) \xrightarrow[nd]{app} P[\tilde{v}/\tilde{x}]}$$

By the substitution assumption, we know that $[\![P[\tilde{v}/\tilde{x}]]\!] = [\![P]\!][\tilde{v}/\tilde{x}]$. The translation is defined, such that $[\![A(\tilde{x}) \triangleq P]\!] = A(\tilde{x}) \triangleq [\![P]\!]$. Thus, the following rule applies

$$\frac{A(\tilde{x}) \triangleq [\![P]\!]}{A(\tilde{v}) \xrightarrow[nd]{app} [\![P]\!][\tilde{v}/\tilde{x}]}$$

(PAR) We assume that the following rule is applicable.

$$\frac{P_1 \xrightarrow[nd]{\alpha} P_1'}{P_1 \mid P_2 \xrightarrow[nd]{\alpha} P_1' \mid P_2}$$

By induction hypothesis, we have that $[\![P_1]\!] \xrightarrow[nd]{\alpha} [\![P_1']\!]$. Since the translation is compositional, the following rule is applicable:

$$\frac{[\![P_1]\!] \xrightarrow[nd]{\alpha} [\![P_1']\!]}{[\![P_1 \mid P_2]\!] \xrightarrow[nd]{\alpha} [\![P_1' \mid P_2]\!]}$$

(NEW) We assume that the following rule is applicable.

$$\frac{P \xrightarrow[nd]{\alpha} P'}{(\nu x)P \xrightarrow[nd]{\alpha} (\nu x)P'}$$

By induction hypothesis, we have that $[\![P]\!] \xrightarrow[nd]{\alpha} [\![P']\!]$. Since the translation is compositional, the following rule is applicable:

$$\frac{[\![P_1]\!] \xrightarrow[nd]{\alpha} [\![P_1']\!]}{[\![(\nu x)P]\!] \xrightarrow[nd]{\alpha} [\![(\nu x)P']\!]}$$

(STRUCT) We assume the following rule is applicable.

$$\frac{P \equiv P_1 \quad P_1 \xrightarrow[nd]{\alpha} P_2 \quad P_2 \equiv Q}{P \xrightarrow[nd]{\alpha} Q}$$

By the claim on the preservation of structural congruence, we have $[\![P]\!] \equiv [\![P_1]\!]$ and $[\![P_2]\!] \equiv [\![Q]\!]$. By induction hypothesis $[\![P_1]\!] \xrightarrow[nd]{\alpha} [\![P_2]\!]$. Thus, the following rule is applicable:

$$\frac{[\![P]\!] \equiv [\![P_1]\!] \quad [\![P_1]\!] \xrightarrow[nd]{\alpha} [\![P_2]\!] \quad [\![P_2]\!] \equiv [\![Q]\!]}{[\![P]\!] \xrightarrow[nd]{\alpha} [\![Q]\!]}$$

(CONV) Suppose that the following rule is applicable.

$$\frac{P \xrightarrow[nd]{app}{}^{*} P' \quad P' \equiv (\nu\tilde{x})\prod_{i=1}^{n} M_i \quad \neg P' \xrightarrow[nd]{err} \bot}{P \Downarrow P'}$$

Since translation preserves structural congruence and application steps, we know that $[\![P]\!] \xrightarrow[nd]{app}{}^{*} [\![P']\!]$. Since it prevents errors, we have $\neg[\![P']\!] \xrightarrow[nd]{err} \bot$. Thus, the following rule is applicable:

$$\frac{[\![P]\!] \xrightarrow[nd]{app}{}^{*} [\![P']\!] \quad [\![P']\!] \equiv (\nu\tilde{x})\prod_{i=1}^{n}[\![M_i]\!] \quad \neg[\![P']\!] \xrightarrow[nd]{err} \bot}{[\![P]\!] \Downarrow [\![P']\!]}$$

(PRIOR) Suppose the following rule is applicable.

$$\frac{P \Downarrow P' \quad P' \xrightarrow[nd]{r} Q \quad \neg\exists r_1 \in R.\exists Q_1.\ r < r_1 \wedge P' \xrightarrow[nd]{r_1} Q_1}{P \rightarrow Q}$$

Since we have treated relations $\Downarrow$ and $\xrightarrow[nd]{r}$ before, we have already shown that $P \Downarrow P'$ yields $[\![P]\!] \Downarrow [\![P']\!]$ and that $P' \xrightarrow[nd]{r} Q$ implies $[\![P']\!] \xrightarrow[nd]{r} [\![Q]\!]$. We can show by contradiction that $\neg\exists r_1 \in R.\exists Q_1.r < r_1 \wedge P \xrightarrow[nd]{r_1} Q_1$ then $\neg\exists r_2.\exists Q_2.r < r_2 \wedge [\![P]\!] \xrightarrow[nd]{r_2} Q_2$. Assume that $\neg\exists r_1 \in R.\exists Q_1.r < r_1 \wedge P \xrightarrow[nd]{r_1}{}' Q_1$, but $\exists r_2.\exists Q_2.r < r_2 \wedge [\![P]\!] \xrightarrow[nd]{r_2} Q_2$. $[\![P]\!] \xrightarrow[nd]{r_2} Q'$ only if rule (COM) applies to $[\![P]\!]$, which is true only if $[\![P]\!] \equiv (\nu\tilde{x})(\ldots \mid x[r_2]!\tilde{y}.P_1 + M_1 \mid x[\lambda y.y]?\tilde{z}.P_2 + M_2 \mid \ldots)$. By the definition of the translation, this is fulfilled only if $P \equiv (\nu\tilde{x})(\ldots \mid x{:}r_2!\tilde{y}.\hat{P}_1 + \hat{M}_1 \mid x?\tilde{z}.\hat{P}_2 + \hat{M}_2 \mid \ldots)$. Thus, $P \xrightarrow[nd]{r_2} Q$ exists, which contradicts with our assumption. Thus, the following rule is applicable

$$\frac{[\![P]\!] \Downarrow [\![P']\!] \quad [\![P']\!] \xrightarrow[nd]{r} [\![Q]\!] \quad \neg\exists r_1 \in R.\exists Q_1.\ r < r_1 \wedge [\![P']\!] \xrightarrow[nd]{r_1} Q_1}{[\![P]\!] \rightarrow [\![Q]\!]}$$

The proof of point 2. is by structural induction on derivations of $P\ \rho\ Q$, under the assumption that $[\![\hat{P}]\!] \equiv P$. We have to consider all rules of the non-deterministic operational semantics of $\pi(\lambda(R)_<)$ and all rules defining the structural congruence, but skip the latter.

(COM) By assumption, we have $[\![\hat{P}]\!] \equiv P$ and $P \xrightarrow[nd]{r} Q$ by applying the following rule:

$$\frac{\pi_1 \Downarrow x[v_1]?\tilde{y} \quad \pi_2 \Downarrow x[v_2]!\tilde{v} \quad v_1 v_2 \Downarrow r \in R \quad |\tilde{v}| = |\tilde{y}|}{P = \pi_1.P_1 + M_1 \mid \pi_2.P_2 + M_2 \xrightarrow[nd]{r} P_1[\tilde{v}/\tilde{y}] \mid P_2 = Q}$$

Inspecting the translation reveals that $\hat{P} \equiv \hat{P}'$ for some process $\hat{P}' = x?\tilde{y}.\hat{P}_1 + \hat{M}_1 \mid x{:}r'!\tilde{v}.\hat{P}_2 + \hat{M}_2$ where $\pi_1 = x[\lambda z.z]?\tilde{y}$, $\pi_2 = x[r']!\tilde{v}$, $[\![\hat{P}_1]\!] \equiv P_1$ and $[\![\hat{P}_2]\!] \equiv P_2$. The prefix equalities yield $v_1 = \lambda z.z$ and $v_2 = r'$. We can deduce $r = r'$ from $v_1 v_2 = (\lambda z.z)r' \Downarrow r'$ by rules (VAL) and (FUN). We define $\hat{Q} = \hat{P}_1[\tilde{v}/\tilde{y}] \mid \hat{P}_2$ so that $[\![\hat{Q}]\!] = Q$ by the substitution claim. Furthermore, rules (COM) and (STRUCT) apply as follows:

$$\frac{\dfrac{|\tilde{y}| = |\tilde{v}|}{\hat{P} \equiv \hat{P}' = x?\tilde{y}.\hat{P}_1 + \hat{M}_1 \mid x{:}r!\tilde{v}.\hat{P}_2 + \hat{M}_2 \xrightarrow[nd]{r} \hat{P}_1[\tilde{v}/\tilde{y}] \mid \hat{P}_2 = \hat{Q} \equiv \hat{Q}}}{\hat{P} \xrightarrow[nd]{r} \hat{Q}}$$

(APP) We assume $[\![\hat{P}]\!] \equiv P$ and that $P \xrightarrow[nd]{app} Q$ is inferred as follows:

$$\frac{[\![A(\tilde{x}) \triangleq P_1]\!]}{P = A(\tilde{v}) \xrightarrow[nd]{app} [\![P_1]\!][\tilde{v}/\tilde{x}] = Q}$$

Since $[\![\hat{P}]\!] \equiv A(\tilde{v})$, the translation yields that $\hat{P} = A(\tilde{v})$. The substitution claim shows that $[\![P_1]\!][\tilde{v}/\tilde{x}] = [\![P_1[\tilde{v}/\tilde{x}]]\!]$. We define $\hat{Q} = P_1[\tilde{v}/\tilde{x}]$ so that $[\![\hat{Q}]\!] = Q$. Furthermore, rule (APP) applies as follows:

$$\frac{A(\tilde{x}) \triangleq P_1}{\hat{P} = A(\tilde{v}) \xrightarrow[nd]{app} P_1[\tilde{v}/\tilde{x}] = \hat{Q}}$$

(PAR) We assume $[\![\hat{P}]\!] \equiv P$ and that $P \xrightarrow[nd]{\alpha} Q$ is obtained as follows:

$$\frac{P_1 \xrightarrow[nd]{\alpha} Q_1}{P = P_1 \mid P_2 \xrightarrow[nd]{\alpha} Q_1 \mid P_2 = Q}$$

Since the translation is compositional, assumption $[\![\hat{P}]\!] \equiv P$ implies the existence of two processes $\hat{P}_1$ and $\hat{P}_2$ such that $\hat{P} \equiv \hat{P}_1 \mid \hat{P}_2$ and $[\![\hat{P}_1]\!] \equiv P_1$ and $[\![\hat{P}_2]\!] \equiv P_2$. The induction hypothesis applied to $P_1 \xrightarrow[nd]{\alpha} Q_1$ shows the existence of a process $\hat{Q}_1$ such that $\hat{P}_1 \xrightarrow[nd]{\alpha} \hat{Q}_1$ and $[\![\hat{Q}_1]\!] \equiv Q_1$. We define $\hat{Q} = \hat{Q}_1 \mid \hat{P}_2$, so that $[\![\hat{Q}]\!] = [\![\hat{Q}_1]\!] \mid [\![\hat{P}_2]\!] \equiv Q_1 \mid P_2 = Q$. Furthermore, we can infer $\hat{P} \xrightarrow[nd]{\alpha} \hat{Q}$ as follows by rules (PAR) and (STRUCT):

$$\frac{\hat{P} \equiv \hat{P}_1 \mid \hat{P}_2 \quad \dfrac{\hat{P}_1 \xrightarrow[nd]{\alpha} \hat{Q}_1}{\hat{P}_1 \mid \hat{P}_2 \xrightarrow[nd]{\alpha} \hat{Q}_1 \mid \hat{P}_2} \quad \hat{Q}_1 \mid \hat{P}_2 \equiv \hat{Q}}{\hat{P} \xrightarrow[nd]{\alpha} \hat{Q}}$$

(NEW) We assume $[\![\hat{P}]\!] \equiv P$ and that $P \xrightarrow[nd]{\alpha} Q$ is obtained as follows:

$$\frac{P_1 \xrightarrow[nd]{\alpha} Q_1}{P = (\nu x)P_1 \xrightarrow[nd]{\alpha} (\nu x)Q_1 = Q}$$

By the definition of the translation, we know that there exists a process $\hat{P}_1$ such that $\hat{P} \equiv (\nu x)\hat{P}_1$ and $P_1 \equiv [\![\hat{P}_1]\!]$. By induction hypothesis, there exists a process $\hat{Q}_1$ with $\hat{P}_1 \xrightarrow[nd]{\alpha} \hat{Q}_1$ and $Q_1 \equiv [\![\hat{Q}_1]\!]$. We define $\hat{Q}$ by $\hat{Q} = (\nu x)\hat{Q}_1$. Hence $[\![\hat{Q}]\!] \equiv Q$ by definition of the translation. Furthermore, we can infer $\hat{P} \xrightarrow[nd]{\alpha} \hat{Q}$ as follows:

$$\frac{\hat{P} \equiv (\nu x)\hat{P}_1 \quad \dfrac{\hat{P}_1 \xrightarrow[nd]{\alpha} \hat{Q}_1}{(\nu x)\hat{P}_1 \xrightarrow[nd]{\alpha} (\nu x)\hat{Q}_1} \quad (\nu x)\hat{Q}_1 \equiv \hat{Q}}{\hat{P} \xrightarrow[nd]{\alpha} \hat{Q}}$$

(STRUCT) We assume $[\![\hat{P}]\!] \equiv P$ and that $P \xrightarrow[nd]{\alpha} Q$ is inferred as follows:

$$\frac{P \equiv P_1 \quad P_1 \xrightarrow[nd]{\alpha} P_2 \quad P_2 \equiv Q}{P \xrightarrow[nd]{\alpha} Q}$$

Since every congruence relation is transitive, we get $[\![\hat{P}]\!] \equiv P_1$. The induction hypothesis applied to $P_1 \xrightarrow[nd]{\alpha} P_2$ thus proves the existence of a process $\hat{P}_2$ such that $\hat{P} \xrightarrow[nd]{\alpha} \hat{P}_2$ and $[\![\hat{P}_2]\!] \equiv P_2$. Transitivity of structural congruence yields $[\![\hat{P}_2]\!] \equiv Q$. We can thus define $\hat{Q} = \hat{P}_2$, so that $\hat{P} \xrightarrow[nd]{\alpha} \hat{Q}$ and $[\![\hat{Q}]\!] \equiv Q$.

(CONV) We first show the following claim by induction on derivation length $l \geq 0$:

Claim. For all $l \in \mathbb{N}_0$, if $[\![\hat{P}]\!] \equiv P$ and $P(\xrightarrow[nd]{app})^l Q_l$ then there exists $\hat{Q}_l$, such that $\hat{P}(\xrightarrow[nd]{app})^l \hat{Q}_l$ and $Q \equiv [\![\hat{Q}]\!]_l$.

Proof. For $l = 0$, the assumption $P(\xrightarrow[nd]{app})^0 Q_0$ is equivalent to $P \equiv Q_0$ by definition. Thus, $[\![\hat{P}]\!] \equiv Q_0$, so that we can define $\hat{Q}_0 = \hat{P}$ in order to obtain $[\![\hat{Q}_0]\!] \equiv Q_0$. For the induction step, let $[\![\hat{P}]\!] \equiv P$ such that $P(\xrightarrow[nd]{app})^l Q_l \xrightarrow[nd]{app} Q_{l+1}$. By induction hypothesis, there exists $\hat{Q}_l$ such that $\hat{P}(\xrightarrow[nd]{app})^l \hat{Q}_l$ and $Q_l \equiv [\![\hat{Q}_l]\!]$. Since we have finished the proof for relation $\xrightarrow[nd]{app}$ already, there exists $\hat{Q}_{l+1}$, such that $\hat{Q}_l \xrightarrow[nd]{app} \hat{Q}_{l+1}$ and $Q_{l+1} \equiv [\![\hat{Q}_{l+1}]\!]$. Clearly $\hat{P}(\xrightarrow[nd]{app})^{l+1} \hat{Q}_{l+1}$.

We next assume $P \equiv [\![\hat{P}]\!]$ and that $P \Downarrow Q$ is inferred by rule (CONV) as follows:

$$\frac{P \xrightarrow[nd]{app}{}^{*} Q \quad Q \equiv (\nu\tilde{x}) \prod_{i=1}^{n} M_i \quad \neg Q \xrightarrow[nd]{err} \bot}{P \Downarrow Q}$$

The claim above proves that there exists $\hat{Q}$ such that $\hat{P} \xrightarrow[nd]{app}{}^{*} \hat{Q}$ and $[\![\hat{Q}]\!] \equiv Q$. The definition of the translation yields that $\hat{Q} \equiv (\nu\tilde{x}) \prod_{i=1}^{n} \hat{M}_i$ for some guarded processes $\hat{M}_i$. Since the translation is error-reflecting, $\neg Q \xrightarrow[nd]{err} \bot$ yields $\neg\hat{Q} \xrightarrow[nd]{err} \bot$. We can thus infer $\hat{P} \Downarrow \hat{Q}$ as follows:

$$\frac{\hat{P} \xrightarrow[nd]{app}{}^{*} \hat{Q} \quad \hat{Q} \equiv (\nu\tilde{x}) \prod_{i=1}^{n} \hat{M}_i \quad \neg\hat{Q} \xrightarrow[nd]{err} \bot}{\hat{P} \Downarrow \hat{Q}}$$

(PRIOR) We assume $P \equiv [\![\hat{P}]\!]$ and that $P \Downarrow Q$ is inferred as follows:

$$\frac{P \Downarrow P_1 \quad P_1 \xrightarrow[nd]{r} Q \quad \neg\exists r_1 \in R.\exists Q_1.\ r < r_1 \wedge P_1 \xrightarrow[nd]{r_1} Q_1}{[\![P]\!] \rightarrow Q}$$

Since we have already proved the result for convergence, we know that there exists a process $\hat{P}_1$, such that $\hat{P} \Downarrow \hat{P}_1$ and $P_1 \equiv [\![\hat{P}_1]\!]$. Hence, there exists $\hat{Q}$ such that $\hat{P}_1 \xrightarrow[nd]{r} \hat{Q}$ and $Q \equiv \hat{Q}$. We next show that $\neg\exists r_1 \in R.\exists\hat{Q}_1.r < r_1 \wedge \hat{P}_1 \xrightarrow[nd]{r_1} \hat{Q}_1$.The proof is by contradiction. Suppose that such an r_1 and $\hat{Q}_1$ exist. For the first part of this theorem we have shown that this implies $[\![\hat{P}_1]\!] \xrightarrow[nd]{r_1} [\![\hat{Q}_1]\!]$. Since $P_1 \equiv [\![\hat{P}_1]\!]$, we can choose a $Q_1 \equiv [\![\hat{Q}_1]\!]$, such that, by rule (STRUCT), we obtain $P_1 \xrightarrow[nd]{r_1} Q_1$ in contradiction to the third hypothesis. □

B.3 Subsection 5.2 (Encoding π@ for Dynamic Compartments)

Lemma 5. *For all constants of variables $v_1, \ldots, v_n, v'_1, \ldots, v'_n$ it is true that:*

1. $eq_n(v_1, \ldots, v_n, v_1, \ldots, v_n) \Downarrow \mathtt{true}$.
2. $eq_n(v_1, \ldots, v_n, v'_1, \ldots, v'_n) \Downarrow \mathtt{false}$ *if* $v_i \neq v'_i$ *for some* $1 \leq i \leq n$.

The proof is straightforward by induction on n. It relies on the definition of conditionals and equality in the big-step evaluator in Figs. 11 and 12.

Proof. (1) The proof is by induction on n. For $n = 0$, we obtain $eq_0() \Downarrow \mathtt{true}$ by definition. For the induction step, we have that:

$$eq_{n+1}(v_1, \ldots, v_{n+1}, v_1, \ldots, v_{n+1}) =_{df}$$
$$\mathtt{if}\ v_{n+1}{=}v_{n+1}\ \mathtt{then}\ eq_n(v_1, \ldots, v_n, v_1 \ldots, v_n)\ \mathtt{else\ false}.$$

By rules ($\textsc{eq}_1$) and ($\textsc{cond}_1$) apply, we obtain $eq_n(v_1, \ldots, v_n, v_1 \ldots, v_n)$. By induction hypothesis:

$$eq_n(v_1, \ldots, v_n, v_1, \ldots, v_n) \Downarrow \texttt{true}$$

(2) The proof is by induction. The case $n = 0$ is trivial, since the hypothesis of the implication is always wrong. For the induction step from n to $n+1$, we assume that $1 \leq i \leq n+1$ such that $v_i \neq v'_i$ and:

$$\begin{aligned} &eq_{n+1}(v_1, \ldots, v_{n+1}, v'_1, \ldots, v_{n+1}) =_{df} \\ &\quad \texttt{if } v_{n+1}{=}v'_{n+1} \texttt{ then } eq_n(v_1, \ldots, v_n, v'_1 \ldots, v'_n) \texttt{ else false.} \end{aligned}$$

Suppose $1 \leq i \leq n$ and $v_{n+1} = v'_{n+1}$. Then, by rules ($\textsc{eq}_1$) and ($\textsc{cond}_1$), we obtain $eq_n(v_1, \ldots, v_n, v'_1 \ldots, v'_n)$, which evaluates to `false` by induction hypothesis. If $i = n+1$, then we know by rules ($\textsc{eq}_2$) and ($\textsc{cond}_2$):

$$\texttt{if } v_{n+1}{=}v'_{n+1} \texttt{ then } eq_n(v_1, \ldots, v_n, v'_1 \ldots, v'_n) \texttt{ else false} \Downarrow \texttt{false}$$

□

Theorem 4 (Operational correspondence).

(a) if $P \to Q$ w.r.t. $\mathcal{D}$, then $[\![P]\!] \to [\![Q]\!]$ w.r.t. $[\![\mathcal{D}]\!]$
(b) if $[\![P]\!] \to Q$ w.r.t $[\![\mathcal{D}]\!]$, then $\exists \hat{Q}$ s.t. $Q \equiv [\![\hat{Q}]\!]$, and $P \to \hat{Q}$ w.r.t $\mathcal{D}$.

Proof. With Theorem 2, we proved the encoding of the π-calculus with priorities in $\pi(\lambda(R)_<)$ correct. The non-deterministic semantics of π@ and the π-calculus with priorities are the same, with only one exception: the communication rule. Therefore, in the following proof, we only consider the communication rule.

(a) We define κ_o and κ_i:

$$\begin{aligned} \kappa_o &= \lambda y_2 \ldots \lambda y_n \lambda r'.\ \texttt{if}\ \ eq_n(x_2, \ldots, x_n, r, y_2, \ldots, y_n, r')\ \texttt{then}\ r \\ \kappa_i &= \lambda e.e x_2 \ldots x_n r \end{aligned}$$

Rule ($\textsc{com}$) yields $P \xrightarrow[nd]{r} P'$ as follows:

$$\frac{|\tilde{y}| = |\tilde{z}|}{P = \tilde{x}{:}r?\tilde{y}.P_1 + M_1 \mid \tilde{x}{:}r!\tilde{z}.P_2 + M_2 \xrightarrow[nd]{r} P_1[\tilde{v}/\tilde{y}] \mid P_2 = P'}$$

where $\tilde{x} = (x_1, \ldots, x_n)$. Thus, $[\![P]\!] = x_1[\kappa_i]?\tilde{y}.[\![P_1]\!] + [\![M_1]\!] \mid x_1[\kappa_o]!\tilde{z}.[\![P_2]\!] + [\![M_2]\!]$. By Lemma 5, we know that $\kappa_i\kappa_o \Downarrow r$, such that the ($\textsc{com}$) rule of $\pi(R, <)$ applies:

$$\frac{x_1[\kappa_i]?\tilde{y} \Downarrow x_1[v_1]?\tilde{y} \quad x_1[\kappa_o]!\tilde{z} \Downarrow x_1[v_2]!\tilde{z} \quad v_1 v_2 \Downarrow r \in R \quad |\tilde{y}| = |\tilde{z}|}{[\![P]\!] \xrightarrow[nd]{r} [\![P_1]\!][\tilde{v}/\tilde{y}] \mid [\![P_2]\!]}$$

By the claim on substitutions, we can show that the reduction result is $[\![P_1[\tilde{v}/\tilde{y}]]\!] \mid [\![P_2]\!] = [\![P']\!]$ as required.

(b) We define κ_o and κ_i:

$$\kappa_o = \lambda y_2 \ldots \lambda y_n \lambda r'.\ \mathtt{if}\ eq_n(x'_2, \ldots, x'_n, \hat{r}', y_2, \ldots, y_n, r')\ \mathtt{then}\ r$$
$$\kappa_i = \lambda e.ex_2 \ldots x_n \hat{r}$$

By assumption, we have $[\![\hat{P}]\!] \equiv P$ and $P \xrightarrow[nd]{r} Q$ by applying the following rule:

$$\frac{\pi_1 \Downarrow x_1[v_1]?\tilde{y} \quad \pi_2 \Downarrow x_1[v_2]!\tilde{v} \quad v_1 v_2 \Downarrow r \in R \quad |\tilde{v}| = |\tilde{y}|}{P = \pi_1.P_1 + M_1 \mid \pi_2.P_2 + M_2 \xrightarrow[nd]{r} P_1[\tilde{v}/\tilde{y}] \mid P_2 = Q}$$

Inspecting the translation reveals that a $\hat{P}'$ exists, s.t. $\hat{P} \equiv \hat{P}'$ and $\hat{P}' = (x_1, x_2 \ldots, x_n){:}\hat{r}?\tilde{y}.\hat{P}_1 + \hat{M}_1 \mid (x_1, x'_2, \ldots, x'_n){:}\hat{r}'!\tilde{z}.\hat{P}_2 + \hat{M}_2$, with $[\![\hat{P}_1]\!] \equiv P_1$, $[\![\hat{P}_2]\!] \equiv P_2$, $\pi_1 = x_1[\kappa_i]?\tilde{y}$, and $\pi_2 = x_1[\kappa_o]!\tilde{z}$. The prefix equalities yield $v_1 = \kappa_i$ and $v_2 = \kappa_o$. By Lemma 5, we know that $x'_2 = x_2 \ldots, x'_n = x_n$, $\hat{r}' = \hat{r}$, and $r = \hat{r}$, since $P \xrightarrow[nd]{r} Q$ by the rule above. We define $\hat{Q} = \hat{P}_1[\tilde{v}/\tilde{y}] \mid \hat{P}_2$ so that $[\![\hat{Q}]\!] = Q$ by the substitution claim. Furthermore, rules (COM) and (STRUCT) apply as follows:

$$\frac{\dfrac{|\tilde{y}| = |\tilde{z}|}{\hat{P} \equiv \hat{P}' = \tilde{x}{:}r?\tilde{y}.\hat{P}_1 + \hat{M}_1 \mid \tilde{x}{:}r!\tilde{z}.\hat{P}_2 + \hat{M}_2 \xrightarrow[nd]{r} \hat{P}_1[\tilde{z}/\tilde{y}] \mid \hat{P}_2 = \hat{Q} \equiv \hat{Q}}}{\hat{P} \xrightarrow[nd]{r} \hat{Q}}$$

□

A Language for Biochemical Systems: Design and Formal Specification

Michael Pedersen and Gordon D. Plotkin

LFCS, School of Informatics, University of Edinburgh

Abstract. This paper introduces a *Language for Biochemical Systems* (LBS) which combines rule-based approaches to modelling with modularity. It is based on the *Calculus of Biochemical Systems* (CBS) which affords modular descriptions of metabolic, signalling and regulatory networks in terms of reactions between modified complexes, occurring concurrently inside a hierarchy of compartments and with possible cross-compartment interactions and transport. Additional features of LBS, targeted towards practical and large-scale applications, include species expressions for manipulating large complexes in a concise manner, parameterised modules with a notion of subtyping for writing reusable modules, and nondeterminism for handling combinatorial explosion. These features are demonstrated through examples. A formal specification of LBS is then given through an abstract syntax and a general semantics which is parametric on a structure pertaining to the specific choice of target semantical objects. Examples of such structures for the specific cases of Petri nets, coloured Petri nets, ODEs and continuous time Markov chains are also given.

Keywords: Large-scale, parametrised modules, subtyping, combinatorial explosion, nondeterminism, Petri nets, coloured Petri nets, ordinary differential equations, continuous time Markov chains.

1 Introduction

Systems biology. Systems biology is a rapidly growing field which seeks a refined and quantitative understanding of organisms, particularly studying how molecular species such as metabolites, proteins and genes interact in cells to form the complex emerging behaviour that living systems exhibit. Such an understanding is for example important for the development of new drugs and to predict the impact of these on an organism. Mathematical modelling plays a key rôle in pursuit of this by facilitating the generation of new knowledge through the cycle of simulation, experimental validation, and model refinement.

Formalisms for systems biology. As our biological knowledge-base increases through rapid improvements of e.g. high-throughput sequencing methods, the models under study also increase in size and complexity. New methods are therefore needed to support the structured development of large models, and also to

C. Priami et al. (Eds.): Trans. on Comput. Syst. Biol. XII, LNBI 5945, pp. 77–145, 2010.

complement simulations with other kinds of analysis. Hence an abundance of formalisms inspired by computer science have been applied to biological modelling over the past decade. These include Petri nets [23] and coloured Petri nets [18]; process calculi such as the stochastic π-calculus [32, 29], the continuous π-calculus [20], Beta binders [33,14], BlenX [12], PEPA [4] and BioPEPA [8]; rule-based languages such as κ [11,10], BioNetGen [13] and BIOCHAM [6]; state-based formalisms such as Statecharts [15]; and more specialised languages such as Bioambients [34], the Brane calculi [5] and P-systems [24] aimed specifically at describing biological compartments and membranes.

Contributions. Some of the above mentioned languages, in particular those from the process calculus family, excel in their support for modularity by allowing large systems to be described in terms of their components. These languages may be difficult for non-specialists, including some biologists, to use and to understand. Other languages, for example from the rule-based family, are more intuitive to use but only allow flat, non-modular descriptions. The contribution of this paper is a *Language for Biochemical Systems* (LBS) which aims at bridging the above gap: it is based on a notion of reaction rules while also being modular.

LBS builds on the *Calculus of Biochemical Systems* (CBS) [31]. Just as CBS, LBS affords modular descriptions of metabolic, signalling and regulatory networks in terms of reactions between modified complexes, occurring concurrently inside a hierarchy of compartments and with possible cross-compartment interactions and transport. It has a compositional semantics, translating programs into semantical objects such as Petri nets, coloured Petri nets, ordinary differential equations (ODEs) and continuous time Markov chains (CTMCs). Petri nets allow a range of established analysis techniques to be used in the biological setting [16], while ODEs and CTMCs enable respectively deterministic and stochastic simulations to be carried out. The compositional semantics of LBS can be exploited in analysis, as previously demonstrated for the case of Petri net flows [25], potentially improving analysis efficiency and enabling the reuse of analysis results.

LBS, unlike CBS, includes support for practical, large-scale applications through three main features, namely: *species expressions*, *parameterised modules* and *nondeterminism*. Species expressions provide a concise way of constructing large complexes and modifying these incrementally, a common scenario in signal transduction pathways. Parameterised modules allow common motifs such as phosphorylation/dephosphorylation cycles or entire MAPK cascades to be modelled once and reused in different contexts. Modules may be parameterised on compartments, rates, and species; species are typed by their component atomic species names and their modification site types; and the typing system includes subtyping and a notion of parametric type. Modules may furthermore output species, thus providing a natural mechanism for connecting related modules. Nondeterminism provides a mechanism for specifying that any one of a given set of species can take place in a reaction, thus going some way towards compact descriptions of combinatorially complex systems.

LBS further includes species and compartment definitions which involve scope and new name generation, and complex species which may span several compartments. LBS also allows arbitrary modification site types for representing e.g. spatial location (real number pairs) or DNA sequences (strings). A suitable choice of modification site type can furthermore capture connectivity in complex species which we anticipate can lead to a translation of LBS to κ and BioNetGen. Finally, LBS allows both mass-action rates and general rate expressions in reactions, and also provides a means of model variation so that a single LBS program may translate to multiple related semantical objects which differ in e.g. their initial conditions.

A compiler from LBS to the *Systems Biology Markup Language* (SBML) [17] has been implemented, allowing existing SBML-compliant tools for simulation and analysis to be exploited; modular compilations to e.g. ODEs are left for future work. The compiler has been validated on a model of the yeast pheromone pathway [19] for which simulation results coincide with published results.

There are two other recently introduced languages which combine the reaction-based approach with modularity, namely Little b [22] and Antimony [35]. Little b is based on Lisp and thus boasts the full power of modularity found in this general-purpose functional programming language. Antimony follows an object-oriented approach rather than the more functional approach of LBS; it also has dedicated features for the composition of genes aimed at applications in synthetic biology. To our knowledge, neither of these languages has a direct counterpart of the LBS species expressions, nondeterminism and subtyping. Furthermore, they do not have a formally defined semantics (barring the standard semantics of Lisp which is not particularly well suited for the biological domain), let alone a compositional one, and are hence not directly amenable to modular analysis.

Paper outline. We start in Section 2 with an informal overview of CBS through basic examples of gene expression and MAPK cascades. In Section 3 we introduce LBS by further examples and demonstrate how specific limitations of CBS can be overcome. We then turn to a formal presentation, starting with an abstract syntax of LBS in Section 4. A general semantics, parameterised on a structure pertaining to specific semantical objects, is given in Section 5. Examples of such structures for translating to Petri nets, coloured Petri nets, ODEs and CTMCs are given in Section 6. Section 7 concludes with a discussion of future directions.

The present version of LBS is a wholesale redesign of a previous version [27], merging patterns and species into one syntactic category with consequent extensive changes to the typing system, and with a number of extensions such as species nondeterminism and model variation.

2 The Calculus of Biochemical Systems

2.1 Located Parallel Reactions

As a first example we consider a basic model of eukaryotic gene expression as illustrated by the informal pictorial diagram in Figure 2.1. A corresponding

Listing 2.1. A CBS program for gene expression.

```
c[
  n[gene + rnap -> gene + rnap + mrna] |
  n[mrna] -> mrna |
  mrna -> prot
]
```

CBS program is shown in Listing 2.1. The program consists of three reactions composed in *parallel* using the operator |, and taking place inside a cytosol *compartment* c. The first reaction models transcription. It is located inside a nested nucleus compartment n and produces mRNA from a gene and an RNA polymerase. The second reaction models transport of mRNA out of the nucleus into the enclosing cytosol compartment, and the third reaction models translation of mRNA into protein. Compartments can thus be used at the level of individual species and at the level of entire programs. In this example we could have omitted the cytosol compartment in which case a default top level compartment would be assumed. Reactants and products can be labelled with stoichiometry, but when omitted, as in this example, a default of 1 is assumed.

As an illustration of the underlying formal semantics, the Petri nets arising from each of the three individual (but located) reactions are shown in Figure 2.2a. Reactions are represented by *transitions* (rectangles), and located species are represented by *places* (circles). Including species location in place names allows for a semantical distinction of e.g. nuclear and cytosolic mRNA.

When considering the reactions together in parallel, the standard interpretation is for the reactions to share and compete for species which have syntactically identical located names. In this example, the first and second reactions hence share the species n[mrna] and the second and third reactions share the species c[mrna]. A Petri net representation of the parallel composition based on this interpretation is shown in Figure 2.2b. This illustrates how located reactions, and

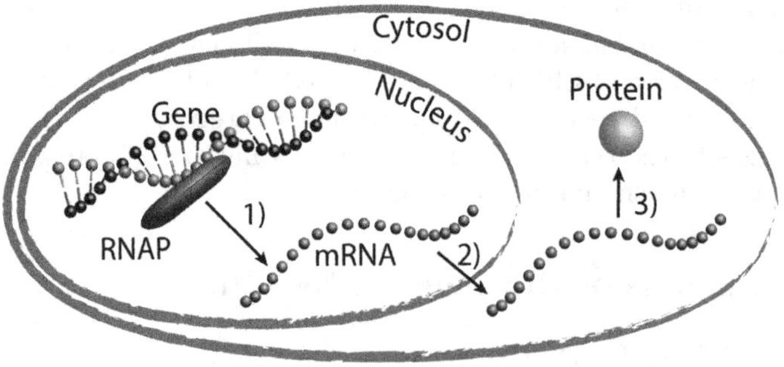

Fig. 2.1. An informal pictorial diagram of eukaryotic gene expression taking place in three steps: 1) transcription, 2) transport and 3) translation

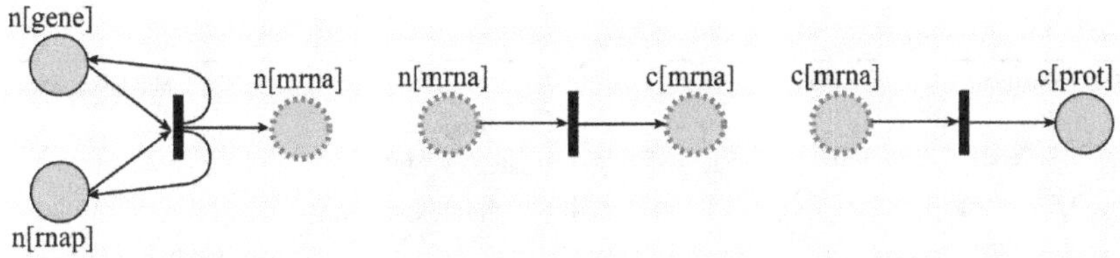

(a) Three Petri nets resulting from individual (but located) reactions. Shared places are highlighted.

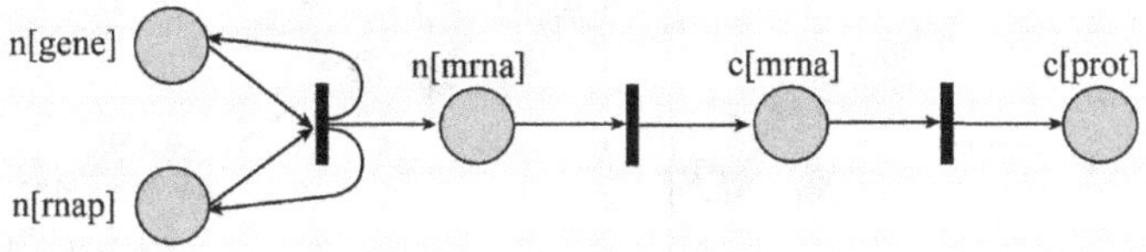

(b) The composition obtained by merging the shared places of the three Petri nets.

Fig. 2.2. Petri net models resulting from the CBS program for gene expression

more generally modules, are composed in CBS and hints at how a compositional semantics in terms of Petri nets can be defined; a formal definition is given in Section 6. Semantically we observe that compartments distribute over parallel compositions and reactant/product sums, and that the parallel composition and sum operators are commutative.

2.2 Modification Sites and Complexes

The next example is based on a scaffolded MAPK cascade from the yeast pheromone pathway [19] and features complex species with modification sites. An informal graphical representation is shown in Figure 2.3 and the corresponding CBS program is shown in Listing 2.2.

The first five reactions in lines $1-12$ model the formation of the scaffold complex and correspond to the left part of Figure 2.3. The last three reactions in lines $14-21$ model the actual MAPK cascade, with each reaction phosphorylating a single atomic species in a complex reactant, and correspond to the right part of Figure 2.3. The scaffold is formed by the atomic species Ste5, Ste20 and Gbg, and the species Fus3, Ste7 and Ste11 serve the MAPK, MAPK2 and MAPK3 rôles, respectively. All species except Ste20 and Gbg have a single modification site, p, which can be either phosphorylated or unphosphorylated, indicated by the assignment of boolean values **tt** and **ff**. For example, Fus3{p=**ff**} represents Fus3 in its unphosphorylated state. Complexes are formed by composing modified primitive species using the *complex formation operator*, -.

Semantically, a complex of modified atomic species can be represented by a Petri net place named by the *multiset* of modified atomic species. Hence the complex formation operator is commutative. More generally, modification sites

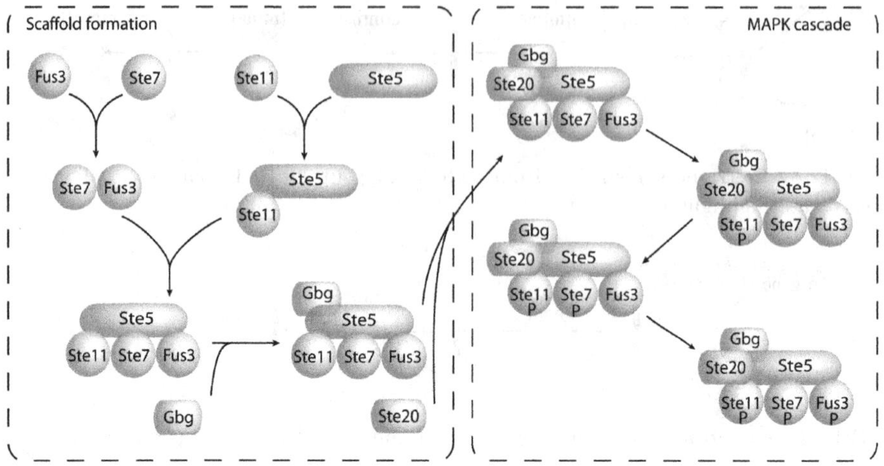

Fig. 2.3. Scaffold formation and a scaffolded MAPK cascade, adapted from [19]

Listing 2.2. A CBS program for a scaffolded MAPK cascade in yeast.

```
Ste5{p=ff} + Ste11{p=ff} -> Ste5{p=ff}-Ste11{p=ff} |

Ste7{p=ff} + Fus3{p=ff} -> Ste7{p=ff}-Fus3{p=ff} |

Ste5{p=ff}-Ste11{p=ff} + Ste7{p=ff}-Fus3{p=ff} ->
    Ste5{p=ff}-Ste11{p=ff}-Ste7{p=ff}-Fus3{p=ff} |

Ste5{p=ff}-Ste11{p=ff}-Ste7{p=ff}-Fus3{p=ff} + Gbg ->
    Ste5{p=ff}-Ste11{p=ff}-Ste7{p=ff}-Fus3{p=ff}-Gbg |

Ste5{p=ff}-Ste11{p=ff}-Ste7{p=ff}-Fus3{p=ff}-Gbg + Ste20 ->
    Ste5{p=ff}-Ste11{p=ff}-Ste7{p=ff}-Fus3{p=ff}-Gbg-Ste20 |

Ste5{p=ff}-Ste11{p=ff}-Ste7{p=ff}-Fus3{p=ff}-Gbg-Ste20 ->
    Ste5{p=ff}-Ste11{p=tt}-Ste7{p=ff}-Fus3{p=ff}-Gbg-Ste20 |

Ste5{p=ff}-Ste11{p=tt}-Ste7{p=ff}-Fus3{p=ff}-Gbg-Ste20 ->
    Ste5{p=ff}-Ste11{p=tt}-Ste7{p=tt}-Fus3{p=ff}-Gbg-Ste20 |

Ste5{p=ff}-Ste11{p=tt}-Ste7{p=tt}-Fus3{p=ff}-Gbg-Ste20 ->
    Ste5{p=ff}-Ste11{p=tt}-Ste7{p=tt}-Fus3{p=tt}-Gbg-Ste20
```

may have arbitrary boolean expressions assigned which may include variables for "matching" multiple physical species, thereby ameliorating the combinatorial explosion problem at the level of species modifications. Semantically, a reaction with such variables can be viewed as generating one concrete reaction for each possible assignment of variables. It is however also possible to handle modification state and variables directly through a semantics in terms of *coloured Petri nets* as is demonstrated in Section 6.

One immediately notices a problem with the CBS program in Listing 2.2: the program is difficult to read due to a high level of *redundancy*. As is common in signalling pathways, some reactions change just a single state of modification in a large complex, yet unaffected parts of the complexes are listed repeatedly. We can improve the situation slightly by introducing an abbreviated notation where an omitted modification site is implicitly assumed to be false, and a modification site with no assignment is assumed to be true, so that for example Ste11{p}−Ste5 is understood as Ste11{p=**tt**}−Ste5{p=**ff**}. We assume this convention in the following. But the underlying problem of redundancy remains and is addressed with dedicated language constructs in LBS.

2.3 Modules

The scaffolded MAPK cascade program is simple in that each reaction represents an autophosphorylation involving only a single reactant. We now shift the focus and consider a larger, unscaffolded cascade model in which the MAPK and MAPK2 proteins each have two phosphorylation sites, and in which each phosphorylation step involves three reactions: binding of kinase and ligand, phosphorylation of bound ligand, and dissociation of phosphorylated ligand and kinase. We furthermore include the corresponding dephosphorylation steps. We choose for this example to adapt a previously published Ras/Raf/MEK/ERK cascade [9]. An informal graphical representation is shown in Figure 2.4, and the corresponding modular CBS program is shown in Listing 2.3.

The new language feature in this program is that of *module definitions* for representing the relevant phosphorylation/dephosphorylation cycles, and the main body of the program in line 47 then consists of five parallel *module invocations*. Such a modular approach simplifies the presentation and should be contrasted with other rule-based approaches using e.g. BIOCHAM where the program would consist of one long, unstructured list of reactions. But as with a previous program, we observe a high degree of redundancy. All five modules have the same structure, consisting of two sets of three reactions for binding, modification and unbinding.

A shorter version of the program could in principle be obtained through appropriate derived forms for enzymatic reactions. But it appears unlikely that a small, fixed set of derived forms can cater for all the variants that a modeller may encounter. We may for example wish to consider variants of the above model in which all the binding reactions are reversible, or in which binding and phosphorylation are combined into a single reaction. Hence we seek to address the

Listing 2.3. A modular CBS program for the Raf/Ras/MEK/ERK MAPK cascade.

```
module rafCycle {
  Ras + Raf -> Ras-Raf |
  Ras-Raf -> Ras-Raf{m} |
  Ras-Raf{m} -> Ras + Raf{m} |
  PP2A1 + Raf{m} -> PP2A1-Raf{m} |
  PP2A1-Raf{m} -> PP2A1-Raf |
  PP2A1-Raf -> PP2A1 + Raf
};

module mekCycle1 {
  Raf{m} + MEK -> Raf{m}-MEK |
  Raf{m}-MEK -> Raf{m}-MEK{S218} |
  Raf{m}-MEK -> Raf{m} + MEK{S218} |
  PP2A2 + MEK{S218} -> PP2A2-MEK{S218} |
  PP2A2-MEK{S218} -> PP2A2-MEK |
  PP2A2-MEK -> PP2A2 + MEK
};

module mekCycle2 {
  Raf{m} + MEK{S218} -> Raf{m}-MEK{S218} |
  Raf{m}-MEK{S218} -> Raf{m}-MEK{S218, S222} |
  Raf{m}-MEK{S218, S222} -> Raf{m} + MEK{S218, S222} |
  PP2A2 + MEK{S218, S222} -> PP2A2-MEK{S218, S222} |
  PP2A2-MEK{S218, S222} -> PP2A2-MEK{S218} |
  PP2A2-MEK{S218} -> PP2A2 + MEK{S218}
};

module erkCycle1 {
  MEK{S218,S222} + ERK -> MEK{S218,S222}-ERK |
  MEK{S218,S222}-ERK -> MEK{S218,S222}-ERK{T185} |
  MEK{S218,S222}-ERK{T185} -> MEK{S218,S222} + ERK{T185} |
  MKP3 + ERK{T185} -> MKP3-ERK{T185} |
  MKP3-ERK{T185} -> MKP3-ERK |
  MKP3-ERK -> MKP3 + ERK
};

module erkCycle2 {
  MEK{S218,S222} + ERK{T185} -> MEK{S218,S222}-ERK{T185} |
  MEK{S218,S222}-ERK{T185}->MEK{S218,S222}-ERK{T185,Y187} |
  MEK{S218,S222}-ERK{T185, Y187} ->
    MEK{S218,S222} + ERK{T185, Y187} |
  MKP3 + ERK{T185,Y187} -> MKP3-ERK{T185,Y187} |
  MKP3-ERK{T185,Y187} -> MKP3-ERK{T185} |
  MKP3-ERK{T185} -> MKP3 + ERK{T185}
};

rafCycle | mekCycle1 | mekCycle2 | erkCycle1 | erkCycle2
```

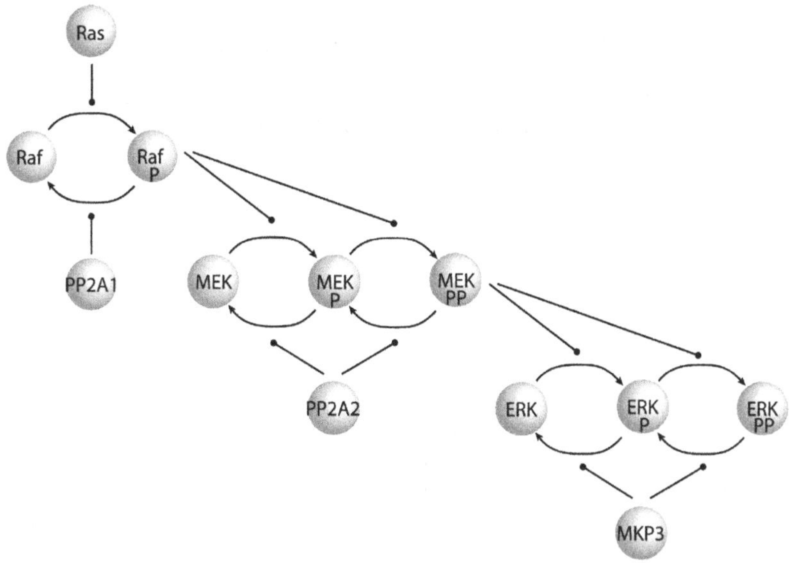

Fig. 2.4. A Ras/Raf/MEK/ERK MAPK cascade represented by five phosphorylation/dephosphorylation cycles. Each phosphorylation and dephosphorylation step covers three underlying reactions for binding, phosphorylation/dephosphorylation, and unbinding.

problem of reusability through language-level support for parameterised modules in LBS.

3 The Language for Biochemical Systems

3.1 New Species and Compartment Definitions

CBS has a static semantics which catches typos by requiring that only species names in a given set are used in programs. In LBS we include both *new species definitions* and *new compartment definitions* directly in the language. New species definitions include a list of modification sites and their type if any, and new compartment definitions include a specification of the parent, if any, and an optional volume. The volume is used when compartments are referred to in rate expressions. The semantics of LBS requires that species are only used with their defined modification types and that compartments are only used inside their defined parents. New species and compartment definitions are demonstrated by the program in Listing 3.1 which is identical to the corresponding CBS program in Listing 2.1 except for the added definitions in the first three lines.

Species *identifiers* such as mrna are here assigned to new species. The text **new**{} is formally a species *expression* which evaluates to a species *value* with no modification sites and with a globally unique name that is used in e.g. the underlying Petri net semantics. Hence the name of the species identifier, mrna, does not in itself hold any identity of a species, and we may bind the same

Listing 3.1. An LBS program for gene expression.

```
spec gene = new{}, rnap = new{}, mrna = new{}, prot = new{};
comp c = new comp; comp n = new comp inside c;

c[
  n[gene + rnap -> gene + rnap + mrna] |
  n [mrna] -> mrna |
  mrna -> prot
]
```

identifier to an entirely different species in another part of the program. This allows different modules, possibly developed by different people, to use the same species identifier for mRNA molecules which are semantically and biologically different, and subsequently combine the modules into a single program without unintended cross-talk. On the other hand, when species *are* intended to be shared between modules, the species should be defined globally or made parameters of modules as we demonstrate next.

3.2 Parameterised Modules

We extend the basic gene expression program to express *two* proteins, prot1 and prot2, from two different genes, gene1 and gene2. We do so by abstracting the gene expression process into a *parameterised module* and invoking the module twice with the relevant parameters. The result is shown in Listing 3.2.

RNAP is defined globally in line 1, meaning that it will be shared between all instances of the module defined in lines 3 – 8. This is biologically meaningful since the same RNAP species is used for transcription independently of the gene in question. The module is parameterised on the nucleus compartment, the gene and the target protein. The body is similar to before, except that the new mRNA species is defined locally. This means that each instance of the module uses semantically distinct mRNA, which again is biologically meaningful. Lines 10 – 13 define the genes and proteins to be expressed together with the relevant compartments, and line 15 is a parallel composition of two module invocations inside the cytosol compartment. We could choose to define the nucleus compartment globally in this particular case, but instead give it as a common parameter in both module invocations in order to illustrate how this can be done in the more general case.

3.3 Species Expressions

New species, species identifiers and complexes are technically considered species expressions. Species identifiers can be bound to any species expressions, not just the new atomic ones, allowing large complexes to be defined once and used

Listing 3.2. A modular LBS program for gene expression instantiated with two genes and two target proteins.

```
spec rnap = new{};

module m(comp nuc; spec gene, prot) {
  spec mrna = new{};
  nuc[gene + rnap -> gene + rnap + mrna] |
  nuc[mrna] -> mrna |
  mrna -> prot
};

spec gene1 = new{}, prot1 = new{};
spec gene2 = new{}, prot2 = new{};
comp c = new comp;
comp n = new comp inside c;

c[ m(n, gene1, prot1) | m(n, gene2, prot2) ]
```

repeatedly. Species expressions also include a construct for *updating* the modification state of atomic species inside a complex. We illustrate this in Listing 3.3 which gives a more concise version of the CBS scaffolded MAPK cascade.

The first two lines consist of new species definitions as before, but now some of the species are defined with a modification site called p of boolean type. Lines 4-8 represent scaffold formation, but now the intermediate complexes are bound to identifiers using the **as** keyword. This *in-line* approach to binding is an abbreviation for binding a species expression to an identifier, then using the identifier in subsequent reactions, so e.g. the program:

```
Ste5 + Ste11 -> Ste5-Ste11 as a; ...
```

is an abbreviation for the program:

```
spec a = Ste5-Ste11;
Ste5 + Ste11 -> Ste5-Ste11 | ...
```

Reactions which have in-line definitions are composed in *sequence*, using the ; operator rather than the parallel one, as the order of such reactions matters since identifiers defined in one reaction can only be used in the following ones.

The species bound to e in line 8 is the full scaffold complex in its unphosphorylated form. Lines 10-12 represent the actual MAPK cascade. In line 10, the complex bound to e becomes the same complex, but updated by changing the phosphorylation state of site p in Ste11 to true. The result is then bound to a new identifier, f. The last two lines follow a similar pattern. When updates are made on atomic species we use an abbreviation and write e.g. Fus3{p} instead of Fus3<Fus3{p}>, although this is not relevant in the above example.

Listing 3.3. An LBS program for a scaffolded MAPK cascade in yeast.

```
spec Fus3 = new{p:bool}, Ste7 = new{p:bool};
spec Ste11 = new{p:bool}, spec Ste5 = new{p:bool};
spec Ste20 = new{}, Gbg = new{};

Ste5 + Ste11 -> Ste5-Ste11 as a;
Ste7 + Fus3  -> Ste7-Fus3  as b;
a + b        -> a-b        as c;
c + Gbg      -> c-Gbg      as d;
d + Ste20 -> d-Ste20 as e;

e -> e<Ste11{p}> as f;
f -> f<Ste7{p}>  as g;
g -> g<Fus3{p}>
```

The reactions in the above LBS program avoid the redundancy which impairs the reactions in the corresponding CBS program in Listing 2.2. This improves readability. It also facilitates the process of program revision since adding e.g. a new phosphorylated site to the definition of Ste5 only involves a subsequent change to the first reaction. Contrast this to the corresponding CBS program in which the same revision requires two changes in each of the eight reactions.

There are two further, perhaps less commonly used, species expression operators which enable complex species to be taken apart. Assuming the definition of g given above, the *selection* expression g.Ste7 results in the atomic species from g identified by Ste7; the *removal* expression g\Ste7 results in the complex species g without Ste7. Hence the reaction g −> g.Ste7 + g\Ste7 represents dissociation of Ste7 from g and could in this case be written explicitly, if more laboriously, as follows:

```
Fus3-Ste7{p}-Ste11{p}-Ste5-Ste20-Gbg ->
  Ste7{p} + Fus3-Ste11{p}-Ste5-Ste20-Gbg
```

However, the species selection and removal operators are needed language constructs, not just notational conveniences. A species identifier used as the target of the selection and removal operators could be a formal parameter, and as a consequence of subtyping (introduced in the next subsection) its complete make-up in terms of atomic species may not generally be known.

If the complex bound to g is a homo-multimer and has several copies of e.g. Ste7, then selection and removal operate on all copies. This convention also applies to the update operator: if the target of an update contains multiple copies of the given species name, all copies are updated accordingly. It is however possible to make distinctions between different copies of the same species in homo-multimers, and we give an example of this when presenting the formal semantics of LBS.

Listing 3.4. A modular LBS program for the Ras/Raf/MEK/ERK signalling cascade.

```
1  module ph( spec k, s:{m} ) {
2        k + s -> k-s |
3        k-s -> k-s{m} |
4        k-s{m} -> k + s{m}
5  };
6
7  module dph( spec p, s:{m} ) {
8        p + s{m} -> p-s{m} |
9        p-s{m} -> p-s |
10       p-s  -> p + s
11 };
12
13 module cycle( spec k, p, s:{m} ) {
14      ph( k, s:{m} )  |  dph( p, s:{m} )
15 };
16
17 spec Ras = new{};
18 spec Raf = new{m:bool};
19 spec MEK = new{S218:bool,S222:bool};
20 spec ERK = new{T185:bool,Y187:bool};
21 spec PP2A1 = new{}, PP2A2 = new{}, MKP3 = new{};
22
23 cycle(Ras, PP2A1, Raf:{m}) |
24 cycle(Raf{m}, PP2A2, MEK:{S218}) |
25 cycle(Raf{m}, PP2A2, MEK{S218}:{S222}) |
26 cycle(MEK{S218,S222}, MKP3, ERK:{T185}) |
27 cycle(MEK{S218,S222}, MKP3, ERK{T185}:{Y187})
```

3.4 Parametric Type and Subtyping

We now return to the unscaffolded MAPK cascade and the issue of reusability. We noted that all the cycle modules used by the CBS program in Listing 2.3 have the same structure, each with two sets of reactions for representing, respectively, phosphorylation and dephosphorylation. The LBS program in listing 3.4 shows how a general, parameterised cycle module can be defined, which in turn relies on two general modules for phosphorylation and dephosphorylation.

The phosphorylation module named ph in lines 1-5 contains three reactions: binding of a kinase k and substrate s, phosphorylation of s in the bound state, and unbinding after phosphorylation. The two species are formal parameters, but in contrast to earlier examples, the formal parameter s has an *annotation* specifying that it must have a modification site m. The dephosphorylation module named dph in lines 7-11 follows a similar structure and is parameterised on a phosphatase p rather than a kinase. The cycle module in lines 13-15 is parameterised on a kinase, a phosphatase and a substrate and invokes the phosphorylation and dephosphorylation modules in parallel. The invocations provide annotations for matching up the modification sites in the actual parameters with the corresponding modification sites in the formal parameters, which in this case is trivial since there is only the single choice, m. Note that there is scope for further abstraction since the phosphorylation and dephosphorylation modules are very similar. In fact they could be abstracted into a single module, but we refrain from doing so for the sake of clarity.

Lines 17-21 define the new species participating in the program and the remaining lines invoke modules for the appropriate cycles. Let us consider the invocation in line 24 in more detail. The first actual parameter, Raf{m}, provides Raf in its phosphorylated state as the kinase, and the second parameter provides PP2A2 as the phosphatase. The third parameter, MEK:{S218}, provides unphosphorylated MEK as the substrate and the annotation {S218} specifies the target site for phosphorylation. This raises two important points. Firstly, the names of modification sites in the actual and formal annotations differ, resulting in a notion of *parametric type*. The underlying semantics maintains a mapping from formal to actual modification site names when evaluating the body of a module. Secondly, there are two possible choices of modification sites to be phosphorylated in the actual parameter, namely S218 and S222. The annotation picks out the former, and the latter then plays no rôle from the perspective of the module. This results in a notion of *subtyping*: any actual parameter will do, as long as it contains at least the sites specified in the annotation and with types that match the corresponding formals. This corresponds to record subtyping in classical programming languages [30]. The module invocation in line 25 is similar but picks out the second site, S222, for phosphorylation, and also specifies that MEK is already phosphorylated on site S218.

In general, parameters may be complexes rather than atomic species. Suppose for example that MEK is in a complex with some other species a in the actual parameter in line 25. This can be written as follows:

```
...
cycle(Raf{m}, PP2A2, MEK{S218}-a:MEK{S222}) |
...
```

This results in an additional layer of subtyping: any actual parameter will do, as long as it contains at least the atomic species in the annotation. The annotation is here extended in order to specify that it is the atomic species MEK rather than a that should be mapped to the substrate. In fact the annotations used in Listing 3.4 are abbreviations of this more general form, so e.g. MEK{S218}:{S222} is an abbreviation of MEK{S218}:MEK{S222}. Similarly, the annotated formal parameter s:{m} in the cycle module abbreviates s:s{m}, and formal annotations may in general contain multiple atomic species.

We end the discussion of parameterised modules with an abstraction of the entire MAPK cascade into a module which is itself reusable. This, together with a module invocation, is shown in Listing 3.5.

Listing 3.5. A general, modular LBS program for the Ras/Raf/MEK/ERK signalling cascade. The species and module definitions from Listing 3.4 are omitted.

```
...
module mapk(
  spec k4, k3:{m}, k2:{m1, m2}, k1:{m1, m2},
       p3, p2, p1) {

  cycle(k4, p3, k3:{m}) |
  cycle(k3{m}, p2, k2:{m1}) |
  cycle(k3{m}, p2, k2{m1}:{m2}) |
  cycle(k2{m1,m2}, p1, k1:{m1}) |
  cycle(k2{m1,m2}, p1, k1{m1}:{m2})
};

mapk(Ras, Raf:{m}, MEK:{S218,S222}, ERK:{T185,Y187},
     PP2A1, PP2A2, MKP3)
```

3.5 Nondeterminism

Nondeterminism for contextual combinatorial explosion. In the previous examples we assumed that species only participate in reactions when they are atomic or when they are in the context of a specific complex as in the scaffolded MAPK cascade. In reality however, atomic species are likely to react in the context of many possible complexes. In the unscaffolded MAPK cascade, Raf may for example continue to function as a kinase for MEK when it is bound to its own kinases and/or phosphatases and when MEK is bound to its own phosphatase. This gives rise to a kind of combinatorial explosion which we call *contextual* and which is difficult to model in CBS under the basic semantics in terms of Petri nets, ODEs and CTMCs. The reason is that species in reactions are interpreted at face value with an "empty context" (except for modification site variables which allow combinatorial explosion at the level of modifications to be handled). In contrast, reaction rules in rule-based languages such as κ and BioNetGen are interpreted "in any context", and can be executed or analysed without generating the full set of empty context reactions which may in some cases be infinite. Languages such as κ and BioNetGen therefore handle combinatorial systems very well.

This could be exploited by giving a translation of LBS into κ or BioNetGen based on a suitable choice of modification site types, but we do not pursue this direction here. Instead we seek a middle ground in which all possible species contexts continue to be specified in reactions, but in a syntactically compact manner through the notion of *nondeterministic* species. Listing 3.6 shows an example of phosphorylation using nondeterministic versions of Raf and MEK.

The **or** operator expresses that either of its operands can take place in reactions where the expression is used. The distinguished species **SNil** is a neutral element under the complex formation operator, i.e. the axiom a−**SNil** = a holds

Listing 3.6. Phosphorylation using nondeterministic species

```
1 spec NRaf = Raf{m}-(SNil or Ras or PP2A1);
2 spec NMEK = MEK-(SNil or PP2A2);
3
4 ph(NRaf, NMEK:MEK{S218})
```

for any species a. The distributivity axiom a−(b **or** c) = a−b **or** a−c also holds for all species a, b and c. Hence line 1 in the above program expands to a choice of three species, namely Raf{m} in isolation or in complex with Ras or PP2A1, and line 2 expands to a choice of two species, namely MEK in isolation or in complex with PP2A2.

A reaction with nondeterministic species semantically gives rise to a number of parallel reactions, one for each possible choice of species. For example, the first reaction k + s −> k−s in the ph module now gives rise to a parallel composition of 6 reactions:

```
Raf{m} + MEK -> Raf{m}-MEK |
Raf{m}-Ras + MEK -> Raf{m}-Ras-MEK |
Raf{m}-PP2A1 + MEK -> Raf{m}-PP2A1-MEK |

Raf{m} + MEK-PP2A2 -> Raf{m}-MEK-PP2A2 |
Raf{m}-Ras + MEK-PP2A2 -> Raf{m}-Ras-MEK-PP2A2 |
Raf{m}-PP2A1 + MEK-PP2A2 -> Raf{m}-PP2A1-MEK-PP2A2 |
```

The two other reactions in the ph module have similar expansions, and the ph module invocation hence results in a total of 18 reactions.

Nondeterminism for species variant combinatorial explosion. The above example demonstrates how nondeterminism can be used to drastically reduce the size of programs in which the combinatorial explosion is contextual. Nondeterminism can also be used to handle combinatorial explosions arising from variants of individual proteins which largely react in the same way. For example, Raf has three variants RafA, RafB and RafC; MEK has two variants MEK1 and MEK2; and ERK has two variants ERK1 and ERK2 [28]. Rule-based languages such as κ and BioNetGen do *not* per se have any dedicated means of handling this source of nondeterminism, but a recent extension of κ provides some level of syntactical support [10]. Indeed this κ extension, together with our need to handle contextual combinatorial explosion, are the two motivating factors for the introduction of nondeterminism into LBS. Listing 3.7 shows how the MAPK cascade module can be used with species variants. In order to be of interest, one would expect that *some* reactions distinguish between the variants, but we omit this aspect in the present example.

Each individual member of the nondeterministic species in lines 16-18 is given a separate annotation at time of definition rather than at time of module invocation. The reason is that the members do not have any common atomic species, and in general they may not have common modification sites either,

Listing 3.7. The MAPK cascade module instantiated with nondeterministic species

```
...
spec Ras   = new{};

spec RafA  = new{m: bool};
spec RafB  = new{m: bool};
spec RafC  = new{m: bool};

spec MEK1  = new{S218 : bool, S222 : bool};
spec MEK2  = new{S218 : bool, S222 : bool};

spec ERK1  = new{T185 : bool, Y187 : bool};
spec ERK2  = new{T185 : bool, Y187 : bool};

spec PP2A1 = new{}, PP2A2 = new{}, MKP3 = new{};

spec NRaf = RafA:{m} or RafB:{m} or RafC:{m};
spec NMEK = MEK1:{S218, S222} or MEK2:{S218, S222};
spec NERK = ERK1:{T185, Y187} or ERK2:{T185, Y187};

mapk(Ras, NRaf, NMEK, NERK, PP2A1, PP2A2, MKP3)
```

so it is necessary to identify the atomic species and sites to be mapped from the corresponding formals on an individual basis; recall here that e.g. RafA:{m} and RafB:{m} abbreviate respectively RafA:RafA{m} and RafB:RafB{m}, so the annotations do indeed differ between different members of the nondeterministic species. For that reason also semantically, annotations are associated with species rather than with module invocations, and the mappings between formals and actuals are maintained locally within individual species rather than globally. Invocation of the mapk module results in 102 reactions as opposed to the 30 reactions in the original program.

The mechanism of nondeterministic selection. An important point about Listing 3.7 is that nondeterministic species are expanded *at the level of reactions* and not at the level of modules. This means that the mapk module invocation is *not* equivalent to a parallel composition of module invocations for each choice of species. Such an interpretation would result in 360 reactions, but $360 - 102 = 258$ of these would be duplicates, effectively adding up the rates of duplicated reactions, which is certainly not what we intend. In this respect LBS has a *call-by-name* semantics. On the other hand, species identifiers are resolved at time of module invocation where actual species parameters are formally evaluated to sets of species values, so in this respect LBS has a *call-by-value* semantics.

An additional subtlety arises in reactions where the same species occurs multiple times as a reactant or product. Consider for example the following:

```
spec a1 = new{}, a2 = new{}, b = new{};
spec a = a1 or a2;
a + a + b -> a-a-b
```

There are two reasonable, but very different, possibilities for expansion of the reaction. The first requires that the same choice for a is made within the scope of the reaction:

```
a1 + a1 + b -> a1-a1-b |
a2 + a2 + b -> a2-a2-b
```

The second possibility allows different copies of the same identifier to take different values, but with the correspondence between the occurrences on the reactant and product sides being preserved:

```
a1 + a1 + b -> a1-a1-b |
a1 + a2 + b -> a2-a2-b |
a2 + a1 + b -> a2-a1-b |
a2 + a2 + b -> a2-a2-b
```

The correct expansion depends on the specific application. Although the first may seem most appropriate in the general case, the second is for example useful for modelling the combinatorial dimerisation of different variants of ErbB receptors [10] (note that two of the resulting reactions are equivalent, effectively duplicating their rates). In order to cater for these different possibilities, LBS has two reaction arrows. The basic reaction arrow, ->, which has been used in the examples so far, results in the first expansion, and we call this a *selection arrow*. The double-headed reaction arrow, ->>, results in the second expansion, and we call this an *L-R equality-preserving arrow*.

In order to give a uniform semantical treatment of nondeterminism, and to enable other expansions than the two described above, LBS has a dedicated **force** operator for forcing nondeterministic choice. For example, the program:

```
spec a = force a1 or a2;
P
```

results in a parallel composition of P with a binding of a to a1 in one parallel component and P with a binding to a2 in the other parallel component. Reactions using either of the two arrows are then derived forms expressed in terms of the force operator and a third *deterministic* reaction arrow, =>, which requires that reactants and products do not contain nondeterministic species. For example, the program:

```
a + a + b -> a-a-b
```

abbreviates the program:

```
spec a = force(a);
spec b = force(b);
a + a + b => a-a-b
```

and the program:

```
a + a + b ->> a-a-b
```

abbreviates the program:

```
spec a1 = force(a);
spec a2 = force(a);
spec b1 = force(b);
a1 + a2 + b1 => a1-a2-b1
```

Nondeterministic species expressions which are not bound to identifiers are not allowed in reactions with any of the three arrows. This is because the implicit forcing is done on identifiers rather than on general species expressions, which allows the identity between different occurrences of e.g. the species a to be preserved after nondeterministic selection.

Limitations. In the examples we have assumed that reaction rates are independent of nondeterministic choices. This assumption appears to be in line with published models of e.g. the EGFR pathway [7], although this may be due to limited knowledge rather than biological reality. We note however that it *is* possible to choose different rates for different members of nondeterministic species based on their state of modification by using conditionals in rate expressions. Even with no distinguishable biological state of modification, one can "cheat" and add a site which serves the sole purpose of identifying members of nondeterministic species. But in the extreme case where all possible combinations of nondeterministic members in reactions give rise to different rates, this approach merely shifts the problem of combinatorial explosion from the reaction level to the reaction rate level. This is a potential practical limitation of nondeterminism.

A related limitation is the inability to constrain the choice of one nondeterministic species based on the choice of another; currently two different nondeterministic species in a reaction gives rise to reactions with *all* possible combinations of choices. This can however be addressed using a similar approach to that described above, where reactions with any incompatible choices are given rates of 0.

Finally, we note that even though nondeterminism can be used to write combinatorial programs in a syntactically compact manner, the resulting semantical objects may be too large for analysis or simulation to be feasible.

3.6 Model Variation

Given an LBS program it is sometimes of interest to vary it in a number of ways and examine the resulting effect on behaviour. In support of a structured approach to variations, LBS has a *variation operator*, ||, which semantically

Listing 3.8. An extension of the MAPK program in Listing 3.7 (not repeated here) with variations for generating one semantical object for each possible initial condition.

```
1  ... |
2  (init RafA 500 || PNil) |
3  (init RafB 500 || PNil) |
4  (init RafC 500 || PNil) |
5
6  (init MEK1 500 || PNil) |
7  (init MEK2 500 || PNil) |
8
9  (init ERK1 500 || PNil) |
10 (init ERK2 500 || PNil)
```

gives the union of its operands, i.e. programs evaluate to *sets* of semantical objects. Listing 3.8 shows how the variation operator can be used to generate a semantical object for each of the given possible initial conditions of species variants in the MAPK cascade program.

Initial condition statements such as **init** RafA 500 are first-class programs and specify a given initial population or concentration for a species. If no initial conditions are specified in a program, the 0 initial population or concentration is assumed for all participating species. The distinguished program **PNil** is a neutral element under parallel composition, i.e. the axiom P | **PNil** = P holds for all programs P; also the distributivity axiom P1 | (P2 || P3) = (P1 | P2) || (P1 | P3) holds for all programs P1, P2 and P3. Hence the parallel composition shown above is a power set construction and expands to a variation composition of all $2^7 = 128$ possible combinations of initial conditions in parallel with the program represented by dots in line 1, in this case the previously defined MAPK cascade.

3.7 Output Species Parameters

Manipulations of large complexes are often spread across multiple modules. Sometimes there is a natural input-output relationship between these modules where a species which is constructed in one module may be the starting point for further manipulation in another. This applies for example to the scaffolded MAPK cascade program in Listing 3.3 which can benefit from a decomposition into two modules, one for scaffold formation, and one for the actual MAPK cascade. The fully formed scaffold in its unphosphorylated state can be considered as an output of the first module and as an input to the second. Although one could simply pass this connecting species as a common parameter to both modules, this would involve the entire scaffold to be written out at the time of module invocation, thus repeating the definitions already given during scaffold formation. In order to avoid this, we introduce the notion of *output species*, and a modular version of the yeast MAPK cascade using this idea is shown in Listing 3.9.

The first two lines define four new species while the remaining two species used in the program are defined locally in the formation module. The formation module

Listing 3.9. A modular LBS program for scaffold formation and a scaffolded MAPK cascade in yeast.

```
1  spec Fus3 = new{p:bool}, Ste7 = new{p:bool};
2  spec Ste11 = new{p:bool}, Ste5 = new {p:bool};
3  
4  module formation(specout e : Fus3-Ste7-Ste11-Ste5) {
5    spec Ste20 = new{}, Gbg = new{};
6    Ste5 + Ste11 -> Ste5-Ste11 as a;
7    Ste7 + Fus3  -> Ste7-Fus3  as b;
8    a + b        -> a-b as c;
9    c + Gbg      -> c-Gbg as d;
10   d + Ste20    -> d-Ste20 as e
11 };
12 
13 module mapk(spec a : k1{m}-k2{m}-k3{m}; specout d : a) {
14   a -> a<k3{m=tt}> as b;
15   b -> b<k2{m=tt}> as c;
16   c -> c<k1{m=tt}> as d
17 };
18 
19 formation(spec link1);
20 mapk(link1 : Fus3{p}-Ste7{p}-Ste11{p}, spec link2);
21 ...
```

has a single formal parameter which specifies that the species e defined in the module body is given as an output, and the associated annotation specifies that the output contains the species Fus3, Ste7, Ste11 and Ste5. In fact the output also contains Ste20 and Gbg, but these are not exposed, which gives rise to a notion of subtyping similar to that of standard species parameters.

The mapk module has a parameter a and also an output species parameter d which is defined in the module body and is specified to contain at least the species of a. In line 19 the formation module is invoked and results in a binding of the identifier link1 to the output scaffold species. In line 20 this is passed on as a parameter to the mapk module which in turn results in a binding of the identifier link2 to the phosphorylated scaffold. In the full model of the yeast pheromone pathway, Ste5 dissociates from the scaffold link2 resulting from the MAPK cascade. We can deduce by inspection of the program that the complex bound to link2 does indeed contain the species Ste5, since link1 contains Ste5 and link2 contains at least the same species as link1. The full program contains five modules which can be connected naturally using this approach [27].

4 The Abstract Syntax of LBS

The previous section gave an informal introduction to the *concrete* syntax and main features of LBS. This section formally defines the *abstract* syntax of LBS which forms the basis of the general and concrete semantics given in the next two sections. The formal definition of the concrete syntax and its mapping into the abstract syntax are omitted, since both can be deduced without surprises from the examples and from the abstract syntax.

In order to achieve our aim of generality, the abstract syntax is parameterised on a set of *modification site types* ρ and *modification site expressions* e_{m}. We divide the language into four main syntactic categories, for compartments, species, programs and definitions, and consider each in turn. But first we introduce the notation used in this section and throughout the paper.

4.1 Notation

We let $\mathbb{R}$ denote the set of *real numbers* and $\mathbb{N}$ denote the set of *natural numbers*. We write $\underline{x}$ for *lists*, $\underline{x}.i$ for the *ith element* (starting from 1) of a list, $|\underline{x}|$ for the *length* of a list and ε for the *empty* list. When a list should be thought of as representing a set, we write $\underset{\sim}{x}$ instead of $\underline{x}$. The set of *indices* of a list $\underline{x}$ is $\{i \mid 1 \leq i \leq |x|\}$. The *sublist* of $\underline{x}$ consisting of the elements at some subset I of the indices of $\underline{x}$ is written $\underline{x}.I$. The *Cartesian product* $\underline{x} \times_\circ \underline{y}$, where $\circ$ is a pairing operator on elements of the respective lists, is the list of length $|\underline{x}| \cdot |\underline{y}|$ s.t. $(\underline{x} \times_\circ \underline{y}).((i-1)|\underline{x}| + j) \overset{\Delta}{\simeq} \underline{x}.i \circ \underline{y}.j$. The *concatenation* of lists $\underline{x}$ and $\underline{y}$ is written $\underline{x}\,\underline{y}$, and the *prefix* and *postfix* of an element a to a list $\underline{x}$ are written $a\underline{x}$ and $\underline{x}a$, respectively.

We write $\{x_i\}_{i \in I}$ for a finite *indexed set* and omit I and/or i and write $\{x_i\}_I$, $\{x_i\}$ or $\{x\}$ when they are understood from the context. The set of *multisets* of a set S is denoted by $MS(S)$ and is defined as the set of total functions from S to the natural numbers, i.e. $MS(S) \overset{\Delta}{\simeq} S \to \mathbb{N}$. We adopt the usual multiset notation and write e.g. $x + 2 \cdot y$ for the multiset containing the element x and two copies of y, and we write $MS(\underline{x})$ for the multiset representation of a list $\underline{x}$. We also use the standard notation $\prod_{i \in I} X_i$ for dependent sets.

We write $x \overset{\Delta}{\simeq} y$ for definitions where x equals y if y is defined, and where x is undefined otherwise. When a notion of well-typedness applies to y, we furthermore write $x \overset{\Delta}{\simeq}_t y$ for definitions where x equals y if y is defined and well-typed, and where x is undefined otherwise.

Partial finite *functions* f are denoted by finite indexed sets of pairs $\{x_i \mapsto y_i\}$ where $f(x_i) = y_i$. The *empty* function is correspondingly denoted by $\emptyset$. The *domain of definition* and *image* of a function f are denoted by $dom(f)$ and $im(f)$, respectively. For functions f and g we define the *update* of f by g, written $f\langle g \rangle$, as follows:

$$f\langle g \rangle(x) \overset{\Delta}{\simeq} \begin{cases} f(x) & \text{if } x \in dom(f) \setminus dom(g) \\ g(x) & \text{if } x \in dom(g) \end{cases}$$

If g consists of a single binding $x \mapsto y$ we write $f\langle x \mapsto y \rangle$ instead of $f\langle \{x \mapsto y\} \rangle$. We specify the type of a partial function f by writing $f(x) = y$ where x and y are given variables ranging over two sets; these sets are then understood to form the domain and image of f.

When an element of a list or an indexed set is referred to without explicit quantification in a semantical definition, the index is assumed to be universally quantified over a set which is understood from the context. Under such circumstances we often omit the index and write e.g. x instead of $\underline{x}.i$. If $\circ$ is an operation on the elements of lists $\underline{x}$ and $\underline{y}$ both of length n, we write $\underline{x \circ y}$ for the list of length n in which the ith element is $\underline{x}.i \circ \underline{y}.i$.

4.2 Compartments

Compartment expressions. The abstract syntax for basic compartment expressions is shown in Table 4.1, where id_c ranges over the set of *compartment identifiers* and $r \in \mathbb{R}$. New compartments are created using the new compartment expression which explicitly records a parent compartment and a volume. In cases where a compartment is used at the top level, the world compartment can be specified as a parent, hence allowing compartment hierarchies to be terminated. A compartment is generally used in multiple contexts by binding it to an identifier at time of creation. The nil compartment functions as a neutral element for the composition of compartment lists. It is paired with a parent compartment, which is necessary for type-checking of compartment hierarchies. Nil compartments can for example be used to decrease the depth of a module hierarchy when passed as parameters to modules.

Table 4.1. The abstract syntax for compartment expressions

e_c ::=	Compartment expression
| **new comp vol** r **inside** e_c	New compartment
| $\top$	World compartment
| id_c	Compartment identifier
| $\mathbf{1}_c$ **in** e_c	Nil compartment

Although the world compartment figures as a general compartment in the abstract syntax, it is only intended for use as a parent of new compartments and of the nil compartment. It is not intended for use in e.g. reactions, and its proper usage is enforced in the semantics for programs. One could enforce this intended usage syntactically by introducing separate production rules for top level and nested new compartment and nil compartment expressions. However, whether or not a compartment features at the global top level of a program is not generally known at time of definition: take for example compartment definitions inside a module, where the parent compartment may be a formal parameter.

Derived compartment expressions. The volume in new compartment expressions may be omitted, in which case a default volume of 1.0 is assumed. The parent compartment in new compartment and nil compartment expressions may also be omitted, in which case the world parent compartment is assumed.

4.3 Species

Modification site expressions. Recall that the abstract syntax for species expressions is parameterised on a set of modification site types ρ and a set of modification site expressions e_m. Since boolean expressions are of widespread practical use as demonstrated in the examples, we assume that the set of modification types contains the boolean type **bool**, and that the set of modification site expressions contains the boolean expressions e_b generated by the grammar in Table 4.2 where x ranges over the set of *variables*. The boolean expressions contain the usual **tt**/**ff** base values and a minimal set of connectives from which the full set of boolean connectives can be defined as derived forms in the usual manner. Variables are used to create species expressions which can match multiple concrete species. We assume that the set of variables is closed by prefixing of underscore-terminated binary strings, i.e. that b_x is a variable for all $b \in \{0,1\}^*$; this is needed to confine variables to their appropriate namespace in the semantics. The type annotation of variables is likewise used for technical convenience in the semantics.

Table 4.2. The abstract syntax for boolean expressions

e_b ::=	BOOLEAN EXPRESSION
| **tt**	TRUE
| **ff**	FALSE
| x : **bool**	TYPED VARIABLE
| e_b **or** e_b' | **not** e_b	BOOLEAN OPERATORS

Species expressions. The abstract syntax for species expressions is shown in Table 4.3, where n_s ranges over the set of *species names*, n_m ranges over the set of *modification site names* and id_s ranges over the set of *species identifiers*. Species names identify atomic species independently of any modification sites, while species identifiers refer to possibly complex species including both the names and modification states of atomic species in the complex. We assume for technical reasons that both the set of species identifiers and the set of binary strings is contained in the set of species names. We assume furthermore, as for variables, that the set of species identifiers is closed by prefixing of underscore-terminated binary strings, i.e. that b_id_s is also a species identifier for all $b \in \{0,1\}^*$.

The grammar distinguishes between species expressions e_s and extended species expressions $e_{\mathrm{s}+}$ which add the new atomic species expression. This is because new species expressions only make sense in the context of definitions, where the resulting new species value can be bound to an identifier. Species bound to an identifier can then be used in multiple contexts and given an initial population through the construct given in the abstract syntax for programs. Technically, separating out the new species expression alleviates the need to consider fresh names in the semantics for the remaining expressions and for certain cases of programs; this significantly simplifies the presentation.

Table 4.3. The abstract syntax for species expressions

$e_{s+} ::= e_s$	EXTENDED SPECIES EXPRESSION
$\mid$ **new** n_s, σ	NEW ATOMIC SPECIES
$e_s ::=$	SPECIES EXPRESSION
$\mid id_c[e_s]$	LOCATED SPECIES
$\mid e_s - e_s'$	COMPOSITE SPECIES
$\mid e_s.\underline{id_c}[n_s]$	SPECIES SELECTION
$\mid e_s \backslash \underline{id_c}[n_s]$	SPECIES REMOVAL
$\mid e_s \langle \underline{id_c}[n_s, \alpha] \rangle$	SPECIES UPDATE
$\mid e_s$ **or** e_s'	SPECIES CHOICE
$\mid e_s : \xi$	SPECIES ANNOTATION
$\mid id_s$	SPECIES IDENTIFIER
$\mid \mathbf{0}_s$	NIL SPECIES
$\xi ::= \underline{\underline{id_c[n_s, \underline{n_m}]}}$	ANNOTATION
$\sigma ::= \{n_m \mapsto \rho\}$	MODIFICATION TYPE
$\alpha ::= \{n_m \mapsto e_m\}$	MODIFICATION ASSIGNMENT

New atomic species are created by specifying a name and a type consisting of a partial finite function from modification site names to modification site types. The modification sites are assigned default expressions appropriate for the corresponding type, e.g. **ff** in the case of the **bool** type. In contrast to new compartment expressions, a new species expression explicitly includes a species name. Often this name is the same as the identifier to which the new species expression is assigned, which is reflected in a derived form of definitions. Although semantically the underlying unique species name will be freshly generated, the specified name is used to identify specific atomic species in subsequent species selection, removal and update expressions. Species names rather than general species expressions are used here for two reasons. First, the update expression updates a specific atomic species in a complex. Second, atomic species names are local to a species, meaning that the same atomic species name in two different species may map to different underlying fresh species names. This is used to cater for nondeterminism in the context of parametric types in species parameters of modules. Similar considerations of nondeterminism apply to compartments, which are only used in species expressions indirectly through compartment identifiers rather than through general compartment expressions.

The species annotations necessary to match the names and sites of actual parameters to those of formal parameters are handled in the abstract syntax for species expressions rather than in the abstract syntax for module invocation in programs. This too is because of nondeterminism where separate annotations may be required for each member of a nondeterministic species expression as demonstrated previously in Listing 3.7.

Derived species expressions. Two derived forms of species expressions allow updates and annotations of atomic species without having to repeat atomic species names. Specifically, the expressions:

$$id_{\mathrm{s}}\{\alpha\} \quad \text{and} \quad id_{\mathrm{s}} : \underline{n_{\mathrm{m}}}$$

abbreviate respectively the expressions:

$$id_{\mathrm{s}}\langle \varepsilon[id_{\mathrm{s}}, \alpha]\rangle \quad \text{and} \quad id_{\mathrm{s}} : \underline{\varepsilon[id_{\mathrm{s}}, \underline{n_{\mathrm{m}}}]}$$

4.4 Programs

Basic programs. The abstract syntax for programs is shown in Table 4.4, where $n \in \mathbb{N}$, id_{p} ranges over the set of *program identifiers* and id_{a} ranges over the set of *algebraic rate function identifiers.* Definitions, ranged over by D, are treated in the next subsection. Module invocations include actual parameters for compartments, species, rates and output species, and as already pointed out, the annotations of actual species parameters necessary to match the formal parameters are handled in the abstract syntax for species expressions.

Reaction rate expressions can either be mass-action rates, given inside braces, or general algebraic rate expressions, given inside square brackets. Algebraic rate expressions include rate constants, compartments and species, where the latter two are interpreted as respectively a volume and a population. Algebraic rate expressions also include a number of basic functions and arithmetic operators which feature regularly in the biological literature; these are inspired by similar features found in BioPEPA. Custom rate functions which are parameterised on compartments, species and algebraic rate expressions can be defined and invoked repeatedly. These also allow the definition of common rate functions for e.g. Michaelis-Menten or Hill kinetics. Conditionals enable different rates to be chosen depending on the state of modification of reactants as recorded by match variables. This mechanism also allows a distinction between different members of a nondeterministic species to be made assuming a convention where an additional site is added in which a value identifies each individual nondeterministic member. Note that mass-action rates are represented by algebraic expressions in the abstract syntax because this allows for a uniform treatment of defined constants and conditionals. Semantically however, mass-action algebraic rate expressions are required to evaluate to constants.

Only the simplest possible reaction is included in the abstract syntax for programs. Species expressions are assumed to be deterministic, requiring any nondeterministic selection to be carried out in advance through the use of the force operator; there are no in-line species definitions; and there are no reversible or enzymatic reactions.

Table 4.4. The abstract syntax for basic programs

$P ::=$	PROGRAM
$\mid \underline{n \cdot e_s} \Rightarrow^{e_r} \underline{n' \cdot e'_s}$ **if** e_b	REACTION
$\mid \mathbf{0}_p$	NIL PROGRAM
$\mid P \mid P'$	PARALLEL COMPOSITION
$\mid P \mid\mid P'$	VARIATION COMPOSITION
$\mid id_c[P]$	LOCATED PROGRAM
$\mid D \,;\, P$	DEFINITION
$\mid id_p(\underline{e_c}; \underline{e_{s+}}; \underline{e_a};$ **out** $\underline{id_s})\ ;\ P$	MODULE INVOCATION
$\mid id_s =$ **force** $e_s\ ;\ P$	NONDETERMINISTIC SELECTION
$\mid$ **init** $e_s = r$	INITIAL POPULATION
$e_r ::=$	RATE EXPRESSION
$\mid \{e_a\}$	MASS-ACTION RATE
$\mid [e_a]$	ALGEBRAIC RATE
$e_a ::=$	ALGEBRAIC RATE EXPRESSION
$\mid r$	CONSTANT
$\mid id_c$	VOLUME
$\mid e_s$	POPULATION
$\mid$ **if** e_b **then** e_a **else** e'_a	CONDITIONAL
$\mid id_a(\underline{e_c}; \underline{e_s}; \underline{e_a})$	FUNCTION INVOCATION
$\mid exp(e_a) \mid log(e_a) \mid sin(e_a) \mid cos(e_a)$	STANDARD FUNCTIONS
$\mid e_a + e'_a \mid e_a - e'_a$	ARITHMETIC OPERATORS
$\mid e_a \times e'_a \mid e_a \,/\, e'_a \mid e_a\hat{\ }e'_a$	

Derived programs. More complicated reactions are generated by the abstract syntax for derived programs in Table 4.5, all of which can be defined in terms of basic programs. The dots in the grammar indicate extension of the grammar for basic programs. Derived programs include the two additional reaction arrows which cater for nondeterministic species and which implicitly force nondeterministic selection in two different manners, as exemplified in section 3.5. Enzymatic reactions are given by a list of enzymes to the left of the tilde symbol. All types of reactions can be reversible with any combination of mass action and general algebraic rate expressions for each of the two directions. Finally, species expressions in derived reactions may contain in-line definitions which go into scope in the sequential program following the reaction.

Further derived forms arise by omitting the enzyme or boolean expression parts of reactions. The absence of an enzyme part is understood as an enzyme part with an empty list of species, and the absence of a boolean expression part

Table 4.5. The abstract syntax for derived programs

$P ::= \ldots$	DERIVED PROGRAM
$\mid \underline{e''_{\mathrm{s}}} \sim \underline{n \cdot e_{\mathrm{s}}}\ A^{e_{\mathrm{r}}}\ \underline{n' \cdot e'_{\mathrm{s}}}\ \mathbf{if}\ e_{\mathrm{b}}\ ;\ P$	GENERAL REACTION
$\mid \underline{e''_{\mathrm{s}}} \sim \underline{n \cdot e_{\mathrm{s}}}\ A_2^{e_{\mathrm{r}}, e'_{\mathrm{r}}}\ \underline{n' \cdot e'_{\mathrm{s}}} \sim \underline{e'''_{\mathrm{s}}}\ \mathbf{if}\ e_{\mathrm{b}}, e'_{\mathrm{b}}\ ;\ P$	GENERAL REVERSIBLE REACTION
$A ::=$	REACTION ARROWS
$\mid \Rightarrow$	DETERMINISTIC ARROW
$\mid \rightarrow$	SELECTION ARROW
$\mid \twoheadrightarrow$	L-R EQUALITY-PRESERVING ARROW
$A_2 ::= \Leftrightarrow \mid \leftrightarrow \mid \twoheadleftarrow\!\!\twoheadrightarrow$	REVERSIBLE REACTION ARROWS
$e_{\mathrm{s}} ::= \ldots$	DERIVED SPECIES EXPRESSIONS
$\mid e_{\mathrm{s}}\ \mathbf{as}\ id_{\mathrm{s}}$	INLINE DEFINITION

is understood as a boolean expression part with the expression **tt**. Stoichiometry in reactions can be omitted, in which case stoichiometry 1 is assumed. Finally, the sequential programs following reactions and module invocations can be omitted when there are no in-line species definitions or output species parameters, respectively. In these cases the nil sequential program is assumed.

4.5 Definitions

Basic definitions. The abstract syntax for definitions is shown in Table 4.6 and should be self-explanatory. Formal species parameters have annotations ξ as defined in the abstract syntax for species expressions. Together with the corresponding annotation of actual species parameters, this is sufficient to construct a mapping that allows use of the species inside the module body.

Derived definitions. There is one important derived form concerning new species definitions. Recall from the abstract syntax for species that a new species

Table 4.6. The abstract syntax for definitions

$D ::=$	DEFINITION
$\mid id_{\mathrm{s}} = e_{\mathrm{s}+}$	SPECIES
$\mid id_{\mathrm{c}} = e_{\mathrm{c}}$	COMPARTMENT
$\mid id_{\mathrm{a}}(\underline{id_{\mathrm{c}}}; \underline{id_{\mathrm{s}} : \xi}; \underline{id_{\mathrm{a}}}) = e_{\mathrm{a}}$	FUNCTION
$\mid id_{\mathrm{p}}(\underline{id_{\mathrm{c}}}; \underline{id_{\mathrm{s}} : \xi}; \underline{id_{\mathrm{a}}}; \mathbf{out}\ \underline{id'_{\mathrm{s}} : e_{\mathrm{s}}}) = P$	MODULE

expression includes a species name. But in most cases this name will be identical to the identifier that the new species expression is bound to. The name can then be omitted, i.e. the expression:

$$id_s = \mathbf{new}\ \sigma$$

abbreviates the expression:

$$id_s = \mathbf{new}\ id_s, \sigma$$

This is the reason that the set of species names is assumed to contain the set of species identifiers.

5 The General Semantics

This section defines a denotational framework for compositionally assigning semantical objects such as Petri nets to LBS programs. Our aim is to abstract away from the specific kind of semantical object under consideration.

Assumptions. We achieve our aim of abstraction by assuming a given structure $(S, |_S, \mathbf{0}_S, R_S, I_S)$ consisting of:

- A set S of *semantical objects* ranged over by $\mathcal{O}$.
- A partial binary *composition function* $|_S$ on semantical objects.
- A distinguished *nil* semantical object $\mathbf{0}_S \in S$.
- A partial *reaction assignment function* of the form $R_S(R, b) = \mathcal{O}$ assigning a semantical object to a given reaction R, named b, in a normal form, defined below (b is used to name e.g. Petri net transitions).
- A partial *initial condition assignment function* of the form $I_S(v_{\text{gns}}, r) = \mathcal{O}$ assigning a semantical object to an initial population or concentration r of species v_{gns} in a ground normal form, defined below.

The last implies that semantical objects have a representation of initial conditions, e.g. an initial marking in the case of a Petri net. Specific examples are given in Section 6.

Recall that the abstract syntax is parameterised on modification site types ρ and modification site expressions e_{m}. We assume the following relations and functions pertaining to these:

- A *typing relation* of the form $e_{\text{m}} : \rho$ giving types to modification site expressions. This is used for determining well-typedness of species expressions.
- A *default expression function* of the form $\mathit{default}(\rho) = e_{\text{m}}$ giving default expressions to types. This is used for assigning expressions to unassigned sites in species expressions.
- A *variable function* of the form $FV(e_{\text{m}}) = \{x_i : \rho_i\}$ giving the set of (typed) variables in a modification site expression. This is used for determining well-typedness of reactions and for computing semantical objects of reactions in some of the concrete semantics.

- An *expression denotation function* of the form $[\![e_{\mathrm{m}}]\!]_{\mathrm{m}}\Gamma_{\mathrm{x}} = v_{\mathrm{m}}$ for evaluating a modification site expression to a value v_{m} in a given set $[\![\rho]\!]_{\mathrm{t}}$ where $e_{\mathrm{m}} : \rho$, given a *variable environment* of the form $\Gamma_{\mathrm{x}}(x : \rho) = v_{\mathrm{m}}$ assigning values $v_{\mathrm{m}} \in [\![\rho]\!]_{\mathrm{t}}$ to typed variables. This is used for computing semantical objects of reactions in some of the concrete semantics.
- An *update function* of the form $e_{\mathrm{m}}\langle e'_{\mathrm{m}}\rangle = e''_{\mathrm{m}}$ for updating one modification site expression with another. This is used in the semantics of species update expressions. While this operation is trivial for e.g. boolean expressions in which the original expression is simply disregarded, we anticipate the addition of other types for which the situation is more subtle.
- A *seal function* of the form $\mathit{seal}(e_{\mathrm{m}}, b) = e'_{\mathrm{m}}$ for confining names in modification site expressions to a *namespace* given by a binary string $b \in \{0,1\}^*$. The namespace is used to avoid capture of e.g. variables in actual species parameters when used inside the body of a module.

In the case where only the boolean modification site type is given, and where the set of modification site expressions is hence the set of boolean expressions, the above functions can be defined as follows:

- $e_{\mathrm{m}} : \mathbf{bool}$ for all e_{m}.

- $\mathit{default}(\rho) \stackrel{\Delta}{\simeq} \mathbf{ff}$.

- $FV(e_{\mathrm{m}})$ is defined inductively as follows:
  - $FV(\mathbf{tt}) \stackrel{\Delta}{\simeq} \emptyset$
  - $FV(\mathbf{ff}) \stackrel{\Delta}{\simeq} \emptyset$
  - $FV(x : \mathbf{bool}) \stackrel{\Delta}{\simeq} \{x : \mathbf{bool}\}$
  - $FV(e_{\mathrm{b}}\ \mathbf{or}\ e'_{\mathrm{b}}) \stackrel{\Delta}{\simeq} FV(e_{\mathrm{b}}) \cup FV(e'_{\mathrm{b}})$
  - $FV(\mathbf{not}\ e_{\mathrm{b}}) \stackrel{\Delta}{\simeq} FV(e_{\mathrm{b}})$

- $[\![e_{\mathrm{b}}]\!]_{\mathrm{m}}\Gamma_{\mathrm{x}} = [\![e_{\mathrm{b}}]\!]_{\mathrm{b}}\Gamma_{\mathrm{x}}$ is defined inductively as follows:
  - $[\![\mathbf{tt}]\!]_{\mathrm{b}}\Gamma_{\mathrm{x}} \stackrel{\Delta}{\simeq} \mathbf{tt}$
  - $[\![\mathbf{ff}]\!]_{\mathrm{b}}\Gamma_{\mathrm{x}} \stackrel{\Delta}{\simeq} \mathbf{ff}$
  - $[\![x : \mathbf{bool}]\!]_{\mathrm{b}}\Gamma_{\mathrm{x}} \stackrel{\Delta}{\simeq} \Gamma_{\mathrm{x}}(x)$
  - $[\![e_{\mathrm{b}}\ \mathbf{or}\ e'_{\mathrm{b}}]\!]_{\mathrm{b}}\Gamma_{\mathrm{x}} \stackrel{\Delta}{\simeq} \begin{cases} \mathbf{tt} & \text{if } [\![e_{\mathrm{b}}]\!]_{\mathrm{b}}\Gamma_{\mathrm{x}} = \mathbf{tt} \text{ or } [\![e_{\mathrm{b}}]\!]_{\mathrm{b}}\Gamma_{\mathrm{x}} = \mathbf{tt} \\ \mathbf{ff} & \text{otherwise} \end{cases}$
  - $[\![\mathbf{not}\ e_{\mathrm{b}}]\!]_{\mathrm{b}}\Gamma_{\mathrm{x}} \stackrel{\Delta}{\simeq} \begin{cases} \mathbf{tt} & \text{if } [\![e_{\mathrm{b}}]\!]_{\mathrm{b}}\Gamma_{\mathrm{x}} = \mathbf{ff} \\ \mathbf{ff} & \text{otherwise} \end{cases}$

- $e_{\mathrm{m}}\langle e'_{\mathrm{m}}\rangle = e'_{\mathrm{m}}$

- $\mathit{seal}(e_{\mathrm{m}}, b)$ is defined inductively as follows:
  - $\mathit{seal}(\mathbf{tt}, b) \stackrel{\Delta}{\simeq} \emptyset$

- $seal(\mathbf{ff}, b) \overset{\Delta}{\simeq} \emptyset$
- $seal(x : \mathbf{bool}, b) \overset{\Delta}{\simeq} b\_x : \mathbf{bool}$
- $seal(e_\mathrm{b} \textbf{ or } e'_\mathrm{b}, b) \overset{\Delta}{\simeq} seal(e_\mathrm{b}, b) \textbf{ or } seal(e'_\mathrm{b}, b)$
- $seal(\textbf{not } e_\mathrm{b}, b) \overset{\Delta}{\simeq} \textbf{not } seal(e_\mathrm{b}, b)$

Overview. As for the abstract syntax, the semantics is presented in four subsections each treating one of the four syntactic categories in detail. An overview of the denotation functions and associated symbols is given in Tables 5.1 and 5.2. The environments are partial finite functions from appropriate sets of identifiers to appropriate sets of values. For the rate function and module environments these values are themselves functions mapping actual parameters to some other appropriate values. The binary string b is a parameter of some of the denotation functions which pass it on to the *seal* and R_S functions. Freshness of b is ensured by appropriate extensions as denotation functions are computed. This follows the approach of CBS, except that in CBS, fresh names are computed bottom-up, whereas we compute them top-down in order to avoid some unpleasant technicalities.

Table 5.1. Denotation functions

Function signature	Denotation of
$[\![e_\mathrm{c}]\!]_\mathrm{c}\Gamma_\mathrm{c}, b = v_\mathrm{c}$	Compartment expressions
$[\![e_\mathrm{s}]\!]_\mathrm{s}\Gamma_\mathrm{c}, \Gamma_\mathrm{s} = \underset{\sim}{v_\mathrm{s}}$	Species expressions
$[\![e_{\mathrm{s}+}]\!]_\mathrm{s}\Gamma_\mathrm{c}, \Gamma_\mathrm{s}, b = \underset{\sim}{v_\mathrm{s}}$	Extended species expressions
$[\![e_\mathrm{a}]\!]_\mathrm{a}\Gamma_\mathrm{c}, \Gamma_\mathrm{s}, \Gamma_\mathrm{a}, \underline{v_\mathrm{c}} = v_\mathrm{a}$	Algebraic rate expressions
$[\![e_\mathrm{m}]\!]_\mathrm{m}\Gamma_\mathrm{x} = v_\mathrm{m}$	Modification site expressions
$[\![P]\!]_\mathrm{p}\Gamma_\mathrm{c}, \Gamma_\mathrm{s}, \Gamma_\mathrm{a}, \Gamma_\mathrm{m}, b, \underline{v_\mathrm{c}} = \{(\mathcal{O}_i, \Gamma_{\mathrm{so}i})\}$	Programs
$[\![D]\!]_\mathrm{d}\Gamma_\mathrm{c}, \Gamma_\mathrm{s}, \Gamma_\mathrm{a}, \Gamma_\mathrm{m}, b = \Gamma'_\mathrm{c}, \Gamma'_\mathrm{s}, \Gamma'_\mathrm{a}, \Gamma'_\mathrm{m}, \Gamma_\mathrm{so}$	Definitions

Table 5.2. Symbols in the denotation function signatures

Symbol	Description
v_c	Compartment value
v_s	Species value
v_a	Algebraic rate value
v_m	Modification site value
$\mathcal{O}$	Semantical object
b	Binary string
Γ_c	Compartment environment
Γ_s	Species environment
Γ_a	Algebraic rate function environment
Γ_x	Variable environment
Γ_m	Module environment
Γ_so	Output species environment

5.1 Compartments

Compartment values. We let n_c range over a given set of *compartment names* which is assumed to include the set of binary strings and contain the nil compartment name $\mathbf{1}_c$. In contrast to compartment identifiers, which are language constructs used for binding compartment values, compartment names are used to uniquely and globally identify a compartment. Compartment values are of the following form:

$v_c ::=$	COMPARTMENT VALUE
$\mid (n_c, r, v_c)$	NESTED COMPARTMENT
$\mid \top$	WORLD COMPARTMENT

We let V_c denote the set of all compartment values generated by this grammar. Parent compartments v_c are recorded as values rather than names, since names do not generally identify a value uniquely. Specifically, $\mathbf{1}_c$ may occur in compartment values with different parents. Compartment volumes r represent the volume of a compartment in the "biological sense" that the volume of a child compartment does not count towards the volume of its enclosing parent.

The denotation function. A *compartment environment* is a partial finite function of the form $\Gamma_c(id_c) = v_c$ mapping compartment identifiers to compartment values. The denotation function for compartment expressions is of the form:

$$[\![e_c]\!]_c \Gamma_c, b = v_c$$

and is defined inductively as follows:

- $[\![\textbf{new comp vol } r \textbf{ inside } e_c]\!]_c \Gamma_c, b \overset{\Delta}{\simeq} (b, r, v_c)$ where
  - $v_c \overset{\Delta}{\simeq} [\![e_c]\!]_c \Gamma_c, 0b$
- $[\![\top]\!]_c \Gamma_c, b \overset{\Delta}{\simeq} \top$
- $[\![\mathbf{1}_c \textbf{ inside } e_c]\!]_c \Gamma_c, b \overset{\Delta}{\simeq} (\mathbf{1}_c, 0.0, v_c)$ where
  - $v_c \overset{\Delta}{\simeq} [\![e_c]\!]_c \Gamma_c, b$
- $[\![id_c]\!]_c \Gamma_c, b \overset{\Delta}{\simeq} \Gamma_c(id_c)$

The denotation function is partial since it is not defined for identifiers which do not have bindings in the given environment. New compartments are named by the binary string argument to the denotation function. The denotation of the parent compartment is computed recursively, but with a 0 prefixed to the fresh name, hence resulting in a new fresh name. Nil compartment values are arbitrarily given the volume 0.0.

Well-typedness of compartment value lists. Compartments generally occur in the context of lists of other compartments, and we are only interested in such lists which respect the hierarchy captured in compartment values. Formally, we say that a *list* $\underline{(n_c, r, v_c)}$ *is well-typed* if $\underline{v_c}.i = \underline{(n_c, r, v_c)}.(i-1)$ for $i \in 2 \ldots |\underline{v_c}|$. Any other lists, including those which contain the world compartment in any other position than possibly the first, are ill-typed.

Compartment lists in turn occur in the context of sets of other compartment lists, and we are only interested in such sets where all compartment lists agree on parent compartments. To formalise this, we define a function of the form $parent(\underline{v_c}) = \{v_{c_i}'\}$ which gives the set of legal parent compartments of a compartment list:

$$parent(\underline{v_c}) \stackrel{\Delta}{\simeq} \begin{cases} V_c & \text{if } \underline{v_c} = \varepsilon \\ \{v_c'\} & \text{if } |\underline{v_c}| > 0 \text{ and } \underline{v_c}.1 = (n_c, r, v_c') \\ \emptyset & \text{if } |\underline{v_c}| > 0 \text{ and } \underline{v_c}.1 = \top \end{cases}$$

In words, the empty list of compartments can be put inside any compartment; a non-empty list of compartments can only be put inside the compartment specified by the first element of the list unless this is the world compartment, in which case it can be put nowhere. Formally, we then say that a *set* $\{\underline{v_{c_i}}\}$ *is well-typed* if all $\underline{v_{c_i}}$ are well-typed and either $parent(\underline{v_{c_i}}) = \emptyset$ for all i or $\bigcap_i parent(\underline{v_{c_i}}) \neq \emptyset$.

The forest structure of well-typed sets of compartment value lists. The motivation for defining well-typedness of sets of compartment value lists is that only physically meaningful compartment hierarchies should be allowed in programs. By this we mean that sets of compartment value lists should form a forest structure, here a directed acyclic graph in which each node has at most one parent.

Observe first that one can obtain a directed graph from a well-typed set of compartment value lists in which nodes are compartment values and edges are determined by left-to-right neighbourhood in lists. Formally, given a well-typed set $\{\underline{v_{c_i}}\}$ we define $G\{\underline{v_{c_i}}\} \stackrel{\Delta}{\simeq} (V, E)$ where

$$V \stackrel{\Delta}{\simeq} \{\underline{v_{c_i}}.j\}$$
$$E \stackrel{\Delta}{\simeq} \{(\underline{v_{c_i}}.j, \underline{v_{c_i}}.(j+1))\}$$

(Recall from our notational convention that the above definitions give indexed sets where i is an index into the set of compartment value lists and j is an index into list positions). We now show that these graphs are indeed forests:

Proposition 5.1. *$G\{\underline{v_{c_i}}\}$ is a forest.*

Proof. By induction in $|\{\underline{v_{c_i}}\}|$. In the following we additionally use a and b to range over compartment values.

- **Basis** ($\{\underline{v_{c_i}}\} = \emptyset$). Holds vacuously.
- **Step** ($\{\underline{v_{c_i}}\} \cup \{\underline{v'_c}\}$).
  Acyclic: by the induction hypothesis, $G\{\underline{v_{c_i}}\}$ is acyclic. Also $G\{\underline{v'_c}\}$ is acyclic, for otherwise $\underline{v'_c}$ would take the form $\underline{v_{c_1}'}\, a\, \underline{v_{c_2}'}\, a\, \underline{v_{c_3}'}$, and it follows from well-typedness that the compartment value a must include itself as an ancestor; this is impossible since compartment values are finite. Suppose towards a contradiction that there is a cycle in $G(\{\underline{v_{c_i}}\} \cup \{\underline{v'_c}\})$. This can then only arise from a branch in $G\{\underline{v_{c_i}}\}$ of the form $\underline{v_{c_1}}\, a\, \underline{v_{c_2}}\, b\, \underline{v_{c_3}}$ and $\underline{v'_c}$ of the form $\underline{v_{c_1}'}\, b\, \underline{v_{c_2}'}\, a\, \underline{v_{c_3}'}$, both of which are well-typed. This means that the compartment value a must include b as an ancestor, and b in turn must include a as an ancestor. Hence a must include itself as an ancestor. But this is impossible since compartment values are finite.
  Max one parent: by the induction hypothesis, each node in $G\{\underline{v_{c_i}}\}$ has at most one parent. Also each node in $G\{\underline{v'_c}\}$ has at most one parent, for otherwise the graph would contain a cycle. Suppose towards a contradiction that there is some node a in $G(\{\underline{v_{c_i}}\} \cup \{\underline{v'_c}\})$ with two parents. This can only arise from a branch in $G\{\underline{v_{c_i}}\}$ of the form $\underline{v_{c_1}}\, b\, a\, \underline{v_{c_2}}$ and $\underline{v'_c}$ of the form $\underline{v_{c_1}'}\, c\, a\, \underline{v_{c_2}'}$ with $b \neq c$. But this is impossible since both lists are well-typed and a can contain only a single parent.

Normal form of compartment value lists. Parent compartments in compartment values are necessary for type-checking in the general semantics, and volumes are necessary in algebraic rate expressions. But from the view of any concrete semantics, we are interested in a normal form of compartment lists in which only compartment names are retained and in which the nil compartments are removed. This *normal form function* is of the form $nf(\underline{v_c}) = \underline{n_c}$ and is defined as $nf(\underline{n_c, r, v_c}) \stackrel{\Delta}{\simeq} \underline{n_c}.I$ where $I \stackrel{\Delta}{\simeq} \{i \mid \underline{n_c}.i \neq \mathbf{1}_c\}$ if $\underline{v_c}$ is a normal form compartment value list and is undefined otherwise. The graph arising from the normal form of a set of compartment value lists is also a forest if all non-nil compartment values have distinct compartment names, i.e. if the same non-nil compartment name does not occur with different parents or with different volumes. This is always the case for graphs in which compartment values arise from compartment expressions.

5.2 Species

Species values. Recall from the abstract syntax for species expressions that ξ is an annotation used to provide a match between actual and formal species parameters. Recall also that ρ ranges over modification site types, that n_m ranges over modification site names, and that n_s ranges over species names. *Species values* are generated by the following grammar where we use $Q \subset_{\text{fin}} \mathbb{N}$ to range over sets of list indices:

$v_{\mathrm{s}} ::= v_{\mathrm{us}}^{\iota:\xi}$	SPECIES VALUE
$v_{\mathrm{us}} ::= \underline{\underline{v_{\mathrm{c}}[n_{\mathrm{s}}, \alpha_{\sigma}]}}$	UNBOXED SPECIES VALUE
$\alpha_{\sigma} ::= \{n_{\mathrm{m}} \mapsto (\rho, e_{\mathrm{m}})\}$	TYPED ASSIGNMENT
$\iota ::= \{\underline{id_{\mathrm{c}}[n_{\mathrm{s}}]} \mapsto (Q, \iota_{\mathrm{m}})\}$	SPECIES INTERFACE
$\iota_{\mathrm{m}} ::= \{n_{\mathrm{m}} \mapsto n'_{\mathrm{m}}\}$	MODIFICATION SITE INTERFACE

An *unboxed species value* represents a possibly complex species by a list of located atomic species, each of which is represented by a name and a *typed assignment* mapping modification site names to pairs of modification types and expressions. *Species values* add annotations and interfaces. Annotations are as in the abstract syntax for species expressions: they are used for selecting the located atomic species and modification sites in an actual species parameter which should be mapped from the corresponding atomic species and modification sites in a formal parameter. *Interfaces* capture this mapping from formals to actuals and can hence be viewed as a product of module invocation. The need for a local mapping from formals to actuals arises from our having nondeterministic species, where different members of the set of values denoted by a nondeterministic actual species parameter may require different mappings from formals to actuals as demonstrated in Listing 3.7.

Interfaces map formal located names to pairs consisting of a set of position indices in the associated unboxed species values and a *modification site interface.* The sets of indices are used to cater for the general case of homo-multimers in which there are multiple instances of some atomic species. Modification site interfaces map formal modification site names to actual modification site names in the unboxed species value. Interfaces may expose only a subset of species indices in an unboxed species value, and for each exposed set of species indices, the associated modification site interface may expose only a subset of the modification sites recorded in the unboxed species value. Hence interfaces give rise to a notion of subtyping. For species values which have not been subjected to module invocations, the interface exposes all atomic species and all modification sites. Interfaces also give rise to a notion of parametric type since they provide means of renaming atomic species and modification site names.

Examples of species values. Examples of some species values arising from Listing 3.9 are shown informally in Figure 5.1 where we let f be the pair (**bool**, **ff**) and t be the pair (**bool**, **tt**). Interfaces are depicted in the top part of each figure, with solid lines representing the mapping from atomic species names to indices, and dotted lines representing the embedded mapping between modification site names. Note that none of these examples are homo-multimers, so interfaces map to singleton sets of indices. Unboxed species values are depicted in the center part of each figure, and annotations are depicted at the bottom.

Figure 5.1a shows a complex species value before it has been subjected to any module invocation, hence all primitive species and their modification sites are exposed by the interface. The species names in the unboxed value are primed, indicating that these are fresh. Figure 5.1b shows the species value after it has

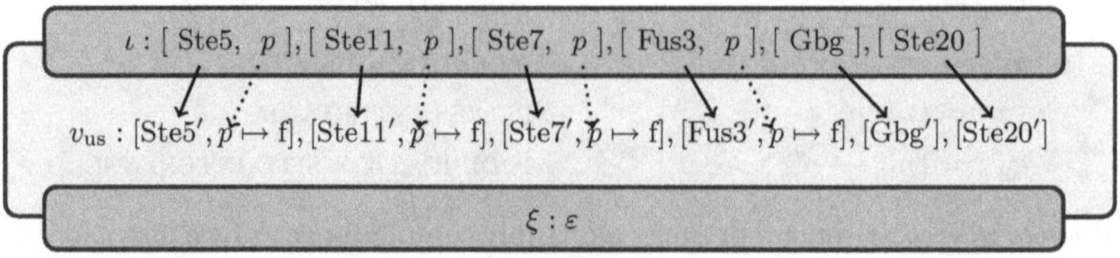

(a) The species value bound to e in the formation module, line 10.

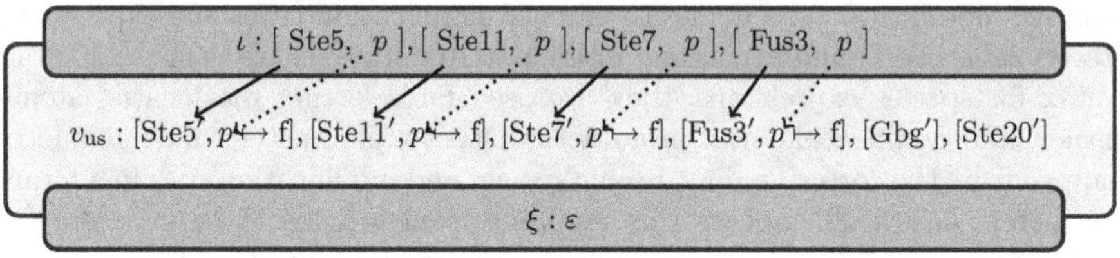

(b) The species value bound to the output species identifier link1 after invocation of the formation module in line 19.

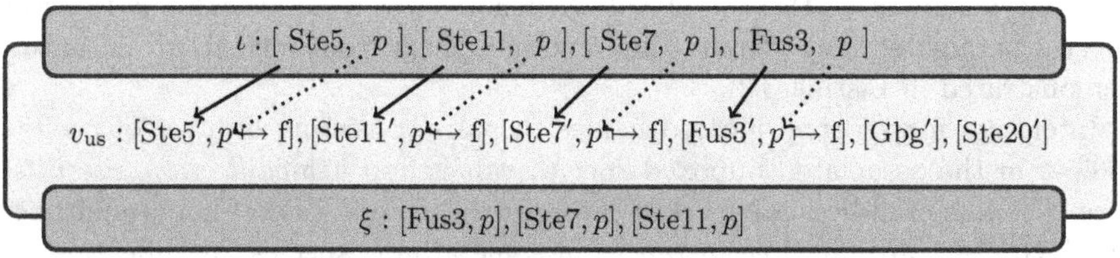

(c) The species value resulting from evaluating the first actual parameter of the mapk module invocation in line 20.

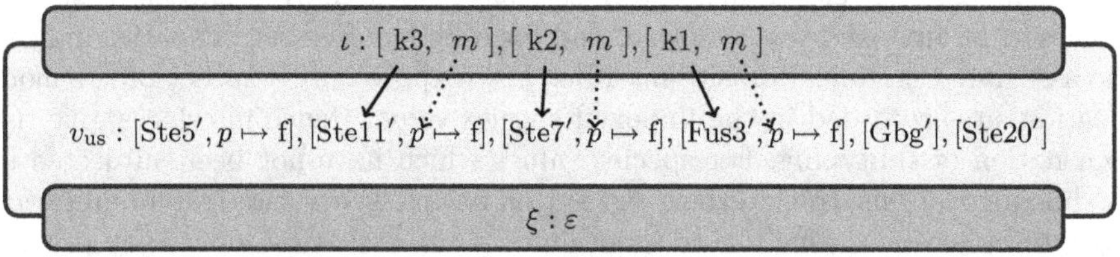

(d) The species value bound to a in the body of the mapk cascade module in line 14.

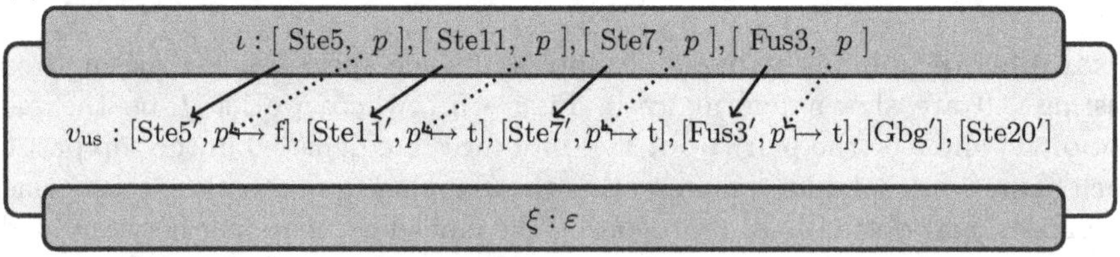

(e) The species value bound to the identifier link2, line 20, after invocation of the mapk module.

Fig. 5.1. Examples of species values from Listing 3.9 represented in an informal graphical notation been output from the formation module where Gbg and Ste20 have been removed from the interface. In Figure 5.1c the annotation of the actual species parameter of the mapk module has been recorded in the species value. Figure 5.1d shows the species value after the interface has been updated based on the annotation in Figure 5.1c and the corresponding formal annotation, $\xi' : [\mathrm{k1}, m], [\mathrm{k2}, m], [\mathrm{k3}, m]$; together these provide a mapping from e.g. k1 to Fus3, which is traced through the interface in Figure 5.1c down to the fourth index of the unboxed species value. The annotation has now served its purpose and is discarded. Finally, figure 5.1d shows the species value where three atomic species have been phosphorylated, and following output from the mapk module, the interface of this species value has been restored to the interface of the original input species value in Figure 5.1d.

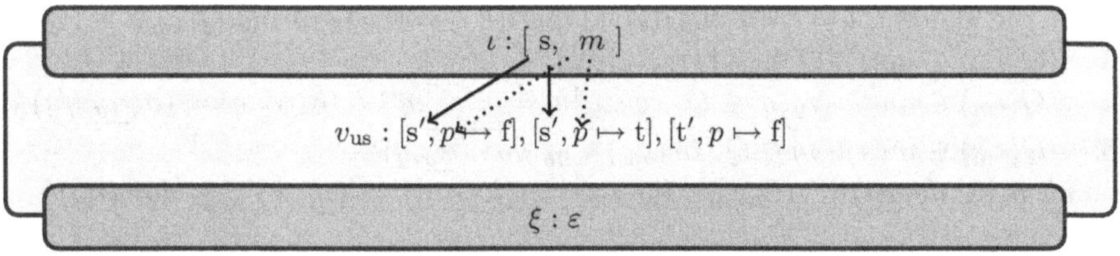

(a) A homomultimer species value with a single atomic species in its interface.

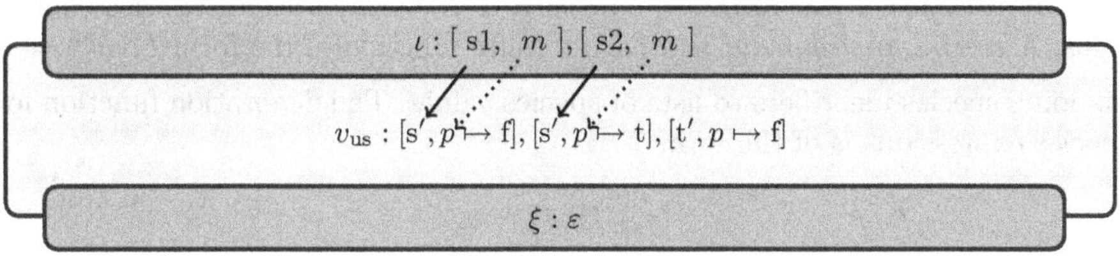

(b) A homomultimer species value with two atomic species in its interface, mapping to different occurrences of the same underlying fresh atomic species.

Fig. 5.2. Examples of homomultimer species values

A smaller example which illustrates how homomultimers can be represented is shown in Figure 5.2a; here the same atomic species name, s, maps to two occurrences of the same underlying fresh species name, s′. An interface may however also map different located names to indices with the same located fresh species names as shown in Figure 5.2b. This allows multiple instances of the same atomic species within a homo-multimer to be distinguished, a capability which a previous version of the language lacked.

Well-typedness of species values. A number of well-typedness conditions apply to species values. For an unboxed species value $\underline{\underline{v_{\mathrm{c}}}}[n_{\mathrm{s}}, \alpha_\sigma]$ we require that the lists of compartment values are well-typed and hence form a forest structure. We also require that assignments respect their associated type. These conditions can be phrased formally as follows:

1. $\{\underline{\underline{v_c}}.i\}$ is a well-typed set of compartment value lists.
2. $\forall(\rho, e_m) \in im(\alpha_\sigma).\ e_m : \rho$

For a species value $v_{us}^{\iota:\xi}$ we furthermore require that the interface maps to 1) non-empty and 2) disjoint sets of indices; that 3) all indices in a set exist in the unboxed species value and 4) contain species with identical located fresh names and modification site names; that 5) the modification site interfaces map to sites which exist in the assignments at the corresponding indices; that 6) the annotation only mentions located species and sites which exist in the interface. These conditions can be summarised formally as follows, where $\underline{\underline{v_c}[n_s, \alpha_\sigma]} = v_{us}$, $\underline{\underline{id_c}[n_s, \underline{n_m}]} = \xi$ and, for $\alpha_\sigma = \{n_m \mapsto (\rho, e_m)\}$, we let $\sigma(\alpha_\sigma) \overset{\Delta}{\simeq} \{n_m \mapsto \rho\}$.

1. $\forall(Q, \iota_m) \in im(\iota).\ |Q| > 0$
2. $\forall l, l' \in dom(\iota).\ l \neq l' \Rightarrow ind(\iota(l)) \cap ind(\iota(l')) = \emptyset$ where $ind(Q, \iota_m) \overset{\Delta}{\simeq} Q$
3. $\forall(Q, \iota_m) \in im(\iota).\ Q \subseteq_{\text{fin}} \{1, \ldots, |v_{us}|\}$
4. $\forall(Q, \iota_m) \in im(\iota).\ \forall q, q' \in Q.\ (\underline{\underline{v_c}[n_s]}.q = \underline{\underline{v_c}[n_s]}.q') \wedge (\sigma(\underline{\alpha_\sigma}.q) = (\sigma(\underline{\alpha_\sigma}.q'))$
5. $\forall(Q, \iota_m) \in im(\iota).\ \forall q \in Q.\ im(\iota_m) \subseteq_{\text{fin}} dom(\underline{\alpha_\sigma}.q)$
6. $\underline{\underline{id_c}[n_s]} \in dom(\iota) \wedge \forall(Q, \iota_m).\ (Q, \iota_m) = \iota(\underline{\underline{id_c}[n_s]}) \Rightarrow \{\underline{n_m}.i\} \subseteq_{\text{fin}} dom(\iota_m)$

The denotation function. We now turn to the semantics for species expressions. A *species environment* is a partial finite function of the form $\Gamma_s(id_s) = \underset{\sim}{v_s}$ mapping species identifiers to lists of species values. The denotation function for species expressions is of the form:

$$[\![e_s]\!]_s \Gamma_c, \Gamma_s = \underset{\sim}{v_s}$$

and is parametric on compartment and species environments. The denotation of a species is a *list* of species values. More than one species value may arise because of nondeterminism, and we use lists rather than sets to cater for output species in module parameters, as will be apparent in the semantics for programs; however, for most purposes we may think of these lists as sets, hence the wavy underline notation.

The definition of the denotation function for species expressions is given in the following. In order to simplify notation, we write Γ instead of Γ_c, Γ_s for cases where the environments are not used. Let us also reiterate the subtle notational convention that given e.g. a list $\underline{v_{us}}$ we write v_{us} for $\underline{v_{us}}.i$, and that i is implicitly assumed to be universally quantified over the indices of $\underline{v_{us}}$ in definitions; see for example the last 3 lines of the first case below.

- $[\![id_c[e_s]]\!]_s \Gamma_c, \Gamma_s \overset{\Delta}{\simeq} \underset{\sim}{v_{us}^{\iota:\xi}}$ where
  - $v_c \overset{\Delta}{\simeq} \Gamma_c(id_c)$
  - $\underset{\sim}{v_{us1}^{\iota_1:\xi_1}} \overset{\Delta}{\simeq} [\![e_s]\!]_s \Gamma_c, \Gamma_s$

- $v_{\mathrm{us}} \overset{\Delta}{\simeq}_t \underline{v_{\mathrm{c}}\underline{v_{\mathrm{c}_1}}[n_{\mathrm{s}}, \alpha_\sigma]}$ where $\underline{\underline{v_{\mathrm{c}_1}}[n_{\mathrm{s}}, \alpha_\sigma]} \overset{\Delta}{\simeq} v_{\mathrm{us}1}$
- $\iota \overset{\Delta}{\simeq} \{id_{\mathrm{c}}\underline{id_{\mathrm{c}_1}}[n_{\mathrm{s}}] \mapsto \iota_1(\underline{id_{\mathrm{c}_1}}[n_{\mathrm{s}}]) \mid \underline{id_{\mathrm{c}_1}}[n_{\mathrm{s}}] \in dom(\iota_1)\}$
- $\xi \overset{\Delta}{\simeq} \underline{\underline{id_{\mathrm{c}}\underline{id_{\mathrm{c}_1}}[n_{\mathrm{s}}, \underline{n_{\mathrm{m}}}]}}$ where $\underline{\underline{\underline{id_{\mathrm{c}_1}}[n_{\mathrm{s}}, \underline{n_{\mathrm{m}}}]}} \overset{\Delta}{\simeq} \xi_1$

– $[\![e_{\mathrm{s}1} - e_{\mathrm{s}2}]\!]_{\mathrm{s}}\Gamma \overset{\Delta}{\simeq}_t \utilde{v_{\mathrm{s}1}} \times_\circ \utilde{v_{\mathrm{s}2}}$ where
  - $\utilde{v_{\mathrm{s}1}} \overset{\Delta}{\simeq} [\![e_{\mathrm{s}1}]\!]_{\mathrm{s}}\Gamma$
  - $\utilde{v_{\mathrm{s}2}} \overset{\Delta}{\simeq} [\![e_{\mathrm{s}2}]\!]_{\mathrm{s}}\Gamma$
  - ${v_{\mathrm{us}1}}^{\iota_1:\xi_1} \circ {v_{\mathrm{us}2}}^{\iota_2:\xi_2} \overset{\Delta}{\simeq} v_{\mathrm{us}}^{\iota:\xi}$ where
    * $v_{\mathrm{us}} \overset{\Delta}{\simeq} v_{\mathrm{us}1}v_{\mathrm{us}2}$
    * $\iota(l) \overset{\Delta}{\simeq} \begin{cases} \iota_1(l) & \text{if } l \in dom(\iota_1) \setminus dom(\iota_2) \\ (A(Q_2), \iota_{\mathrm{m}2}) & \text{if } l \in dom(\iota_2) \setminus dom(\iota_1) \wedge \\ & (Q_2, \iota_{\mathrm{m}2}) = \iota_2(l) \\ (Q_1 \cup A(Q_2), \iota_{\mathrm{m}}) & \text{if } l \in dom(\iota_1) \cap dom(\iota_2) \wedge \\ & (Q_1, \iota_{\mathrm{m}1}) = \iota_1(l) \wedge (Q_2, \iota_{\mathrm{m}2}) = \iota_2(l) \wedge \\ & \iota_{\mathrm{m}} = \iota_{\mathrm{m}1} = \iota_{\mathrm{m}2} \end{cases}$
      where $A(Q) = \{q + |v_{\mathrm{us}1}| \mid q \in Q\}$
    * $\underline{\xi} \overset{\Delta}{\simeq} \underline{\xi_1}\,\underline{\xi_2}$

– $[\![e_{\mathrm{s}}.\underline{id_{\mathrm{c}}}[n_{\mathrm{s}}]]\!]_{\mathrm{s}}\Gamma \overset{\Delta}{\simeq} \utilde{v_{\mathrm{us}}^{\iota:\xi}}$ where
  - $\utilde{{v_{\mathrm{us}1}}^{\iota_1:\xi_1}} \overset{\Delta}{\simeq} [\![e_{\mathrm{s}}]\!]_{\mathrm{s}}\Gamma$
  - $(Q, \iota_{\mathrm{m}}) \overset{\Delta}{\simeq} \iota_1(\underline{id_{\mathrm{c}}}[n_{\mathrm{s}}])$
  - $v_{\mathrm{us}} \overset{\Delta}{\simeq} v_{\mathrm{us}1}.Q$
  - $\iota \overset{\Delta}{\simeq} \{\underline{id_{\mathrm{c}}}[n_{\mathrm{s}}] \mapsto (\{1 \ldots |Q|\}, \iota_{\mathrm{m}})\}$
  - $\xi \overset{\Delta}{\simeq} \xi_1.\{q \mid \exists \underline{n_{\mathrm{m}}}.\ \xi_1.q = (\underline{id_{\mathrm{c}}}[n_{\mathrm{s}}, \underline{n_{\mathrm{m}}}])\}$

– $[\![e_{\mathrm{s}} \backslash \underline{id_{\mathrm{c}}}[n_{\mathrm{s}}]]\!]_{\mathrm{s}}\Gamma \overset{\Delta}{\simeq} \utilde{v_{\mathrm{us}}^{\iota:\xi}}$ where
  - $\utilde{{v_{\mathrm{us}1}}^{\iota_1:\xi_1}} \overset{\Delta}{\simeq} [\![e_{\mathrm{s}}]\!]_{\mathrm{s}}\Gamma$
  - $(Q, \iota_{\mathrm{m}}) \overset{\Delta}{\simeq} \iota_1(\underline{id_{\mathrm{c}}}[n_{\mathrm{s}}])$
  - $v_{\mathrm{us}} \overset{\Delta}{\simeq} v_{\mathrm{us}1}.(\{1 \ldots |v_{\mathrm{us}1}|\} \setminus Q)$
  - $\iota \overset{\Delta}{\simeq} \{l \mapsto (A(Q'), \iota_{\mathrm{m}}) \mid l \in dom(\iota_1) \setminus \{\underline{id_{\mathrm{c}}}[n_{\mathrm{s}}]\} \wedge (Q', \iota_{\mathrm{m}}) = \iota_1(l)\}$
    where
    * $A(Q') \overset{\Delta}{\simeq} \{q' - |\{q \in Q \mid q \leq q'\}| \mid q' \in Q'\}$
  - $\xi \overset{\Delta}{\simeq} \xi_1.\{q \mid \neg\exists \underline{n_{\mathrm{m}}}.\ \xi_1.q = (\underline{id_{\mathrm{c}}}[n_{\mathrm{s}}, \underline{n_{\mathrm{m}}}])\}$

– $[\![e_s\langle \underline{id_c}[n_s, \alpha]\rangle]\!]_s\Gamma \stackrel{\Delta}{\simeq}_t \underset{\sim}{\underline{\underline{v_c}}[n_s', \underline{\alpha_\sigma}'']^{\iota:\xi}}$ where

- $\underset{\sim}{\underline{\underline{v_c}}[n_s', \underline{\alpha_\sigma}']^{\iota:\xi}} \stackrel{\Delta}{\simeq} [\![e_s]\!]_s\Gamma$
- $(Q, \iota_m) \stackrel{\Delta}{\simeq} \iota(\underline{id_c}[n_s])$
- $\underline{\alpha_\sigma}''.q \stackrel{\Delta}{\simeq} \begin{cases} \underline{\alpha_\sigma}'.q\langle \alpha \circ (\iota_m{}^{-1})\rangle & \text{if } q \in Q \\ \underline{\alpha_\sigma}'.q & \text{otherwise} \end{cases}$

  where

  * $\alpha_\sigma'\langle\alpha'\rangle(n_m) \stackrel{\Delta}{\simeq} \begin{cases} \alpha_\sigma'(n_m)\langle\alpha'(n_m)\rangle & \text{if } n_m \in dom(\alpha_\sigma') \cap dom(\alpha') \\ \alpha_\sigma'(n_m) & \text{if } n_m \in dom(\alpha_\sigma') \setminus dom(\alpha') \end{cases}$

  for all α_σ', α' and n_m.

– $[\![e_{s1} \textbf{ or } e_{s2}]\!]_s\Gamma \stackrel{\Delta}{\simeq} \underset{\sim}{v_{s1}}\ \underset{\sim}{v_{s2}}$ where

- $\underset{\sim}{v_{s1}} \stackrel{\Delta}{\simeq} [\![e_{s1}]\!]_s\Gamma$
- $\underset{\sim}{v_{s2}} \stackrel{\Delta}{\simeq} [\![e_{s2}]\!]_s\Gamma$

– $[\![e_s : \xi]\!]_s\Gamma \stackrel{\Delta}{\simeq}_t \underset{\sim}{v_{us}^{\iota:\xi}}$ where

- $\underset{\sim}{v_{us}^{\iota:\xi_1}} \stackrel{\Delta}{\simeq} [\![e_s]\!]_s\Gamma$

– $[\![id_s]\!]_s\Gamma_c, \Gamma_s \stackrel{\Delta}{\simeq} \Gamma_s(id_s)$

– $[\![\mathbf{0}_s]\!]_s\Gamma \stackrel{\Delta}{\simeq} \underset{\sim}{\varepsilon}$

The denotation function is partial because some species expressions do not result in lists of well-typed species values or in environments which are functions, or because some operations are undefined for some of the intermediate objects which arise. Given suitable environments, we say that a species expression is *well-typed* if its denotation is defined.

Explanation of the denotation function. In the case of located species, the compartment value assigned to the compartment identifier is looked up in the compartment environment and the species value denoting the nested species expression is obtained recursively. The denotation of the located species expression is obtained from this species value by adding the compartment value to the left of every located atomic species in the unboxed species value, by adding the compartment identifier to the left of each list of compartment identifiers in the domain of the interface, and by likewise adding the compartment identifier to each list of compartments in the annotation. Note that the interface records the compartment identifier rather than the compartment value. The expression is well-typed when the compartment identifier is defined in the given environment and when the resulting sets of compartment value lists are well-typed.

The denotation of a composite species expression is given by a Cartesian product of the denotations of the two operands. The corresponding pairing operation on species values concatenates the two unboxed species values and composes the interfaces in a manner that reflects this concatenation. The composed interface essentially maps located names to the union of indices given by the individual interfaces, with the twist that indices from the second interface are increased by the length of the first unboxed species value. This adjustment of indices is handled by an auxiliary function A. Annotations are composed simply by list concatenation. The resulting species value is well-typed when the two components agree on atomic species names and modification site interfaces for any common members of their interface.

In the case of species selection, the resulting unboxed species value is obtained by selecting the indices determined by the interface of the target species on the given located name. The resulting interface maps the given located name to a set of consecutive indices together with the original modification site interface. The resulting annotation is obtained by selecting for those entries which contain the given located name. The expression is well-typed when the located name is in the domain of the interface, which is necessary for the list selection operations to be well-defined. The case of species removal follows a similar idea, although special care is needed for the appropriate adjustment of indices using an auxiliary function A.

In the case of species update, the interface and annotation of the denotation of the operand are preserved, and an updated unboxed species is obtained by updating the assignments at the indices with the given located name. The modification site names in the given update expression are first renamed by function composition with the inverse of the modification site interface, and the result is given as an argument to an update operation on typed assignments which is defined as an auxiliary function. Here, the assignment to modification sites which are not mentioned in the update are preserved. For sites which are both in the original assignment and in the update, the expression update function (which is a parameter of the general semantics) is used; the update function is assumed extended to pairs of modification site types and expressions. The expression is well-typed if the located species name is in the domain of the target species interface, if the domain of the update is in the domain of the relevant modification site interface, and if the update respects the relevant species types.

The remaining cases are simpler. For nondeterministic species expressions, the species value lists obtained from the denotations of the operands are simply concatenated. The species annotation expression replaces the annotation in the denotation of the nested species expression with a new annotation. For this to be well-typed, the new annotation must mention only located names and sites which exist in the domain of the interface of the operand, thus ensuring that condition 6 of well-typedness for species values is satisfied. In the case of species identifier expressions, the corresponding value is simply looked up in the species environment which must be defined for the given identifier to be well-typed. Finally, the nil species evaluates to a singleton list containing just the empty species value.

Extended species expressions. The denotation function for extended species expressions is of the form:

$$[\![e_{s+}]\!]_s \Gamma_c, \Gamma_s, b = \underset{\sim}{v_s}$$

It is parametric on compartment and species environments, and also on a binary string b used to create fresh names for new species. Here is the definition:

- $[\![e_s]\!]_{s+}\Gamma_c, \Gamma_s, b \overset{\Delta}{\simeq} [\![e_s]\!]_s\Gamma_c, \Gamma_s$
- $[\![\mathbf{new}\ n_s, \sigma]\!]_{s+}\Gamma_c, \Gamma_s, b \overset{\Delta}{\simeq} \underset{\sim}{\varepsilon[b, \alpha_\sigma]}^{\iota:\varepsilon}$ where
  - $\alpha_\sigma \overset{\Delta}{\simeq} \{n_m \mapsto (\rho, \mathit{default}(\rho)) \mid n_m \in \mathit{dom}(\sigma) \wedge \rho = \sigma(n_m)\}$
  - $\iota \overset{\Delta}{\simeq} \{\varepsilon[n_s] \mapsto (\{1\}, \{n_{mi} \mapsto n_{mi}\})$ where $\{n_{mi} \mapsto \rho_i\} \overset{\Delta}{\simeq} \sigma$

A new species expression evaluates to a singleton unboxed species with a fresh species name given by the binary string parameter to the denotation function, together with a typed assignment which extends the given type with default modification expressions. The interface simply maps the given species name in the empty list of compartment identifiers to the first and only index of the unboxed species value, together with the identity interface on modification site names.

Normal form species values. Interfaces, annotations and parent definitions in compartment values are needed for determining well-typedness and for making module invocation work, as detailed in the next subsection, but they are not needed for normal form reactions. All that is needed here is a *normal form*:

$$v_{ns} ::= \underline{\underline{n_c}[n_s, \alpha_\sigma]}$$

of species values. The *normal form function* then takes the form $\mathit{nf}(v_s, \underline{v_c}) = v_{ns}$ where $\underline{v_c}$ is a list of compartment values which is required to go all the way from the world compartment down to the compartment enclosing the species value. The function is used in the semantics for programs and is applied to species values in a located reaction. In the following definition we use the normal form function for compartment value lists defined in the previous subsection:

$$\mathit{nf}(\underline{\underline{v'_c}[n_s, \alpha_\sigma]}^{\iota,\xi}, \underline{v_c}) \overset{\Delta}{\simeq} \underline{\mathit{nf}(\underline{v_c}\,\underline{v'_c})[n_s, \alpha_\sigma]}$$

The function is defined only when $\underline{v_c}\,\underline{v'_c}$ is a well-typed list of compartment values. Although retaining the full list of compartment names is unnecessary for some concrete semantics such as Petri nets, this information may be relevant in other cases. For example, it allows the compartment forest structure of programs to be obtained through the general semantical framework by defining an appropriate concrete semantics for representing and composing forests.

Ground normal form species values. We introduce one further *ground normal form* of species values:

$$v_{\text{gns}} ::= \underline{\underline{n_{\text{c}}}}[n_{\text{s}}, \{n_{\text{m}} \mapsto (\rho, v_{\text{m}})\}]$$

Here modification sites map to pairs of modification types and values, rather than to pairs of modification types and expressions with variables. Ground normal form species values are used in the semantics for programs in the case of initial conditions; indeed, as described above, the general semantics is parameterised on a function of the form $I_S(v_{\text{gns}}, r) = \mathcal{O}$ for assigning semantical objects to initial populations or concentrations of ground normal form species values. Recall also that the general semantics is parameterised on a function of the form $[\![e_{\text{m}}]\!]_{\text{m}} \Gamma_{\text{x}} = v_{\text{m}}$ which assigns values to modification site expressions given a variable environment. This function can be extended in an evident manner to a function from normal form species values to ground normal form species values; this is needed when defining concrete semantics.

Further functions on species values. We define two functions on species values which are required for the semantics of module invocations. For the first function, the intuition is that one can update an interface ι_1 given the associated annotation ξ_1 together with a second matching annotation ξ_2 obtained from a corresponding formal species parameter. The updated interface is then used to access the species within the body of the module. An example of this is shown in the transition from the species value in Figure 5.1c to that in Figure 5.1d. Formally, the function takes the form $close(v_{\text{s}1}, \xi_2) = v_{\text{s}2}$ and is defined as follows:

$$close(v_{\text{us}}^{\iota_1:\xi_1}, \xi_2) \stackrel{\Delta}{\simeq} v_{\text{us}}^{\iota_2:\varepsilon}$$

where, for $\underline{\underline{id_{\text{c}}[n_{\text{s}}, \underline{n_{\text{m}}}]_1}} \stackrel{\Delta}{\simeq} \xi_1$ and $\underline{\underline{id_{\text{c}}[n_{\text{s}}, \underline{n_{\text{m}}}]_2}} \stackrel{\Delta}{\simeq} \xi_2$,

$$\iota_2 \stackrel{\Delta}{\simeq} \{\underline{\underline{id_{\text{c}}[n_{\text{s}}]_2}}.i \mapsto (Q, \{\underline{\underline{n_{\text{m}2}}}.i.j \mapsto \iota_{\text{m}}(\underline{\underline{n_{\text{m}1}}}.i.j)\}) \mid (Q, \iota_{\text{m}}) = \iota_1(\underline{\underline{id_{\text{c}}[n_{\text{s}}]_1}}.i)\}$$

The function is only defined if the lists ξ and ξ' have the same length, and if all of the embedded lists $\underline{\underline{n_{\text{m}1}}}.i$ and $\underline{\underline{n_{\text{m}2}}}.i$ also have the same length.

The second function enables one species value to take on the interface and annotation of another species value. This is needed in the semantics for output species in programs, and an example is shown in the transition of the species value in Figure 5.1a to that in Figure 5.1b, and from the species value in Figure 5.1d to that in Figure 5.1e. Formally, we first need two supporting functions. The first gives the located name of an unboxed species value at a given index, and the second counts the number of previous occurrences of the located species name at a given index:

$$l^q(v_{\text{us}}) \stackrel{\Delta}{\simeq} \underline{id_{\text{c}}}[n_{\text{s}}] \text{ where } \exists \alpha_\sigma .\ \underline{id_{\text{c}}}[n_{\text{s}}, \alpha_\sigma] = v_{\text{us}}.q$$

$$c^q(v_{\text{us}}) \stackrel{\Delta}{\simeq} |\{q' \mid l^q(v_{\text{us}}) = l^{q'}(v_{\text{us}}) \wedge q' < q\}|$$

The function of interest then takes the form $adapt(v_{s1}, v_{s2}) = v_{s3}$ and is defined as follows:

$$adapt(v_{us1}^{\iota_1:\xi_1}, v_{us2}^{\iota_2:\xi_2}) \stackrel{\Delta}{\simeq}_t v_{us1}^{\iota_3:\xi_2}$$

where, for all $l \in dom(\iota_2)$:

$$\begin{aligned}(Q_2, \iota_{m2}) &\stackrel{\Delta}{\simeq} \iota_2(l)\\ Q_3 &\stackrel{\Delta}{\simeq} \{q \mid \exists q_2 \in Q_2.\ l^q(v_{us1}) = l^{q_2}(v_{us2}) \wedge c^q(v_{us1}) = c^{q_2}(v_{us2})\}\\ \iota_3(l) &\stackrel{\Delta}{\simeq} (Q_3, \iota_{m2})\end{aligned}$$

Here the new interface ι_3 maps a located species name l to the indices in v_{us1} which have the same located fresh name l' as the indices in v_{us2} mapped from l by ι_2. If there are multiple choices of such indices, the index which results in the same number of previous occurrences of the fresh name l' in the two unboxed species values is chosen. The function is defined only when the resulting species value is well-typed, which is the case whenever the resulting sets of indices are non-empty and modification site interfaces map to sites which exist in the resulting unboxed species value.

Species value design choices. We end the treatment of species values with some remarks about possible alternative representations. First note that we choose to include modification types in species values. However, since new species values are always created with fresh names and there are no expressions which allow modification sites to change type, species values with identical names also have identical modification types. Hence it would also be possible to maintain modification types separately from species values as indeed was done in a previous version of the language [27], with the benefit of reduced redundancy. But this approach would have the downside of cluttering the presentation of the semantics with an additional environment needing to be maintained.

Alternative representations of species interfaces are also possible. We choose for example to include compartment identifiers in the renaming of interfaces, which allows compartments to differ between different members of a nondeterministic species in the same way that atomic species may differ. But interfaces could instead provide local mappings for only species and modification site names, and require compartments to be evaluated externally. This would ensure that two species with the same location in their interface are indeed in the same location. Such guarantees cannot be made when location is included in the interface.

Another design choice involves the relatively relaxed conditions on interfaces. For example, an interface may map a located name to indices in which the compartment structure is completely different, and different located names may map to indices with the same species names. The latter allows the elements of a homomultimer to be distinguished within the same species as demonstrated by the earlier example in Figure 5.2b. This is the reason why unboxed species values are lists rather than multisets. We also anticipate that the ordering of atomic species within a complex may be significant for future concrete semantics, although the Petri net, ODE and CTMC concrete semantics disregard the ordering.

Annotations are maintained explicitly in species values. This incurs some overhead in the semantics for species expressions since all operations must take annotations into account. Alternatively, the interface could be represented by lists and the annotation could be captured by an appropriate ordering and restriction of the interface. This would give a more compact semantics at the cost of reduced transparency. It is however impossible for another reason pertaining to output species: these can adopt the interface of actual species values at time of module invocation using the *adapt* function, so interfaces of actual species parameters must be preserved.

5.3 Programs

Normal form reactions. Recall that the general semantics is parameterised on a structure $(S, |_S, \mathbf{0}_S, R_S, I_S)$ and that R_S is a function of the form $R_S(R, b) = \mathcal{O}$ assigning a semantical object to a reaction R, named b, in a suitable normal form. More precisely, R takes the form:

$$\underline{n \cdot v_{\mathrm{ns}}} \Rightarrow^{v_{\mathrm{r}}} \underline{n' \cdot v'_{\mathrm{ns}}} \ \textbf{if}\ e_{\mathrm{b}}$$

where v_{ns} and v'_{ns} are normal form species values as defined in the semantics for species and v_{r} is a *rate value*, i.e. a rate expression in which species expressions have been evaluated to their normal forms, compartment expressions have been replaced by their resulting volumes, and rate function invocations have been evaluated. Rate values and algebraic rate values are generated by the grammar in Table 5.3.

The denotation function for algebraic rate expressions is of the form:

$$[\![e_{\mathrm{a}}]\!]_{\mathrm{a}} \Gamma_{\mathrm{c}}, \Gamma_{\mathrm{s}}, \Gamma_{\mathrm{a}}, \underline{v_{\mathrm{c}}} = v_{\mathrm{a}}$$

Here Γ_{a} is an *algebraic rate function environment* of the form $\Gamma_{\mathrm{a}}(id_{\mathrm{a}}) = f$ where f in turn is a function of the form $f(\underline{v_{\mathrm{c}}}, \underset{\sim}{\underline{v_{\mathrm{s}}}}, \underline{v_{\mathrm{a}}}, \underline{v'_{\mathrm{c}}}) = v_{\mathrm{a}}$ mapping actual parameters, together with a list $\underline{v'_{\mathrm{c}}}$ of parent compartment values at time of invocation, to algebraic rate values. The denotation function is defined below, but with some standard cases for functions and arithmetic operators omitted. We adopt a convention here and throughout where any parameters of a denotational function that are not explicitly used by a given case are represented by Γ; in the second case below, for example, Γ hence represents the parameters Γ_{s}, Γ_{a} and $\underline{v_{\mathrm{c}}}$.

- $[\![r]\!]_{\mathrm{a}} \Gamma \overset{\Delta}{\simeq} r$
- $[\![id_{\mathrm{c}}]\!]_{\mathrm{a}} \Gamma, \Gamma_{\mathrm{c}} \overset{\Delta}{\simeq} r$ where
  - $(n_{\mathrm{c}}, r, v_{\mathrm{c}}) \overset{\Delta}{\simeq} \Gamma_{\mathrm{c}}(id_{\mathrm{c}})$

Table 5.3. The abstract syntax for rate values

$v_{\mathrm{r}} ::=$	Rate value
$\mid \{v_{\mathrm{a}}\}$	Mass-action rate
$\mid [v_{\mathrm{a}}]$	Algebraic rate
$v_{\mathrm{a}} ::=$	Algebraic rate value
$\mid r$	Constant
$\mid v_{\mathrm{ns}}$	Population
$\mid \textbf{if } e_{\mathrm{b}} \textbf{ then } v_{\mathrm{a}} \textbf{ else } v'_{\mathrm{a}}$	Conditional
$\mid exp(v_{\mathrm{a}}) \mid log(v_{\mathrm{a}}) \mid sin(v_{\mathrm{a}}) \mid cos(v_{\mathrm{a}})$	Functions
$\mid v_{\mathrm{a}} + v'_{\mathrm{a}} \mid v_{\mathrm{a}} \text{ - } v'_{\mathrm{a}}$	Arithmetic operators
$\mid v_{\mathrm{a}} \times v'_{\mathrm{a}} \mid v_{\mathrm{a}} \,/\, v'_{\mathrm{a}} \mid v_{\mathrm{a}} \,\hat{}\, v'_{\mathrm{a}}$	

– $[\![e_{\mathrm{s}}]\!]_{\mathrm{a}}\Gamma_{\mathrm{c}}, \Gamma_{\mathrm{s}}, \Gamma_{\mathrm{a}}, \underline{v_{\mathrm{c}}} \stackrel{\Delta}{\simeq} nf(\underset{\sim}{v_{\mathrm{s}}}.1, \underline{v_{\mathrm{c}}})$ where
  - $\underset{\sim}{v_{\mathrm{s}}} \stackrel{\Delta}{\simeq} [\![e_{\mathrm{s}}]\!]_{\mathrm{s}}\Gamma_{\mathrm{c}}, \Gamma_{\mathrm{s}}$

  if $|\underset{\sim}{v_{\mathrm{s}}}| = 1$

– $[\![id_{\mathrm{a}}(\underline{id_{\mathrm{c}}}; \underline{e_{\mathrm{s}}}; \underline{e_{\mathrm{a}}})]\!]_{\mathrm{a}}\Gamma_{\mathrm{c}}, \Gamma_{\mathrm{s}}, \Gamma_{\mathrm{a}}, \underline{v'_{\mathrm{c}}} \stackrel{\Delta}{\simeq} \Gamma_{\mathrm{a}}(id_{\mathrm{a}})(\underline{v_{\mathrm{c}}}, \underline{\underset{\sim}{v_{\mathrm{s}}}}, \underline{v_{\mathrm{a}}}, \underline{v'_{\mathrm{c}}})$ where
  - $v_{\mathrm{c}} \stackrel{\Delta}{\simeq} \Gamma_{\mathrm{c}}(id_{\mathrm{c}})$
  - $\underset{\sim}{v_{\mathrm{s}}} \stackrel{\Delta}{\simeq} [\![e_{\mathrm{s}}]\!]_{\mathrm{s}}\Gamma_{\mathrm{c}}, \Gamma_{\mathrm{s}}$
  - $v_{\mathrm{a}} \stackrel{\Delta}{\simeq} [\![e_{\mathrm{a}}]\!]_{\mathrm{a}}\Gamma_{\mathrm{c}}, \Gamma_{\mathrm{s}}, \Gamma_{\mathrm{a}}, \underline{v'_{\mathrm{c}}}$

  if $|\underset{\sim}{v_{\mathrm{s}}}| = 1$

– $[\![\textbf{if } e_{\mathrm{b}} \textbf{ then } e_{\mathrm{a}} \textbf{ else } e'_{\mathrm{a}}]\!]_{\mathrm{a}}\Gamma \stackrel{\Delta}{\simeq} \textbf{if } e_{\mathrm{b}} \textbf{ then } [\![e_{\mathrm{a}}]\!]_{\mathrm{a}}\Gamma \textbf{ else } [\![e'_{\mathrm{a}}]\!]_{\mathrm{a}}\Gamma$

– $[\![exp(e_{\mathrm{a}})]\!]_{\mathrm{a}}\Gamma \stackrel{\Delta}{\simeq} exp([\![e_{\mathrm{a}}]\!]_{\mathrm{a}}\Gamma)$

– $[\![e_{\mathrm{a}} + e'_{\mathrm{a}}]\!]_{\mathrm{a}}\Gamma \stackrel{\Delta}{\simeq} [\![e_{\mathrm{a}}]\!]_{\mathrm{a}}\Gamma + [\![e'_{\mathrm{a}}]\!]_{\mathrm{a}}\Gamma$

The case of compartments is only defined for non-world compartment expressions, reflecting our convention that the world compartment should only be used as a parent in definitions of new compartments. The case of species is only defined when the species expression does not contain nondeterminism because any nondeterministic choice is forced in the appropriate derived forms of reactions. In the case of algebraic rate function invocation, the semantic function is looked up in the algebraic rate function environment and applied to the actual parameters after these have been evaluated.

The remaining cases simply evaluate components recursively. Note in particular that conditionals are preserved in rate values, since a full evaluation requires an assignment to variables. As for normal form species expressions, this assignment is left as a concern for the concrete semantics because certain semantical objects, such as coloured Petri nets, have their own distinct way of handing variables.

The denotation function for basic programs. The denotation function for basic programs is of the form:

$$[\![P]\!]_{\mathrm{p}} \Gamma_{\mathrm{c}}, \Gamma_{\mathrm{s}}, \Gamma_{\mathrm{a}}, \Gamma_{\mathrm{m}}, b, \underline{v_{\mathrm{c}}} = \{(\mathcal{O}_i, \Gamma_{\mathrm{so}i})\}$$

Here Γ_{m} is a *module environment* of the form $\Gamma_{\mathrm{m}}(id_{\mathrm{p}}) = g$ where g in turn is a function of the form $g(\underline{v_{\mathrm{c}}}, \underset{\sim}{v_{\mathrm{s}}}, \underline{v_{\mathrm{a}}}, \underline{id_{\mathrm{s}}}, b, \underline{v_{\mathrm{c}}}) = \{(\mathcal{O}_i, \Gamma_{\mathrm{so}i})\}$ mapping actual parameters to a set of pairs of semantical objects and *output species environments* $\Gamma_{\mathrm{so}i}$ which take the same form as species environments. Note that we obtain a *set* of semantical objects and output species environments in order to account for variation composition. The output species environments allows the formal output species, defined inside a module, to become available in the program following module invocation where they are bound to the corresponding actual output species identifiers. Note also that g is parameterised on a fresh name b and a list $\underline{v_{\mathrm{c}}}$ of parent compartments. The latter is because parent compartments for a module are determined dynamically rather than statically.

The denotation function is defined below and relies on the function

$$\delta_n^m \;\stackrel{\Delta}{\simeq}\; \{0\}^{n-1}\,1\,\{0\}^{m-n}$$

for constructing a binary string of length m with zeros everywhere except in the nth position which holds a one.

- $[\![\underline{n \cdot e_{\mathrm{s}}} \Rightarrow^{e_{\mathrm{r}}} \underline{n' \cdot e'_{\mathrm{s}}} \textbf{ if } e_{\mathrm{b}}]\!]_{\mathrm{p}} \Gamma_{\mathrm{c}}, \Gamma_{\mathrm{s}}, \Gamma_{\mathrm{a}}, \Gamma_{\mathrm{m}}, b, \underline{v_{\mathrm{c}}} \stackrel{\Delta}{\simeq} \{(\mathcal{O}, \emptyset)\}$ where
  - $v_{\mathrm{r}} \stackrel{\Delta}{\simeq} [\![e_{\mathrm{r}}]\!]_{\mathrm{a}} \Gamma_{\mathrm{c}}, \Gamma_{\mathrm{s}}, \Gamma_{\mathrm{a}}, \underline{v_{\mathrm{c}}}$
  - $\underset{\sim}{v_{\mathrm{s}}} \stackrel{\Delta}{\simeq} [\![e_{\mathrm{s}}]\!]_{\mathrm{s}} \Gamma_{\mathrm{c}}, \Gamma_{\mathrm{s}}$
  - $\underset{\sim}{v'_{\mathrm{s}}} \stackrel{\Delta}{\simeq} [\![e'_{\mathrm{s}}]\!]_{\mathrm{s}} \Gamma_{\mathrm{c}}, \Gamma_{\mathrm{s}}$
  - $v_{\mathrm{ns}} \stackrel{\Delta}{\simeq} nf(\underset{\sim}{v_{\mathrm{s}}}.1, \underline{v_{\mathrm{c}}})$
  - $v'_{\mathrm{ns}} \stackrel{\Delta}{\simeq} nf(\underset{\sim}{v'_{\mathrm{s}}}.1, \underline{v_{\mathrm{c}}})$
  - $\mathcal{O} \stackrel{\Delta}{\simeq} R_S(\underline{n \cdot v_{\mathrm{ns}}} \Rightarrow^{v_{\mathrm{r}}} \underline{n' \cdot v'_{\mathrm{ns}}} \textbf{ if } e_{\mathrm{b}}, b)$
  if $|\underset{\sim}{v_{\mathrm{s}}}| = |\underset{\sim}{v'_{\mathrm{s}}}| = 1$
- $[\![\mathbf{0}_{\mathrm{P}}]\!]_{\mathrm{p}} \Gamma \stackrel{\Delta}{\simeq} \{(\mathbf{0}_S, \emptyset)\}$
- $[\![P \mid P']\!]_{\mathrm{p}} \Gamma, b \stackrel{\Delta}{\simeq} \{(\mathcal{O}_i |_S \mathcal{O}'_j, \Gamma_{\mathrm{so}i} \langle \Gamma'_{\mathrm{so}j} \rangle)\}$ where
  - $\{(\mathcal{O}_i, \Gamma_{\mathrm{so}i})\} \stackrel{\Delta}{\simeq} [\![P]\!]_{\mathrm{p}} \Gamma, 0b$
  - $\{(\mathcal{O}'_j, \Gamma'_{\mathrm{so}j})\} \stackrel{\Delta}{\simeq} [\![P']\!]_{\mathrm{p}} \Gamma, 1b$

- $\llbracket P \mid\mid P' \rrbracket_{\mathrm{p}} \Gamma, b \overset{\Delta}{\simeq} \{(\mathcal{O}_i, \Gamma_{\mathrm{so}i})\} \cup \{(\mathcal{O}'_j, \Gamma'_{\mathrm{so}j})\}$ where
  - $\{(\mathcal{O}_i, \Gamma_{\mathrm{so}i})\} \overset{\Delta}{\simeq} \llbracket P \rrbracket_{\mathrm{p}} \Gamma, 0b$
  - $\{(\mathcal{O}'_j, \Gamma'_{\mathrm{so}j})\} \overset{\Delta}{\simeq} \llbracket P' \rrbracket_{\mathrm{p}} \Gamma, 1b$

- $\llbracket id_{\mathrm{c}}[P] \rrbracket_{\mathrm{p}} \Gamma, \Gamma_{\mathrm{c}}, \underline{v_{\mathrm{c}}} \overset{\Delta}{\simeq} \llbracket P \rrbracket_{\mathrm{p}} \Gamma, \Gamma_{\mathrm{c}}, v'_{\mathrm{c}}\underline{v_{\mathrm{c}}}$ where
  - $v'_{\mathrm{c}} \overset{\Delta}{\simeq} \Gamma_{\mathrm{c}}(id_{\mathrm{c}})$

- $\llbracket id_{\mathrm{p}}(\underline{e_{\mathrm{c}}}; \underline{e_{\mathrm{s+}}}; \underline{e_{\mathrm{a}}}; \textbf{out } \underline{id_{\mathrm{s}}}); P \rrbracket_{\mathrm{p}} \Gamma_{\mathrm{c}}, \Gamma_{\mathrm{s}}, \Gamma_{\mathrm{a}}, \Gamma_{\mathrm{m}}, b, \underline{v_{\mathrm{c}}} \overset{\Delta}{\simeq} \{(\mathcal{O}_i |_S \mathcal{O}'_{j_i}, \Gamma_{\mathrm{so}i}\langle \Gamma'_{\mathrm{so}j_i} \rangle)\}$
  where
  - $m \overset{\Delta}{\simeq} |\underline{e_{\mathrm{c}}}| + |\underline{e_{\mathrm{s+}}}| + 2$
  - $b_i \overset{\Delta}{\simeq} \delta^m_i$
  - $b'_j \overset{\Delta}{\simeq} \delta^m_{|\underline{e_{\mathrm{s+}}}|+j}$
  - $b^1 \overset{\Delta}{\simeq} \delta^m_{|\underline{e_{\mathrm{s+}}}|+|\underline{e_{\mathrm{c}}}|+1}$
  - $b^2 \overset{\Delta}{\simeq} \delta^m_{|\underline{e_{\mathrm{s+}}}|+|\underline{e_{\mathrm{c}}}|+2}$
  - $\underline{\underset{\sim}{v}_{\mathrm{s}}}.i \overset{\Delta}{\simeq} \llbracket \underline{e_{\mathrm{s+}}}.i \rrbracket_{\mathrm{s}} \Gamma_{\mathrm{c}}, \Gamma_{\mathrm{s}}, b_i b,$
  - $\underline{v_{\mathrm{c}}}.j \overset{\Delta}{\simeq} \llbracket \underline{e_{\mathrm{c}}}.j \rrbracket_{\mathrm{c}} \Gamma_{\mathrm{c}}, b'_j b$
  - $\underline{v_{\mathrm{a}}}.k \overset{\Delta}{\simeq} \llbracket \underline{e_{\mathrm{a}}}.k \rrbracket_{\mathrm{a}} \Gamma_{\mathrm{c}}, \Gamma_{\mathrm{s}}, \Gamma_{\mathrm{a}}, \underline{v_{\mathrm{c}}}$
  - $\{(\mathcal{O}_i, \Gamma_{\mathrm{so}i})\} \overset{\Delta}{\simeq} \Gamma_{\mathrm{m}}(id_{\mathrm{p}})(\underline{v_{\mathrm{c}}}, \underline{\underset{\sim}{v}_{\mathrm{s}}}, \underline{v_{\mathrm{a}}}, \underline{id_{\mathrm{s}}}, \underline{v_{\mathrm{c}}}, b^1 b)$
  - $\{\{(\mathcal{O}'_j, \Gamma'_{\mathrm{so}j})\}_i\} \overset{\Delta}{\simeq} \llbracket P \rrbracket_{\mathrm{p}} \Gamma_{\mathrm{c}}, \Gamma_{\mathrm{s}}\langle \Gamma_{\mathrm{so}i} \rangle, \Gamma_{\mathrm{a}}, \Gamma_{\mathrm{m}}, b^2 b$

- $\llbracket D \,;\, P \rrbracket_{\mathrm{p}} \Gamma, b, \underline{v_{\mathrm{c}}} \overset{\Delta}{\simeq} \{(\mathcal{O}_i, \Gamma'_{\mathrm{so}}\langle \Gamma_{\mathrm{so}i} \rangle)\}$ where
  - $\Gamma', \Gamma'_{\mathrm{so}} \overset{\Delta}{\simeq} \llbracket D \rrbracket_{\mathrm{d}} \Gamma, 0b$
  - $\{(\mathcal{O}_i, \Gamma_{\mathrm{so}i})\} \overset{\Delta}{\simeq} \llbracket P \rrbracket_{\mathrm{p}} \Gamma', 1b, \underline{v_{\mathrm{c}}}$

- $\llbracket id_{\mathrm{s}} = \textbf{force}(e_{\mathrm{s}}) \;;\; P \rrbracket_{\mathrm{p}} \Gamma_{\mathrm{c}}, \Gamma_{\mathrm{s}}, \Gamma_{\mathrm{a}}, \Gamma_{\mathrm{m}}, b, \underline{v_{\mathrm{c}}} \overset{\Delta}{\simeq} \{(\underline{\mathcal{O}_{j_1}}.1 \,|_S \ldots |_S\, \underline{\mathcal{O}_{j_m}}.m, \emptyset)\}$
  where
  - $\underset{\sim}{v}_{\mathrm{s}} \overset{\Delta}{\simeq} \llbracket e_{\mathrm{s}} \rrbracket_{\mathrm{s}} \Gamma_{\mathrm{c}}, \Gamma_{\mathrm{s}}$
  - $\underline{\{(\mathcal{O}_j, \Gamma_{\mathrm{so}j})\}}.i \overset{\Delta}{\simeq} \llbracket P \rrbracket_{\mathrm{p}} \Gamma_{\mathrm{c}}, \Gamma_{\mathrm{s}}\langle id_{\mathrm{s}} \mapsto \underset{\sim}{v}_{\mathrm{s}}.i \rangle, \Gamma_{\mathrm{a}}, \Gamma_{\mathrm{m}}, \delta_i^{|\underset{\sim}{v}_{\mathrm{s}}|} b, \underline{v_{\mathrm{c}}}$

- $\llbracket \textbf{init } e_{\mathrm{s}} = r \rrbracket_{\mathrm{p}} \Gamma_{\mathrm{c}}, \Gamma_{\mathrm{s}}, \Gamma_{\mathrm{a}}, \Gamma_{\mathrm{m}}, b, \underline{v_{\mathrm{c}}} \overset{\Delta}{\simeq} \{(\mathcal{O}, \emptyset)\}$ where
  - $\underset{\sim}{v}_{\mathrm{s}} \overset{\Delta}{\simeq} \llbracket e_{\mathrm{s}} \rrbracket_{\mathrm{s}} \Gamma_{\mathrm{c}}, \Gamma_{\mathrm{s}}$
  - $v_{\mathrm{ns}} \overset{\Delta}{\simeq} nf(\underset{\sim}{v}_{\mathrm{s}}.1, \underline{v_{\mathrm{c}}})$
  - $\mathcal{O} \overset{\Delta}{\simeq} I_S(v_{\mathrm{ns}}, r)$

  if $|\underset{\sim}{v}_{\mathrm{s}}| = 1$

We furthermore define $[\![P]\!]_{\mathrm{P}} = [\![P]\!]_{\mathrm{P}}\emptyset, \emptyset, \emptyset, \emptyset, \epsilon, \top$ for programs which constitute a complete model, i.e. which have no free identifiers.

Explanation of the denotation function. The case of reactions relies on the given concrete semantic function for assigning a semantical object to the reaction evaluated to its normal form. This normal form reaction is in turn obtained by evaluating the species expressions to their normal forms, which involves completing the compartment hierarchy in species values, and by evaluating rate expressions to rate values. The latter assumes the denotation function for algebraic rate expressions to be extended to rate expressions in an evident manner. There is one explicit condition for well-typedness, namely that species expressions must be deterministic, i.e. evaluate to singleton lists of species values. There is also the implicit condition that the concrete semantic function must be defined for the computed normal form reaction, which may e.g. fail if non-mass-action rates are used in a CTMC semantics.

The denotation of the nil program is simply the singleton set with the nil semantical object and the empty output species environment.

The denotation of a parallel composition is the pairwise composition of all semantical objects in the denotations of the operands, together with the pairwise update of output species environments from the first component with those of the second. The fresh name prefixes are extended appropriately. This case is well-typed when the composition operation, which is a parameter of the general semantic function, is defined.

The denotation of a variation composition is similar to that of parallel composition but results in a union of semantical objects rather than a Cartesian product.

In the case of located programs, the compartment identifier is looked up in the compartment environment and appended to the list of compartment values used to compute the denotation of the nested program. The denotation is defined when the compartment identifier is in the given compartment environment and when the resulting list of compartment values is well-typed.

The case of module invocation evaluates the actual parameters and passes the resulting values as parameters to the function denoting the module as given by the module environment. This function takes two additional parameters, namely the parent compartments at time of invocation and a fresh name string. From the function we obtain a set of semantical objects together with output species environments with bindings for the actual output species parameters. The sequential program is then evaluated in the species environment updated with the appropriate bindings for the output species. The result is the set of all pairwise compositions of semantical objects from the module and from the sequential program, together with the pairwise update of output species environments from the module with those from the sequential program. Hence the sequential program is treated as a parallel program with respect to semantical objects.

Special care must be taken to ensure the proper extension of fresh name strings for evaluating compartment expressions, species expressions, the module body and the sequential program. The crucial characteristic of these strings is that none is a postfix of another, ensuring that there is no way of extending one string to match another. So far we have achieved this in the semantics of binary operators by prefixing respectively a 0 and a 1 to the fresh name string. But here we are faced with lists of expressions to be evaluated. We then achieve the desired property by letting all prefixes be of length $|\underline{e_\mathrm{s}}| + |\underline{e_{\mathrm{s}+}}| + 2$, where the plus two term accounts for the module body and for the sequential program. For the ith compartment expression we choose a prefix in which the ith symbol is 1 and the remaining symbols are 0s, and a similar construction is used for the remaining prefixes. The denotation function for module invocation is defined when the module identifier is in the given environment and the associated function is defined for the given arguments.

The case of definitions relies on the denotation function for definitions to obtain an updated collection of environments in which the sequential program following the definition is evaluated.

The case of nondeterministic selection evaluates the species expression, and for each resulting species value, it evaluates the sequential program. As for module invocation, special care is needed to ensure that the fresh name strings are extended appropriately. The resulting set of semantical objects consists of all possible compositions of semantical objects associated with each species value, and is hence effectively a Cartesian product. The output species environments resulting from repeated evaluation of the sequential program are disregarded, since there does not appear to be any meaningful way to reconcile them. They all have the same domain, but generally differ in their images, since each is a result of evaluating the same sequential program with different bindings for the forced species.

Finally, the case of initial population or concentration definitions evaluates the given species expression, obtains the corresponding normal form based on the current parent compartments, and uses the concrete semantic function to obtain a semantical object.

Derived programs. Next we define the denotation of derived forms in terms of basic programs. We start by considering in-line species definitions which intuitively give rise to a sequence of standard species definitions followed by the reaction in which the in-line definitions have been removed and by the given sequential program; an example of this is given in Section 3.3. The formal presentation relies on an auxiliary *definition extraction function* of the form $[\![e_\mathrm{s}]\!]_\mathrm{ds} = e'_\mathrm{s}, \underline{D}$ where e_s is a derived species expression as defined in the abstract syntax for derived programs, e'_s is a basic species expression and $\underline{D}$ is a list of extracted species definitions. Selected cases of the definition are shown below; the remaining cases are similar:

- $[\![id_\mathrm{c}[e_\mathrm{s}]]\!]_\mathrm{ds} \stackrel{\Delta}{\simeq} id_\mathrm{c}[e'_\mathrm{s}], \underline{D}$ where
  - $e'_\mathrm{s}, \underline{D} \stackrel{\Delta}{\simeq} [\![e_\mathrm{s}]\!]_\mathrm{ds}$

- $[\![e_{\rm s} - e'_{\rm s}]\!]_{\rm ds} \stackrel{\Delta}{\simeq} e''_{\rm s} - e'''_{\rm s}, \underline{D}\,\underline{D'}$ where
  - $e''_{\rm s}, \underline{D} \stackrel{\Delta}{\simeq} [\![e_{\rm s}]\!]_{\rm ds}$
  - $e'''_{\rm s}, \underline{D'} \stackrel{\Delta}{\simeq} [\![e'_{\rm s}]\!]_{\rm ds}$

- $[\![id_{\rm s}]\!]_{\rm ds} \stackrel{\Delta}{\simeq} id_{\rm s}$

- $[\![e_{\rm s}\ \textbf{as}\ id_{\rm s}]\!]_{\rm ds} \stackrel{\Delta}{\simeq} e'_{\rm s}, (id_{\rm s} = e'_{\rm s})\underline{D}$ where
  - $e'_{\rm s}, \underline{D} \stackrel{\Delta}{\simeq} [\![e_{\rm s}]\!]_{\rm ds}$

The definition of L-R-equality preserving arrows, outlined informally in Section 3.5, relies on an auxiliary *linearisation function* for renaming identifiers in species expressions in a linear manner. Informally, the renaming is such that all species identifiers in the reactants become distinct, all species identifiers in the products become distinct, and the ith occurrences of a given identifier in the original reactants and products are given the same name.

The linearisation function is of the form $lin(e_{\rm s}, M) \stackrel{\Delta}{\simeq} e'_{\rm s}, M'$ where M and M' are multisets of species identifiers. Two key cases of the definition are given below where, for $i \in \mathbb{N}$, $bs(i)$ is the binary string representation of i:

- $lin(id_{\rm s}, M) \stackrel{\Delta}{\simeq} bs(M(id_{\rm s}))\_id_{\rm s}, M + id_{\rm s}$

- $lin(e_{\rm s} - e'_{\rm s}, M) \stackrel{\Delta}{\simeq} e''_{\rm s} - e'''_{\rm s}, M''$ where
  - $e''_{\rm s}, M' \stackrel{\Delta}{\simeq} lin(e_{\rm s}, M)$
  - $e'''_{\rm s}, M'' \stackrel{\Delta}{\simeq} lin(e'_{\rm s}, M')$

The base case prefixes an identifier with the binary string representation of the number of the identifier's previous occurrences, and adds the identifier to the multiset. The case of complex formation evaluates the first expression in the given multiset, resulting in a new multiset in which the second expression is evaluated. The remaining cases which are not shown here simply evaluate components recursively in the original multisets.

We extend the linearisation function to the form $lin(\underline{e_{\rm s}}, M) = \underline{e'_{\rm s}}, M'$ in order to rename species identifiers in reactant and product lists:

- $lin(\varepsilon, M) \stackrel{\Delta}{\simeq} \varepsilon, M$

- $lin(e_{\rm s}\underline{e'_{\rm s}}, M) \stackrel{\Delta}{\simeq} e''_{\rm s}\underline{e'''_{\rm s}}, M''$ where
  - $e''_{\rm s}, M' \stackrel{\Delta}{\simeq} lin(e_{\rm s}, M)$
  - $\underline{e'''_{\rm s}}, M'' \stackrel{\Delta}{\simeq} lin(\underline{e'_{\rm s}}, M')$

We also extend it to the form $lina(e_{\rm a}, M) = \underline{e'_{\rm a}}, M'$ for algebraic rate expressions with selected cases defined as follows:

- $lina(r, M) \stackrel{\Delta}{\simeq} r, M$

– $\mathit{lina}(e_s, M) \stackrel{\Delta}{\simeq} \mathit{lin}(e_s, M)$

– $\mathit{lina}(\textbf{if } e_b \textbf{ then } e_a \textbf{ else } e'_a) \stackrel{\Delta}{\simeq} \textbf{if } e_b \textbf{ then } e''_a \textbf{ else } e'''_a, M''$ where
  - $e''_a, M' \stackrel{\Delta}{\simeq} \mathit{lina}(e_a, M)$
  - $e'''_a, M'' \stackrel{\Delta}{\simeq} \mathit{lina}(e'_a, M')$

– $\mathit{lina}(e_a + e'_a, M) \stackrel{\Delta}{\simeq} e''_a + e'''_a, M''$ where
  - $e''_a, M' \stackrel{\Delta}{\simeq} \mathit{lina}(e_a, M)$
  - $e'''_a, M'' \stackrel{\Delta}{\simeq} \mathit{lina}(e'_a, M)$

The derived forms are then defined by a denotation function of the form:

$$[\![P]\!]_{\mathrm{dp}} = P'$$

where P is a derived form program and P' is a basic program. In the following, we assume a function of the form $\odot\underline{D}; P = P'$ which, given a list $\underline{D}$ of definitions and a program P, gives a program P' in which the definitions in $\underline{D}$ have been composed sequentially following the order of $\underline{D}$ and have scope P. We assume a function of the form $\mathit{order}(\mathcal{D}) = \underline{D}$ which orders a set $\mathcal{D}$ of definitions in some arbitrary but definite order. The function FS gives the set of species identifiers in a species expression and is defined along standard lines.

– $[\![\underline{e''_s} \sim \underline{n \cdot e_s}\ A_2^{e_r, e'_r}\ \underline{n' \cdot e'_s} \sim \underline{e'''_s} \textbf{ if } e_b, e'_b\ ; P]\!]_{\mathrm{dp}} \stackrel{\Delta}{\simeq}$

$$[\![\underline{e''_s} \sim \underline{n \cdot e_s}\ A(A_2)^{e_r}\ \underline{n' \cdot e'_s} \textbf{ if } e_b\ ; \mathbf{0}_{\mathrm{p}}]\!]_{\mathrm{dp}} \mid$$
$$[\![\underline{e'''_s} \sim \underline{n' \cdot e'_s}\ A(A_2)^{e'_r}\ \underline{n \cdot e_s} \textbf{ if } e'_b\ ; P]\!]_{\mathrm{dp}}$$

where
  - $A(\Leftrightarrow) \stackrel{\Delta}{\simeq} \Rightarrow$
  - $A(\leftrightarrow) \stackrel{\Delta}{\simeq} \rightarrow$
  - $A(\twoheadleftarrow\!\!\twoheadrightarrow) \stackrel{\Delta}{\simeq} \twoheadrightarrow$

– $[\![\underline{e''_s} \sim \underline{n \cdot e_s}\ A^{e_r}\ \underline{n' \cdot e'_s} \textbf{ if } e_b\ ; P]\!]_{\mathrm{dp}} \stackrel{\Delta}{\simeq}$

$$[\![\underline{1 \cdot e''_s}\ \underline{n \cdot e_s}\ A^{e_r}\ \underline{1 \cdot e''_s}\ \underline{n' \cdot e'_s} \textbf{ if } e_b\ ; P]\!]_{\mathrm{dp}}$$

– $[\![\underline{n \cdot e_s}\ A^{e_r}\ \underline{n' \cdot e'_s} \textbf{ if } e_b\ ; P]\!]_{\mathrm{dp}} \stackrel{\Delta}{\simeq} \odot\underline{D}\,\underline{D'}; ([\![\underline{n \cdot e''_s}\ A^{e_r}\ \underline{n' \cdot e'''_s} \textbf{ if } e_b]\!]_{\mathrm{dp}} \mid P)$
where
  - $\underline{e''_s}, \underline{D} = [\![\underline{e_s}]\!]_{\mathrm{ds}}$
  - $\underline{e'''_s}, \underline{D'} = [\![\underline{e'_s}]\!]_{\mathrm{ds}}$

– $[\![\underline{n \cdot e_s}\ \rightarrow^{e_r}\ \underline{n' \cdot e'_s} \textbf{ if } e_b]\!]_{\mathrm{dp}} \stackrel{\Delta}{\simeq}$

$$\odot\mathit{order}\{\mathit{id}_s = \textbf{force } \mathit{id}_s \mid \mathit{id}_s \in \mathit{FS}(\underline{e_s}, \underline{e'_s})\}\ ;\ \underline{n \cdot e_s}\ \Rightarrow^{e_r}\ \underline{n' \cdot e'_s} \textbf{ if } e_b$$

– $[\![\underline{n \cdot e_s}\ \twoheadrightarrow^{e_r}\ \underline{n' \cdot e'_s} \textbf{ if } e_b]\!]_{\mathrm{dp}} \stackrel{\Delta}{\simeq}$

$$\odot\mathit{order}\{\mathit{bs}(i)\_\mathit{id}_s = \textbf{force } \mathit{id}_s \mid \mathit{id}_s \in \mathit{dom}(M) \wedge i \in M(\mathit{id}_s)\}\ ;$$
$$\underline{n \cdot e''_s}\ \Rightarrow^{e'_r}\ \underline{n' \cdot e'''_s} \textbf{ if } e_b$$

where

- $e'_{\mathrm{r}}, M' \overset{\Delta}{\simeq} \mathit{lin}(e_{\mathrm{r}}, \emptyset)$
- $\underline{e''_{\mathrm{s}}}, M'' \overset{\Delta}{\simeq} \mathit{lin}(\underline{e_{\mathrm{s}}}, \emptyset)$
- $\underline{e'''_{\mathrm{s}}}, M''' \overset{\Delta}{\simeq} \mathit{lin}(\underline{e'_{\mathrm{s}}}, \emptyset)$
- $M(id_{\mathrm{s}}) \overset{\Delta}{\simeq} \{1 \ldots max(M'(id_{\mathrm{s}}), M''(id_{\mathrm{s}}), M'''(id_{\mathrm{s}}))\}$

The first case defines a reversible reaction as the parallel compositions of the two reactions, one for each direction. The second case defines an enzymatic reaction as a non-enzymatic reaction in which the enzymes are included in both reactants and products and hence do not get consumed. Other choices, following e.g. Michaelis-Menten kinetics, could also be made here.

The third case defines a reaction with in-line definitions as a reaction where definitions have been extracted and put in scope of both the reaction and the following program which is composed in parallel. Although conceptually simple, we note that there are some artificial cases which may not expand to the expected result, specifically when the same identifier is defined multiple times in a reaction as in e.g. a–b **as** a + a–b **as** a => a–b; here all occurrences of a will be bound to the same expression, namely that resulting from the second in-line definition.

The last two cases define nondeterministic reaction arrows in terms of the force operator and the deterministic reaction arrow. Note that nondeterministic species must be bound to identifiers in reactions in order to preserve the relationship between identical nondeterministic species in reactants and products. Reactions with explicit nondeterminism are therefore ill-typed. Note also that for reactions with the L-R equality-preserving arrow, a given identifier should generally have the same number of occurrences in the reactants and products to obtain meaningful results, although this condition is not explicitly enforced. In particular, reactions such as 2 s –>> s–s and s + s –>> s–s are *not* equivalent according to the above definition of derived forms.

The order of evaluation of derived forms is significant. Specifically, in-line species definitions are expanded before nondeterministic selection. This ensures that e.g. the program s + t –> s–t **as** a; P expands correctly, i.e. to:

```
spec a = s-t;
(s = force s;
 t = force t;
 s + t => s-t
) | P
```

rather than to:

```
spec a = s-t;
s = force s;
t = force t;
(s + t => s-t | P)
```

5.4 Definitions

The denotation function. The denotation function for definitions updates the environments with bindings for a given definition. It takes the following form:

$$[\![D]\!]_{\mathrm{d}}\Gamma_{\mathrm{c}},\Gamma_{\mathrm{s}},\Gamma_{\mathrm{a}},\Gamma_{\mathrm{m}},b=\Gamma'_{\mathrm{c}},\Gamma'_{\mathrm{s}},\Gamma'_{\mathrm{a}},\Gamma'_{\mathrm{m}},\Gamma_{\mathrm{so}}$$

The output species environment is *created* by the denotation function and is always empty except for the case of species definitions, where it captures the binding for the defined species. This is in contrast to the other environments which are *updated* by the denotation function. Here is the definition:

- $[\![id_{\mathrm{s}}=e_{\mathrm{s+}}]\!]_{\mathrm{d}}\Gamma_{\mathrm{c}},\Gamma_{\mathrm{s}},\Gamma_{\mathrm{a}},\Gamma_{\mathrm{m}},b\stackrel{\Delta}{\simeq}\Gamma_{\mathrm{c}},\Gamma_{\mathrm{s}}\langle id_{\mathrm{s}}\mapsto\underset{\sim}{v_{\mathrm{s}}}\rangle,\Gamma_{\mathrm{a}},\Gamma_{\mathrm{m}},\{id_{\mathrm{s}}\mapsto\underset{\sim}{v_{\mathrm{s}}}\}$ where
  - $\underset{\sim}{v_{\mathrm{s}}}\stackrel{\Delta}{\simeq}[\![e_{\mathrm{s+}}]\!]_{\mathrm{s+}}\Gamma_{\mathrm{c}},\Gamma_{\mathrm{s}},b$
- $[\![id_{\mathrm{c}}=e_{\mathrm{c}}]\!]_{\mathrm{d}}\Gamma,\Gamma_{\mathrm{c}}\stackrel{\Delta}{\simeq}\Gamma,\Gamma_{\mathrm{c}}\langle id_{\mathrm{c}}\mapsto v_{\mathrm{c}}\rangle,\emptyset$ where
  - $v_{\mathrm{c}}\stackrel{\Delta}{\simeq}[\![e_{\mathrm{c}}]\!]_{\mathrm{c}}\Gamma_{\mathrm{c}},b$
- $[\![id_{\mathrm{a}}(\underline{id_{\mathrm{c}}};\underline{id_{\mathrm{s}}:\xi};\underline{id_{\mathrm{a}}})=e_{\mathrm{a}}]\!]_{\mathrm{d}}\Gamma_{\mathrm{c}},\Gamma_{\mathrm{s}},\Gamma_{\mathrm{a}},\Gamma_{\mathrm{m}},b\stackrel{\Delta}{\simeq}\Gamma_{\mathrm{c}},\Gamma_{\mathrm{s}},\Gamma_{\mathrm{a}}\langle id_{\mathrm{a}}\mapsto f\rangle,\Gamma_{\mathrm{m}},\emptyset$ where
  - $f(\underline{v_{\mathrm{c}}},\underline{\underset{\sim}{v_{\mathrm{s}}}},\underline{v_{\mathrm{a}}},\underline{v_{\mathrm{c}}}')\stackrel{\Delta}{\simeq}[\![e_{\mathrm{a}}]\!]_{\mathrm{a}}\Gamma'_{\mathrm{c}},\Gamma'_{\mathrm{s}},\Gamma'_{\mathrm{a}},\underline{v_{\mathrm{c}}}'$
  - $\Gamma'_{\mathrm{c}}\stackrel{\Delta}{\simeq}\Gamma_{\mathrm{c}}\langle\{\underline{id_{\mathrm{c}}}\mapsto\underline{v_{\mathrm{c}}}\}\rangle$
  - $\Gamma'_{\mathrm{s}}\stackrel{\Delta}{\simeq}\Gamma_{\mathrm{s}}\langle\{\underline{id_{\mathrm{s}}}\mapsto close(\underset{\sim}{v_{\mathrm{s}}},\xi)\}\rangle$
  - $\Gamma'_{\mathrm{a}}\stackrel{\Delta}{\simeq}\Gamma_{\mathrm{a}}\langle\{\underline{id_{\mathrm{a}}}\mapsto\underline{v_{\mathrm{a}}}\}\rangle$
- $[\![id_{\mathrm{p}}(\underline{id_{\mathrm{c}}};\underline{id_{\mathrm{s}}:\xi};\underline{id_{\mathrm{a}}};\ \mathbf{out}\ \underline{id'_{\mathrm{s}}:e'_{\mathrm{s}}})=P]\!]_{\mathrm{d}}\Gamma_{\mathrm{c}},\Gamma_{\mathrm{s}},\Gamma_{\mathrm{a}},\Gamma_{\mathrm{m}},b\stackrel{\Delta}{\simeq}$
  $\Gamma_{\mathrm{c}},\Gamma_{\mathrm{s}},\Gamma_{\mathrm{a}},\Gamma_{\mathrm{m}}\langle id_{\mathrm{p}}\mapsto g\rangle,\emptyset$

  where
  - $g(\underline{v_{\mathrm{c}}},\underline{\underset{\sim}{v_{\mathrm{s}}}},\underline{v_{\mathrm{a}}},\underline{id''_{\mathrm{s}}},b',\underline{v_{\mathrm{c}}}')\stackrel{\Delta}{\simeq}\{(\mathcal{O}_i,\Gamma_{\mathrm{so}i})\}$
  - $\Gamma'_{\mathrm{c}}\stackrel{\Delta}{\simeq}\Gamma_{\mathrm{c}}\langle\{\underline{id_{\mathrm{c}}}\mapsto\underline{v_{\mathrm{c}}}\}\rangle$
  - $\Gamma'_{\mathrm{s}}\stackrel{\Delta}{\simeq}\Gamma_{\mathrm{s}}\langle\{\underline{id_{\mathrm{s}}}\mapsto seal(close(\underset{\sim}{v_{\mathrm{s}}},\xi),b)\}\rangle$
  - $\Gamma''_{\mathrm{s}}\stackrel{\Delta}{\simeq}\Gamma_{\mathrm{s}}\langle\{\underline{id_{\mathrm{s}}}\mapsto\underline{\underset{\sim}{v_{\mathrm{s}}}}\}\rangle$
  - $\Gamma'_{\mathrm{a}}\stackrel{\Delta}{\simeq}\Gamma_{\mathrm{a}}\langle\{\underline{id_{\mathrm{a}}}\mapsto seal(\underline{v_{\mathrm{a}}},b)\}\rangle$
  - $\{(\mathcal{O}_i,\Gamma'_{\mathrm{so}i})\}\stackrel{\Delta}{\simeq}[\![P]\!]_{\mathrm{p}}\Gamma'_{\mathrm{c}},\Gamma'_{\mathrm{s}},\Gamma'_{\mathrm{a}},\Gamma_{\mathrm{m}},b',\underline{v_{\mathrm{c}}}'$
  - $\Gamma_{\mathrm{so}i}\stackrel{\Delta}{\simeq}\{\underline{id''_{\mathrm{s}}}\mapsto adapt(\underline{\underset{\sim}{v'_{\mathrm{s}}}},\underline{\underset{\approx}{v''_{\mathrm{s}}}})\}$ where
    - $\underline{\underset{\sim}{v'_{\mathrm{s}}}}\stackrel{\Delta}{\simeq}\Gamma'_{\mathrm{so}i}(\underline{id'_{\mathrm{s}}})$
    - $\underline{\underset{\approx}{v''_{\mathrm{s}}}},\emptyset=[\![\underline{e'_{\mathrm{s}}}]\!]_{\mathrm{s}}\Gamma'_{\mathrm{c}},\Gamma''_{\mathrm{s}}$

Explanation of the denotation function. The cases for species and compartment definitions are straightforward since they rely on the respective denotation functions. The case of rate function definitions updates the rate function environment with a new binding to a function f from actual parameters and parent compartments to an algebraic rate value. This algebraic rate value is computed in the environments at time of definition updated with bindings for the actual parameters, and with the parent compartments at time of invocation. The interfaces of the actual species parameters are updated based on the annotations of the corresponding formal parameters using the *close* function defined in Subsection 5.2 which here is assumed extended to lists of species values. The function f is only defined when the number of actual and formal parameters match, and when the species interface updates result in well-typed species values.

The case of module definitions updates the module environment with a new binding to a function from actual parameters, a fresh name string and parent compartments, to a set of semantical objects and species output environments. The semantical objects are computed in the environments at time of definition updated with bindings for actual parameters, and with the fresh name string and parent compartments at time of invocation. As for algebraic rate expressions, the interfaces of actual species values are updated. But an additional step is taken to confine species values to a namespace given by the fresh name string at time of definition, ensuring that e.g. variables in actual parameters are not captured inside the module. This is done using the *seal* function on modification site expressions, which is given as a parameter of the general semantics; we assume this to be extended appropriately to lists of species values and also to algebraic rate values. Finally, the resulting output species environment is given by a mapping from actual output species identifiers to the values of the corresponding formal output species identifiers as recorded in the output species environment of the body, but with interfaces updated using the *adapt* function defined in Subsection 5.2. The function is assumed extended to pairs of species value lists of the same length. Hence the updates are carried out in a pair-wise manner by matching up corresponding positions in the lists constituting the nondeterministic species values; this is the reason for having nondeterministic species represented by lists rather than sets.

6 Some Concrete Semantics

Practical applications of LBS require specific choices of concrete semantics to be made, and any questions of language expressiveness must also be addressed in the context of a specific concrete semantics. This section therefore gives four examples of concrete semantics, namely: basic Petri nets; coloured Petri nets; ordinary differential equations; and continuous time Markov chains. These follow the ideas in [31], but are adapted to adhere to the general semantics of LBS.

6.1 Preliminaries

The general semantics preserves variables in species modification sites because variables can be exploited by some concrete semantics. But for other concrete semantics this is not the case, and we can instead parameterise the general semantic function on a structure $(S, |_S, \mathbf{0}_S, G_S, I_S)$ which is the same as before, except that G_S is a function assigning semantical objects to named *ground normal form reactions.* These are normal form reactions in which expressions have been appropriately evaluated based on a variable environment: species values have been evaluated to ground normal form species values as defined previously; rate values have been evaluated to obtain *ground rate values* defined below; and reaction conditionals are omitted because reactions with conditionals which evaluate to **ff** are simply discarded. We therefore start by defining the general assignment R_S of semantical objects to named normal form reactions in terms of an assignment G_S to named ground normal form reactions, allowing a concrete semantics to be defined in terms of either of these.

Ground normal form reactions. *Ground algebraic rate values* differ from algebraic rate values in that species values are replaced by ground normal form species values and conditionals are not included. Ground rate values contain ground algebraic rate values rather than algebraic rate values, and for the mass-action case, these must be constants. The formal definition is given by the grammar in Table 6.1.

Table 6.1. The abstract syntax for ground rate values

$v_{\mathrm{gr}} ::=$	GROUND RATE VALUE
$\mid \{r\}$	MASS-ACTION RATE CONSTANT
$\mid [v_{\mathrm{ga}}]$	GROUND ALGEBRAIC RATE VALUE
$v_{\mathrm{ga}} ::=$	GROUND ALGEBRAIC RATE VALUE
$\mid r$	CONSTANT
$\mid v_{\mathrm{gns}}$	POPULATION
$\mid exp(v_{\mathrm{ga}}) \mid log(v_{\mathrm{ga}}) \mid sin(v_{\mathrm{ga}}) \mid cos(v_{\mathrm{ga}})$	FUNCTIONS
$\mid v_{\mathrm{ga}} + v'_{\mathrm{ga}} \mid v_{\mathrm{ga}}$ - v'_{ga}	ARITHMETIC OPERATORS
$\mid v_{\mathrm{ga}} \times v'_{\mathrm{ga}} \mid v_{\mathrm{ga}} / v'_{\mathrm{ga}} \mid v_{\mathrm{ga}}$^$v'_{\mathrm{ga}}$	

A denotation function of the form $[\![v_{\mathrm{a}}]\!]_{\mathrm{a}}\Gamma_{\mathrm{x}} = v_{\mathrm{ga}}$ assigning ground algebraic rate values to algebraic rate values, given a variable environment, is defined below. Only selected cases for functions and arithmetic operators are shown since the remaining cases are similar.

- $[\![r]\!]_{\mathrm{a}}\Gamma_{\mathrm{x}} \stackrel{\Delta}{\simeq} r$
- $[\![v_{\mathrm{ns}}]\!]_{\mathrm{a}}\Gamma_{\mathrm{x}} \stackrel{\Delta}{\simeq} [\![v_{\mathrm{ns}}]\!]_{\mathrm{m}}\Gamma_{\mathrm{x}}$

– $[\![\textbf{if } e_\mathrm{b} \textbf{ then } v_\mathrm{a} \textbf{ else } v'_\mathrm{a}]\!]_\mathrm{a}\Gamma_\mathrm{x} \stackrel{\Delta}{\simeq} \begin{cases} [\![v_\mathrm{a}]\!]_\mathrm{a}\Gamma_\mathrm{x} \text{ if } [\![e_\mathrm{b}]\!]_\mathrm{b}\Gamma_\mathrm{x} = \mathbf{tt} \\ [\![v'_\mathrm{a}]\!]_\mathrm{a}\Gamma_\mathrm{x} \text{ otherwise} \end{cases}$

– $[\![\mathit{exp}(v_\mathrm{a})]\!]_\mathrm{a}\Gamma_\mathrm{x} \stackrel{\Delta}{\simeq} \mathit{exp}([\![v_\mathrm{a}]\!]_\mathrm{a}\Gamma_\mathrm{x})$

– $[\![v_\mathrm{a} + v'_\mathrm{a}]\!]_\mathrm{a}\Gamma_\mathrm{x} \stackrel{\Delta}{\simeq} [\![v_\mathrm{a}]\!]_\mathrm{a}\Gamma_\mathrm{x} + [\![v'_\mathrm{a}]\!]_\mathrm{a}\Gamma_\mathrm{x}$

In the case of normal form species values we assume the denotation function for modification site expressions extended in an evident manner.

Ground normal form reactions are then of the form:

$$G ::= \underline{n \cdot v_\mathrm{gns}} \Rightarrow^{v_\mathrm{gr}} \underline{n' \cdot v'_\mathrm{gns}}$$

The general semantics in terms of ground normal form reactions. The idea in the following construction is to obtain a ground normal form reaction for each possible variable environment associated with a normal form reaction, then get the semantical object of each ground normal form reaction, and finally apply the parallel composition operator to these objects. We therefore start by defining a function of the form $R_S(R, b, \Gamma_\mathrm{x}) = \mathcal{O}$ assigning a semantical object $\mathcal{O}$ to a normal form reaction R, named b, given a variable environment Γ_x:

$$R_S(\underline{n \cdot v_\mathrm{ns}} \Rightarrow^{v_\mathrm{r}} \underline{n' \cdot v'_\mathrm{ns}} \textbf{ if } e_\mathrm{b}, b, \Gamma_\mathrm{x}) \stackrel{\Delta}{\simeq}$$
$$\begin{cases} G_S(\underline{n \cdot [\![v_\mathrm{ns}]\!]_\mathrm{m}\Gamma_\mathrm{x}} \Rightarrow^{[\![v_\mathrm{r}]\!]_\mathrm{a}\Gamma_\mathrm{x}} \underline{n' \cdot [\![v'_\mathrm{ns}]\!]_\mathrm{m}\Gamma_\mathrm{x}}, b) & \text{if } [\![e_\mathrm{b}]\!]_\mathrm{b}\Gamma_\mathrm{x} = \mathbf{tt} \\ \mathbf{0}_S & \text{otherwise} \end{cases}$$

If the conditional evaluates to **ff**, the reaction is assigned the nil object, and otherwise the assignment relies on the function G_S for assigning a semantical object to the ground normal form of the reaction. Again we assume the denotation function on modification site expressions to be extended to normal form species values in an evident manner. We also assume the denotation function for ground algebraic rate values to be extended to ground rate values in an evident manner; note that this function is only defined when ground algebraic rate values which are used as mass-action rates evaluate to constants.

The set of all variable environments associated with a normal form reaction is defined as follows, using the standard notation for dependent sets:

$$\mathit{Val}(R) \stackrel{\Delta}{\simeq} \prod_{(x:\rho) \in FV(R)} [\![\rho]\!]_\mathrm{t}$$

We here assume the variable function FV on modification site expressions to be extended to reactions in an evident manner. Observe that variable environments are restricted to only assign values of given types to variables, and that for finite types, we get a finite set of variable environments.

In order to construct appropriate binary strings for naming reactions, we assume an arbitrary but fixed total ordering $\leq$ on variable environments Γ_x.

In practise this can for example be obtained from a lexicographical ordering on variables together with a suitable ordering on values. We assume an operator Ⓢ which gives the parallel composition in some definite order of its operands. Recall also that the function δ_i^m gives a binary string of length m with 0s everywhere except for the ith entry. The assignment R_S can then be defined in terms of G_S as follows:

$$R_S(R, b) \stackrel{\Delta}{\simeq}$$
$$Ⓢ\{R_S(R, \delta_i^m b, \Gamma_{\mathrm{x}}) \mid \Gamma_{\mathrm{x}} \in \mathit{Val}(R) \wedge i = |\{\Gamma'_{\mathrm{x}} \in \mathit{Val}(R) \mid \Gamma'_{\mathrm{x}} \leq \Gamma_{\mathrm{x}}\}| \wedge m = |\mathit{Val}(R)|\}$$

6.2 A Petri Net Semantics

Petri nets. We already encountered a graphical representation of a Petri net in Figure 2.2. *Places*, depicted as circles, represent species, and *transitions*, depicted as rectangles, represent reactions. In the figure we considered only atomic species with no modification sites, but in the more general case, a separate place is used to represent each modification state of a complex species. *Flow functions*, depicted as arcs between places and transitions, are used to identify the reactants and products of a reaction. In the general case, arcs are labelled with integers representing stoichiometry. Finally, a *marking* defines the state of a Petri net by the number of *tokens*, representing individual molecules, contained in each place. For our purposes, tokens are multiset versions of ground normal form species values.

The formal definition of our Petri nets is given below, where $V_{\mathbf{gns}}$ is the set of all ground normal form species values $v_{\mathbf{gns}}$.

Definition 1. *An LBS-*Petri net *P is a tuple $(S, T, F^{in}, F^{out}, M^0)$ where*

- $S \subset_{fin} \{MS(v_{gns}) \mid v_{gns} \in V_{gns}\}$ *is the set of* places.
- $T \subset_{fin} \{0,1\}^*$ *is the set of* transitions.
- $F^{in}, F^{out} : T \times S \to \mathbb{N}$ *are the* flow-in *and* flow-out *functions, respectively.*
- $M^0 \in MS(S)$ *is the* initial marking

Recall in the above definition that $MS(\underline{x})$ gives the multiset representation of a list $\underline{x}$. The set of places hence contains multiset-representations of normal form species values, reflecting that the ordering of atomic species within normal form species values is insignificant, i.e. that the complex formation operator is commutative. Transitions are binary strings since these are used to name reactions in the general semantics. We use the notation S_P to refer to the places S of Petri net P, and similarly for the other Petri net elements. The set of all Petri nets is denoted by $\mathcal{P}$.

The qualitative semantics of Petri nets. The qualitative semantics determines how the marking of a Petri net changes over discrete time. Informally, this can be illustrated by playing the *token game*: a transition can fire whenever all its input places contain at least the number of tokens specified by the corresponding arc weights; when a transition fires, the number of tokens specified by arc weights are consumed from the input places and added to the output places.

Formally, the set of all *markings* of a Petri net is the set of multisets of places:

$$\mathcal{M}(P) \overset{\Delta}{\simeq} MS(S_P)$$

The behaviour of a Petri net is defined in terms of a transition relation which captures all possible moves in the token game.

Definition 2. *Let P be a Petri net, let $X \in MS(T_P)$ and let $M, N \in \mathcal{M}(P)$. Then define $M \xrightarrow{X} N$ iff*

1. $M \geq \sum_{t \in X} F_P^{in}(t)$
2. $N = M + \sum_{t \in X} F_P^{out}(t) - F_P^{in}(t)$

Note that a flow function applied to only one argument, a transition, is interpreted as a function on places, here a marking. The arithmetic operations and relations are understood to be extended to markings in the expected way, e.g. $M \geq M'$ iff $M(s) \geq M'(s)$ for all s. Condition 1 hence states that the marking M must have sufficient tokens for transitions in X to fire, and condition 2 states that N is the marking resulting from firing the transitions from X in marking M.

The concrete Petri net semantics of LBS

Definition 3. *The concrete semantics for LBS in terms of Petri nets is given by the tuple $(\mathcal{P}, |_\mathcal{P}, \mathbf{0}_\mathcal{P}, G_\mathcal{P}, I_\mathcal{P})$ where*

- $P_1 |_\mathcal{P} P_2 \overset{\Delta}{\simeq} P$ *where*
  - $S_P \overset{\Delta}{\simeq} S_{P_1} \cup S_{P_2}$
  - $T_P \overset{\Delta}{\simeq} T_{P_1} \cup T_{P_2}$
  - $F_P^{io}(t,s) \overset{\Delta}{\simeq} \begin{cases} F_{P_1}^{io}(t,s) & \text{if } t \in T_{P_1} \wedge s \in S_{P_1} \\ F_{P_2}^{io}(t,s) & \text{if } t \in T_{P_2} \wedge s \in S_{P_2} \\ 0 & \text{otherwise} \end{cases} \quad \text{for } io \in \{in, out\}$
  - $M_P^0 \overset{\Delta}{\simeq} M_{P_1}^0 + M_{P_2}^0$

  if $T_{P_1} \cap T_{P_2} = \emptyset$
- $\mathbf{0}_P \overset{\Delta}{\simeq} (\emptyset, \emptyset, \emptyset, \emptyset, \emptyset)$
- $G_\mathcal{P}(\underline{n \cdot v_{gns}} \Rightarrow^{v_{gr}} \underline{n' \cdot v'_{gns}}, t) \overset{\Delta}{\simeq} P$ *where*
  - $S_P \overset{\Delta}{\simeq} \{MS(\underline{v_{gns}}.i)\} \cup \{MS(\underline{v'_{gns}}.j)\}$
  - $T_P \overset{\Delta}{\simeq} \{t\}$
  - $F_P^{in}(t,s) \overset{\Delta}{\simeq} \sum_{MS(\underline{v_{gns}}.i)=s} \underline{n}.i$

- $F^{out}_P(t, s) \stackrel{\Delta}{\simeq} \sum_{MS(\underline{v'_{gns}}.j)=s} \underline{n'}.j$
- $M^0_P \stackrel{\Delta}{\simeq} \emptyset$

– $I_P(v_{gns}, n) \stackrel{\Delta}{\simeq} (\{MS(v_{gns})\}, \emptyset, \emptyset, \emptyset, \{MS(v_{gns}) \mapsto n\})$

The function I_P is only defined for natural-numbered initial populations, not for real-numbered initial concentrations, because markings in Petri nets are discrete. The parallel composition operator is only defined for Petri nets with disjoint sets of transitions. The transition sets of two Petri nets resulting from the general semantics are however always disjoint because reactions have fresh names. This is in contrast to CBS where a bottom-up approach is taken: the semantics for parallel composition renames transitions before composition.

6.3 A Coloured Petri Net Semantics

Coloured Petri nets (CPNs) allow a single place to represent a species in any of its possible states of modification. CPNs hence allow for a compact description of models and can potentially lead to more efficient simulation and analysis, and in contrast to standard Petri nets, they are capable of representing species with infinite modification site types such as strings.

Coloured Petri nets. Places in CPNs are assigned *types* (or *colour*), and tokens are structured values of the type assigned to the place in which they reside. In our case, the type of a place is given by a multiset of located atomic species names and their modification site types, hence representing a complex species independently of its state of modification. As for standard Petri nets, tokens are multiset versions of ground normal form species values. But in contrast to standard Petri nets, arcs are equipped with multiset representations of normal form species values which are not necessarily ground. This enables a transition to operate selectively on species in a given state of modification, or indeed to ignore the state of certain sites. *Boolean guards* with variables allow transitions to assert further control over tokens.

We give a definition of coloured Petri nets which is tailored to our needs and which avoids some details of the standard definition [18]. For example, the standard definition distinguishes between place names and place types, but for our purposes a place is identified uniquely by its type. Our definition can however be recast in standard terms, as would be necessary for exploiting existing CPN tools.

Formally, we define a *species type* τ as follows:

$$\tau ::= \sum_i \underline{n_{c_i}}[n_{si}, \sigma_i]$$

and we let *Types* be the set of all species types. We define a function of the form $type(v_{ns}) = \tau$ giving the type of a normal form species value:

$$type(\underline{n_c[n_s, \alpha_\sigma]}) \stackrel{\Delta}{\simeq} \sum_i \underline{n_{c_i}}[n_{si}, \sigma_i]$$

where σ_i is α_i in which each pair of the image has been projected to the type component. We assume a similar definition for a function of the form $type(v_{\text{gns}}) = \tau$ for ground normal form species values.

The formal definition of our coloured Petri nets is given below, where $E_{\mathbf{bool}}$ is set of boolean expressions e_{b}.

Definition 4. *An LBS-*coloured Petri net *C is a tuple $(S, T, F^{in}, F^{out}, B, M^0)$ where*

- $S \subset_{fin}$ *Types is a finite set of* places.
- $T \subset_{fin} \{0,1\}^*$ *is a finite set of* transitions.
- $F^{in}, F^{out} : \prod_{(t,\tau)\in T\times S} MS(\{MS(v_{ns}) \mid type(v_{ns}) = \tau\})$ *are the* flow-in *and* flow-out *functions, respectively.*
- $B : T \to E_{\mathbf{bool}}$ *is the* transition guard *function.*
- $M^0 : \prod_{\tau\in S} MS(\{MS(v_{gns}) \mid type(v_{gns}) = \tau\})$ *is the* initial marking.

As for basic Petri nets, we use the notation S_C to refer to the places S of a coloured Petri net C, and similarly for the other elements. The set of all coloured Petri nets is denoted by $\mathcal{C}$.

The qualitative semantics of coloured Petri nets. The set of all *markings* $\mathcal{M}(C)$ of a coloured Petri net C is defined as follows:

$$\mathcal{M}(C) \stackrel{\Delta}{\simeq} \prod_{\tau\in S_C} MS(\{MS(v_{\text{gns}}) \mid type(v_{\text{gns}}) = \tau\})$$

We furthermore let

$$VE_X \stackrel{\Delta}{\simeq} \{\Gamma_{\text{x}} \mid dom(\Gamma_{\text{x}}) = X\}$$

be the set of variable environments with domain X, and we let

$$FV(t, C) \stackrel{\Delta}{\simeq} FV(F_C^{in}(t)) \cup FV(F_C^{out}(t)) \cup FV(B_C(t))$$

be the set of typed variables associated with a transition t in CPN C; here FV is assumed extended in an evident manner. The behaviour of a coloured Petri net is defined in terms of a transition relation as follows.

Definition 5. *Let C be a coloured Petri net, let $X \in MS(\prod_{t\in T_C} VE_{FV(t,C)})$ and let $M, N \in \mathcal{M}(C)$. Then define $M \xrightarrow{X} N$ iff*

1. $M \geq \sum_{(t,\Gamma_x)\in X} [\![F_C^{in}(t)]\!]_m \Gamma_x$
2. $N = M + \sum_{(t,\Gamma_x)\in X} [\![F_C^{out}(t)]\!]_m \Gamma_x - [\![F_C^{in}(t)]\!]_m \Gamma_x$
3. $\bigwedge_{(t,\Gamma_x)\in X} [\![B_C(t)]\!]_m \Gamma_x = \mathbf{tt}$

Recall that the modification site denotation function is a parameter of the species semantics, and in the above definition we assume this function to be extended from modification site expressions to normal form species values and to markings in an evident manner. A flow function applied to a transition is here interpreted as a marking, i.e. a mapping from places to multisets, and the multiset operations are assumed to be appropriately extended. Conditions 1 and 2 then correspond to conditions 1 and 2 in the qualitative semantics of standard Petri nets. Condition 3 states that the guards of all fired transitions must evaluate to **tt**.

The concrete coloured Petri net semantics of LBS

Definition 6. *The concrete semantics for LBS in terms of coloured Petri nets is given by the tuple* $(\mathcal{C}, |_C, \mathbf{0}_C, R_C, I_C)$ *where*

- $C_1 \,|_C\, C_2 \stackrel{\Delta}{\simeq} C$ *where*
  - $S_C \stackrel{\Delta}{\simeq} S_{C_1} \cup S_{C_2}$
  - $T_C \stackrel{\Delta}{\simeq} T_{C_1} \cup T_{C_2}$
  - $F_C^{io}(t,\tau) \stackrel{\Delta}{\simeq} \begin{cases} F_{C_1}^{io}(t,\tau) & \textit{if } t \in T_{C_1} \wedge \tau \in S_{C_1} \\ F_{C_2}^{io}(t,\tau) & \textit{if } t \in T_{C_2} \wedge \tau \in S_{C_2} \\ \emptyset & \textit{otherwise} \end{cases}$ *for* $io \in \{in, out\}$
  - $B_C(t) \stackrel{\Delta}{\simeq} \begin{cases} B_{C_1}(t) & \textit{if } t \in T_{C_1} \\ B_{C_2}(t) & \textit{if } t \in T_{C_2} \end{cases}$
  - $M_C^0 \stackrel{\Delta}{\simeq} M_{C_1}^0 + M_{C_2}^0$

  if $T_{C_1} \cap T_{C_2} = \emptyset$
- $\mathbf{0}_C \stackrel{\Delta}{\simeq} (\emptyset,\emptyset,\emptyset,\emptyset,\emptyset,\emptyset)$
- $R_C(\underline{n \cdot v_{ns}} \Rightarrow^{v_r} \underline{n' \cdot v'_{ns}} \textbf{ if } e_b, t) \stackrel{\Delta}{\simeq} C$ *where*
  - $S_C \stackrel{\Delta}{\simeq} \{type(\underline{v_{ns}}.i)\} \cup \{type(\underline{v'_{ns}}.j)\}$
  - $T_C \stackrel{\Delta}{\simeq} \{t\}$
  - $F_C^{in}(t,\tau) \stackrel{\Delta}{\simeq} \sum_{type(\underline{v_{ns}}.i)=\tau} \underline{n}.i \cdot MS(\underline{v_{ns}}.i)$
  - $F_C^{out}(t,\tau) \stackrel{\Delta}{\simeq} \sum_{type(\underline{v'_{ns}}.j)=\tau} \underline{n'}.j \cdot MS(\underline{v'_{ns}}.j)$
  - $B_C(t) \stackrel{\Delta}{\simeq} e_b$
  - $M_C^0 \stackrel{\Delta}{\simeq} \emptyset$
- $I_C(v_{ns}, n) \stackrel{\Delta}{\simeq} (\{type(v_{ns})\},\emptyset,\emptyset,\emptyset,\emptyset,\{MS(v_{ns}) \mapsto n\})$

The definition is similar to that for standard Petri nets, but differs in the inclusion of guards and in the definition of the reaction and initial condition functions where species types and normal form values are used rather than ground values.

6.4 An ODE Semantics

The Petri net semantics presented above are *qualitative* in that they do not take reaction rates into account. In this section we give a *quantitative* semantics in terms of *ordinary differential equations* (ODEs). ODEs are *continuous* since they define system dynamics in terms of species concentrations. They are also *deterministic* since they, given initial conditions, uniquely determine the state of a system at any point of time in terms of species concentrations.

Ordinary differential equations. A set of ODEs specifies how the concentration $[s_i]$ of a species s_i changes over time and is traditionally written in the following notation:

$$\begin{aligned} \mathrm{d}[s_1] &= p_1 \\ &\vdots \\ \mathrm{d}[s_n] &= p_n \end{aligned}$$

where the p_i are real polynomials over $[s_i]$. The initial conditions of a set of ODEs are specified by the concentration of each species at time 0.

Formally, let $(Pol(X), +, \cdot)$ be the ring of real polynomials over variables in the set X. We then define the structure of ODEs with initial conditions as follows:

Definition 7. *A structure D of LBS-*ODEs with initial conditions *is given by a tuple (X, P, I) where*

- $X \subset_{fin} \{MS(v_{gns}) \mid v_{gns} \in V_{gns}\}$ *is the set of* variables.
- $P : X \to Pol(X)$ *is the assignment of* polynomials *to variables.*
- $I : X \to \mathbb{R}$ *is the* initial condition.

The set of all structures of ODEs with initial conditions is denoted by $\mathcal{D}$, and we denote e.g. X in D by X_D. Although non-linear ODEs cannot generally be solved in closed form, numerical integration methods are available and described in standard text books [1].

The ordinary differential equation semantics of LBS. Given two total functions $f_1 : X_1 \to Y$ and $f_2 : X_2 \to Y$ with a binary operator $+$ on the elements of Y, we define $f_1 + f_2 : X_1 \cup X_2 \to Y$ as follows:

$$(f_1 + f_2)(x) \overset{\Delta}{\simeq} \begin{cases} f_1(x) & \text{if } x \in X_1 \setminus X_2 \\ f_2(x) & \text{if } x \in X_2 \setminus X_1 \\ f_1(x) + f_2(x) & \text{if } x \in X_1 \cap X_2 \end{cases}$$

The semantics of LBS in terms of ODEs is defined below.

Definition 8. *The concrete semantics for LBS in terms of ODEs is given by the tuple $(\mathcal{D}, |_{\mathcal{D}}, \mathbf{0}_{\mathcal{D}}, G_{\mathcal{D}}, I_{\mathcal{D}})$ where*

- $D_1 \,|_{\mathcal{D}}\, D_2 \overset{\Delta}{\simeq} D$ *where*
  - $X_D \overset{\Delta}{\simeq} X_{D_1} \cup X_{D_2}$
  - $P_D \overset{\Delta}{\simeq} P_{D_1} + P_{D_2}$
  - $I_D \overset{\Delta}{\simeq} I_{D_1} + I_{D_2}$
- $\mathbf{0}_{\mathcal{D}} \overset{\Delta}{\simeq} (\emptyset, \emptyset, \emptyset)$

– $G_{\mathcal{D}}(\underline{n \cdot v_{gns}} \Rightarrow^{v_{gr}} \underline{n' \cdot v'_{gns}}, b) \stackrel{\Delta}{\simeq} D$ *where*
  - $X_D \stackrel{\Delta}{\simeq} \{MS(\underline{v_{gns}}.i)\} \cup \{MS(\underline{v'_{gns}}.j)\}$
  - $P_D(s) \stackrel{\Delta}{\simeq} \begin{cases} (N(s) - M(s)) \cdot r \cdot \prod_i (MS(\underline{v_{gns}}.i))^{\underline{n}.i} & \text{if } v_{gr} = \{r\} \\ v_{ga} & \text{if } v_{gr} = [v_{ga}] \end{cases}$

    where
    - $M(s) \stackrel{\Delta}{\simeq} \sum_{MS(\underline{v_{gns}}.i)=s} \underline{n}.i$
    - $N(s) \stackrel{\Delta}{\simeq} \sum_{MS(\underline{v'_{gns}}.j)=s} \underline{n'}.j$
  - $I_D(s) = 0$.

– $I_{\mathcal{V}}(v_{gns}, r) \stackrel{\Delta}{\simeq} (\{s\}, \{s \mapsto 0\}, \{s \mapsto r\})$ *where*
  - $s \stackrel{\Delta}{\simeq} MS(v_{gns})$

In the case of reactions, rate expressions are constructed from mass-action rate constants in the standard way [36]. Note that the assignment to reactions is only defined when mass-action rates are constants.

6.5 A CTMC Semantics

We now give another quantitative semantics in terms of *continuous time Markov chains* (CTMCs). In contrast to ODEs, CTMCs are *discrete* since they describe the system state in terms of species populations rather than concentrations, and they give rise to *stochastic* behaviour.

Continuous time Markov chains. The state of a CTMC corresponds to a marking of a Petri net and is hence given by a multiset of ground normal form species values in their multiset form. State transitions are described directly in terms of a transition rate matrix. Here is the formal definition:

Definition 9. *An LBS-*continuous time Markov chain with initial state *V is a tuple (X, Q, I) where*

1. *$X \subset_{fin} MS(\{MS(v_{gns}) \mid v_{gns} \in V_{gns}\})$ is the set of* states.
2. *$Q : X^2 \to \mathbb{R}$ is the* transition rate matrix *satisfying*
   (a) $Q(M, N) \geq 0$ for all $M, N \in X$ with $M \neq N$.
   (b) $Q(M, M) = -\sum_{M \neq N} Q(M, N)$.
3. *$I \in X$ is the* initial state.

The set of all CTMCs with initial state is denoted by $\mathcal{V}$, and we denote e.g. X in V by X_V. We refer to the literature [36] for further details on CTMCs and their associated simulation methods.

The continuous time Markov chain semantics of LBS

Definition 10. *The concrete semantics for LBS in terms of CTMCs is given by the tuple* $(\mathcal{V}, |_{\mathcal{V}}, \mathbf{0}_{\mathcal{V}}, G_{\mathcal{V}}, I_{\mathcal{V}})$ *where*

- $V_1 |_{\mathcal{V}} V_2 \overset{\Delta}{\simeq} V$ *where*
  - $X_V \overset{\Delta}{\simeq} \{M + N \mid M \in X_{V_1} \wedge N \in X_{V_2}\}$
  - $Q_V \overset{\Delta}{\simeq} Q_{V_1} + Q_{V_2}$
  - $I_V \overset{\Delta}{\simeq} I_{V_1} + I_{V_2}$
- $\mathbf{0}_{\mathcal{V}} \overset{\Delta}{\simeq} (\emptyset, \emptyset, \emptyset)$
- $G_{\mathcal{V}}(\underline{n \cdot v_{gns}} \Rightarrow^{\{r\}} \underline{n' \cdot v'_{gns}}, t) \overset{\Delta}{\simeq} V$ *where*
  - $X_V \overset{\Delta}{\simeq} MS(\{MS(\underline{v_{gns}}.i)\} \cup \{MS(\underline{v'_{gns}}.j)\})$
  - $Q_V(M', N') \overset{\Delta}{\simeq} \begin{cases} r\binom{M'}{M} & \textit{if } (M, N) \preceq (M', N') \wedge M \neq N \\ -r\binom{M'}{M} & \textit{if } M' = N' \wedge M \neq N \wedge M' \geq M \\ 0 & \textit{otherwise} \end{cases}$

    where
    - $M(s) \overset{\Delta}{\simeq} \sum_{MS(\underline{v_{gns}}.i)=s} \underline{n}.i$
    - $N(s) \overset{\Delta}{\simeq} \sum_{MS(\underline{v'_{gns}}.j)=s} \underline{n'}.j$
    - $\binom{M'}{M} \overset{\Delta}{\simeq} \prod_{s \in dom(M')} \binom{M'(s)}{M(s)}$
    - $(M, N) \preceq (M', N')$ *iff* $\exists L.\ M' = M + L \wedge N' = N + L$
  - $I_V(s) = 0$
- $I_{\mathcal{V}}(v_{gns}, n) \overset{\Delta}{\simeq} V$ *where*
  - $X_V \overset{\Delta}{\simeq} MS(\{MS(v_{gns})\})$
  - $Q_V(M, N) \overset{\Delta}{\simeq} 0$
  - $I_V(\{MS(v_{gns})\}) = n$

In the case of reactions, $\binom{x}{y}$ is the binomial coefficient and state transition rates are constructed from mass-action constants in the standard way [36]. Note that the reaction assignment is only defined when mass-action rates are used and, as in the ODE semantics, mass-action rates must be constants.

7 Future Directions

Combinatorial explosion. Whether the explicit modelling of "empty context" reactions (i.e. where species are listed in fully specified complexes) is desirable or not depends on the particular system under study. Systems which exhibit low or moderate levels of combinatorial complexity are amenable to modelling in LBS and can benefit from its simplicity and the large body of tools and techniques available for the analysis and simulation of Petri nets, ODEs and CTMCs. Although nondeterminism provides means of handling moderately combinatorial systems in a compact manner, one would like a more refined approach to nondeterminism in order to control which members of nondeterministic species interact in reactions.

Systems which are characterised by high levels of combinatorial complexity may be better modelled in e.g. rule-based languages such as κ and BioNetGen. As mentioned in the introduction, there is however scope for supporting κ and BioNetGen within the general framework of LBS through an appropriate choice of concrete semantics and modification site types. This may furthermore open possibilities for exploiting modularity in the κ analysis methods.

Graphical representations. Although LBS has been designed with ease of use in mind, it is still a textual language that may not be easily accessible to some biologists. Graphical representations of LBS models can ameliorate this problem. Specifically, tools for visualising LBS programs and, conversely, for generating LBS programs from visual diagrams, would be useful. These tools might follow the *Systems Biology Graphical Notation* (SBGN) [21].

A static semantics. The denotation functions impose certain constraints on their arguments. The resulting notion of well-typedness is a dynamical one: the denotation of a module is a function, and whether or not this function is defined for a given set of actual parameters is determined by applying the function to these parameters. This approach falls short in two respects. Firstly, the function may not be defined for any inputs at all. In this case it is the module definition that should be reported as ill-typed, rather than the module invocation. Secondly, well-typedness of a module invocation should be determined based on the actual parameters and an appropriate interface of the module, rather than by attempting to translate the module implementation under a given set of actual parameters. It may be that these limitations can be addressed through a dedicated type system along the lines of a previous version of LBS [27].

Synthetic biology. Whether natural biological system are amenable to extensive modular decomposition remains an open question that must be addressed through further modelling exercises. In the setting of *synthetic* biology, however, systems are *designed* rather than *modelled*, so it should be possible to exploit modularity fully there. Languages with dedicated features for synthetic biology are starting to emerge, including the previously mentioned Antimony language as well as GenoCad [2, 3] and the language for *Genetic Engineering of Cells* (GEC) [26]. GEC has a notion of parallel and located reactions which serve as constraints for deducing appropriate genetic parts for constructing genes, and it also has a basic notion of modularity. It may then be of interest to add some of the more advanced features of LBS for e.g. structured species and subtyping into an extension of GEC, or conversely to add the support of GEC for genetic design into an extension of LBS.

Acknowledgements. The authors would like to thank Vincent Danos, Stuart Moodie and Nicolas Oury for useful discussions. This work was supported by Microsoft Research through its European PhD Scholarship Programme and by a Royal Society-Wolfson Award. The second author is grateful for the support from The Centre for Systems Biology at Edinburgh, a Centre for Integrative Systems Biology funded by BBSRC and EPSRC, reference BB/D019621/1. Part of the research was carried out at the Microsoft Silicon Valley Research Center.

References

1. Blanchard, P., Devaney, R.L., Hall, G.R.: Differential Equations. Brooks/Cole (2002)
2. Cai, Y., Hartnett, B., Gustafsson, C., Peccoud, J.: A syntactic model to design and verify synthetic genetic constructs derived from standard biological parts. Bioinformatics 23(20), 2760–2767 (2007)
3. Cai, Y., Lux, M.W., Adam, L., Peccoud, J.: Modeling structure-function relationships in synthetic DNA sequences using attribute grammars. PLoS Comput. Biol. 5(10), e1000529 (2009)
4. Calder, M., Gilmore, S., Hillston, J.: Modelling the influence of RKIP on the ERK signalling pathway using the stochastic process algebra PEPA. Trans. on Comput. Syst. Biol. VII 4230, 1–23 (2006)
5. Cardelli, L.: Brane calculi. In: Danos, V., Schächter, V. (eds.) CMSB 2004. LNCS (LNBI), vol. 3082, pp. 257–280. Springer, Heidelberg (2005)
6. Chabrier-Rivier, N., Fages, F., Soliman, S.: The biochemical abstract machine BIOCHAM. In: Danos, V., Schächter, V. (eds.) CMSB 2004. LNCS (LNBI), vol. 3082, pp. 172–191. Springer, Heidelberg (2005)
7. Chen, W.W., Schoeberl, B., Jasper, P.J., Niepel, M., Nielsen, U.B., Lauffenburger, D.A., Sorger, P.K.: Input-output behavior of ErbB signaling pathways as revealed by a mass action model trained against dynamic data. Mol. Syst. Biol. 5(239) (2009)
8. Ciocchetta, F., Hillston, J.: Bio-PEPA: An extension of the process algebra PEPA for biochemical networks. Electron. Notes Theor. Comput. Sci. 194(3), 103–117 (2008)
9. Danos, V.: Agile modelling of cellular signalling. In: Computation in Modern Science and Engineering, vol. 2, Part A 963, pp. 611–614 (2007)
10. Danos, V., Feret, J., Fontana, W., Harmer, R., Krivine, J.: Rule-based modelling and model perturbation. TCSB 5750(11), 116–137 (2009)
11. Danos, V., Feret, J., Fontana, W., Harmer, R., Krivine, J.: Rule-based modelling of cellular signalling. In: Caires, L., Vasconcelos, V.T. (eds.) CONCUR 2007. LNCS, vol. 4703, pp. 17–41. Springer, Heidelberg (2007)

12. Dematté, L., Priami, C., Romanel, A.: Modelling and simulation of biological processes in BlenX. SIGMETRICS Performance Evaluation Review 35(4), 32–39 (2008)
13. Faeder, J.R., Blinov, M.L., Hlavacek, W.S.: Graphical rule-based representation of signal-transduction networks. In: Liebrock, L.M. (ed.) Proc. 2005 ACM Symp. Appl. Computing, pp. 133–140. ACM Press, New York (2005)
14. Guerriero, M.L., Heath, J.K., Priami, C.: An automated translation from a narrative language for biological modelling into process algebra. In: Calder, M., Gilmore, S. (eds.) CMSB 2007. LNCS (LNBI), vol. 4695, pp. 136–151. Springer, Heidelberg (2007)
15. Harel, D.: Statecharts: A visual formalism for complex systems. Sci. Comput. Program. 8(3), 231–274 (1987)
16. Heiner, M., Gilbert, D., Donaldson, R.: Petri nets for systems and synthetic biology. In: Bernardo, M., Degano, P., Zavattaro, G. (eds.) SFM 2008. LNCS, vol. 5016, pp. 215–264. Springer, Heidelberg (2008)
17. Hucka, M., et al.: The systems biology markup language (SBML): a medium for representation and exchange of biochemical network models. Bioinformatics 19(4), 524–531 (2003)
18. Jensen, K.: Coloured Petri Nets: Basic Concepts, Analysis Methods and Practical Use, vol. 1. Springer, Heidelberg (1992)
19. Kofahl, B., Klipp, E.: Modelling the dynamics of the yeast pheromone pathway. Yeast 21(10), 831–850 (2004)
20. Kwiatkowski, M., Stark, I.: The continuous π-calculus: a process algebra for biochemical modelling. In: Heiner, M., Uhrmacher, A.M. (eds.) CMSB 2008. LNCS (LNBI), vol. 5307, pp. 103–122. Springer, Heidelberg (2008)
21. Le Novère, N., et al.: The systems biology graphical notation. Nature Biotechnology 27, 735–741 (2009)
22. Mallavarapu, A., Thomson, M., Ullian, B., Gunawardena, J.: Programming with models: modularity and abstraction provide powerful capabilities for systems biology. J. R. Soc. Interface (2008)
23. Murata, T.: Petri nets: properties, analysis and applications. Proceedings of the IEEE 77(4), 541–580 (1989)
24. Paun, G., Rozenberg, G.: A guide to membrane computing. Theor. Comput. Sci. 287(1), 73–100 (2002)
25. Pedersen, M.: Compositional definitions of minimal flows in Petri nets. In: Heiner, M., Uhrmacher, A.M. (eds.) CMSB 2008. LNCS (LNBI), vol. 5307, pp. 288–307. Springer, Heidelberg (2008)
26. Pedersen, M., Phillips, A.: Towards programming languages for genetic engineering of living cells. J. R. Soc. Interface special issue (2009)
27. Pedersen, M., Plotkin, G.: A Language for Biochemical Systems. In: Heiner, M., Uhrmacher, A.M. (eds.) CMSB 2008. LNCS (LNBI), vol. 5307, pp. 63–82. Springer, Heidelberg (2008)
28. Peyssonnaux, C., Eychène, A.: The Raf/MEK/ERK pathway: new concepts of activation. Biol. Cell. 93(1-2), 53–62 (2001)
29. Phillips, A., Cardelli, L., Castagna, G.: A graphical representation for biological processes in the stochastic pi-calculus. In: Priami, C., Ingólfsdóttir, A., Mishra, B., Riis Nielson, H. (eds.) Transactions on Computational Systems Biology VII. LNCS (LNBI), vol. 4230, pp. 123–152. Springer, Heidelberg (2006)

30. Pierce, B.C.: Types and Programming Languages. MIT Press, Cambridge (2002)
31. Plotkin, G.: A calculus of biochemical systems (in preparation)
32. Priami, C.: Stochastic pi-calculus. The Computer Journal 38(7), 578–589 (1995)
33. Priami, C., Quaglia, P.: Beta binders for biological interactions. In: Danos, V., Schächter, V. (eds.) CMSB 2004. LNCS (LNBI), vol. 3082, pp. 20–33. Springer, Heidelberg (2005)
34. Regev, A., Paninab, E.M., Silverman, W., Cardelli, L., Shapiro, E.: BioAmbients: an abstraction for biological compartments. Theor. Comput. Sci. 325(1), 141–167 (2004)
35. Smith, L.P., Bergmann, F.T., Chandran, D., Sauro, H.M.: Antimony: a modular model definition language. Bioinformatics 25(18), 2452–2454 (2009)
36. Wilkinson, D.J.: Stochastic Modelling for Systems Biology. Chapman & Hall/CRC, Boca Raton (2006)

Mechanistic Insights into Metabolic Disturbance during Type-2 Diabetes and Obesity Using Qualitative Networks

Antje Beyer[1], Peter Thomason[2], Xinzhong Li[2], James Scott[2], and Jasmin Fisher[3]

[1] Department of Genetics, University of Cambridge, Cambridge, UK
[2] National Heart and Lung Institute, Imperial College London, UK
[3] Microsoft Research, Cambridge, UK
jasmin.fisher@microsoft.com

Abstract. In many complex biological processes quantitative data is scarce, which makes it problematic to create accurate quantitative models of the system under study. In this work, we suggest that the Qualitative Networks (QNs) framework is an appropriate approach for modeling biological networks when only little quantitative data is available. Using QNs we model a metabolic network related to fat metabolism, which plays an important role in type-2 diabetes and obesity. The model is based on gene expression data of the regulatory network of a key transcription factor Mlxipl. Our model reproduces the experimental data and allows *in-silico* testing of new hypotheses. Specifically, the QN framework allows to predict new modes of interactions between components within the network. Furthermore, we demonstrate the value of the QNs approach in directing future experiments and its potential to facilitate our understanding of the modeled system.

Keywords: computational modeling, Qualitative Networks, metabolic pathways, obesity, type-2 diabetes, Mlxipl.

1 Introduction

One of the key objectives of systems biology is to understand the behavior of cellular signaling- and transcription-networks. High-throughput techniques enable the accumulation of large amounts of data relating to the behavior of these networks. However, we need to establish the ways and methodologies to organize this data into a coherent whole that sheds light on the underlying biological system. In recent years, different modeling approaches have been suggested and used to this end. For example, differential equations and Boolean Networks [3,5,8,16,17]. It is clear that for different networks and different kinds of data, different approaches may need to be taken. Many modeling techniques require accurate quantitative data in order to calibrate and set the values of certain "constant rate" parameters. Our main interest is the case where such accurate

C. Priami et al. (Eds.): Trans. on Comput. Syst. Biol. XII, LNBI 5945, pp. 146–162, 2010.

data is missing but still there is a need to create formal executable models that help analyze and understand large amounts of qualitative data.

In this paper we advocate the usage of the *Qualitative Networks* framework [12] as a means for modeling metabolic and transcriptional networks. Qualitative Networks are an extension of Boolean Networks. The network associates variables that range over finite (and usually very small) domains with each of the substances of the network. All variables are updated synchronously in discrete computation steps that are intended to abstract the continuous deterministic evolvement of models described by differential equations. In every step of the system a variable may either increase by one, decrease by one, or remain unchanged. The changes of variables are determined by so called *target functions* that given the current state of the network (i.e., the value assignment to each one of the variables) compute the target value for each substance. This substance then changes in increments / decrements of one until reaching this value (assuming that the value of the target function does not change). Qualitative Networks have several characteristics that arise from their underlying semantics. The system can be in one of a finite number of states (value assignments to all the variables) and as the system is deterministic the execution of the system inevitably terminates in a loop.

There are mainly two analysis techniques that are applied to Qualitative Networks. The first is the execution of a network from a given initial state. In such a simulation, the changes in values of substances are followed over time from a given initial state and mimics specific evolutions of the system. In order to compare such simulations to evolutions of the underlying biological system several measurements of the level of substances in the biological system over a short period of time need to be available. While this information is very useful for debugging and understanding the way the model works, such accurate experimental data is very hard to come by and is almost impossible to find. A second analysis technique is to find the set of stable states (or stable loops) of the model. These are the states in which the model can remain looping forever. Usually, these are computed for a large number of initial conditions simultaneously. These states relate to the stable states of the biological system and their analysis can relate to the values of substances in the stable states of the system. Experimentally, these values are easier to come by with as they correspond to measurements in a single time point and in constant conditions that are easier to maintain. Technically, we use symbolic state-space analysis techniques [1] in order to efficiently and practically compute the set of stable states of large qualitative Networks. We then compare the values of variables in these stable states against data extracted from experimental data.

We illustrate the suitability of Qualitative Networks to modeling of metabolic networks in the case where quantitative data is not available. We show that even with restricted amounts of data, Qualitative Networks can still suggest useful hypotheses that lead to new avenues to explore experimentally and interesting insights. This is demonstrated through an extensive case study of a metabolic and transcriptional interaction network involved in obesity and type-2 diabetes mellitus (T2D). In both diseases, metabolic and inflammatory pathways play important roles [19] as their dysregulation can lead to insulin resistance which is a characteristic symptom of obesity and T2D [10,11,15]. Nowadays, T2D affects over 110 million people worldwide and, as well as obesity, it highly increases the risk of cardiovascular disease, blindness, amputation and kidney failure [9,10]. With cardiovascular disease being a major cause of mortality, T2D and obesity pose a significant threat to global health [9,19].

Our model is based on unpublished data by Scott and colleagues [15]. The data was obtained in a study looking at macrophage infiltration and gene expression profiles in insulin-responsive tissues of C57Bl6 mice which are susceptible to these metabolic diseases. To determine the onset of inflammation and metabolic disturbance in adipose tissue, mice were fed a high saturated fat diet. The data includes gene expression measurements in white adipose tissue in mice, performed at two days, eight days, three weeks, and 15 weeks after the beginning of a fat-feeding process. We concentrate on the measurements at eight days and 15 weeks as understanding the differences between the gene expression levels in these time points could help understand the difference between processes operating during early onset of T2D and obesity (8 days) and the chronic disease stage (15 weeks) [15]. For the computational model, we used gene expression data obtained from white adipose tissue in the gonadal region of male mice. Scott and colleagues selected genes based on epididymal fat weight after 8 days of fat feeding with fat as 60% of the caloric intake (cf. [15]). Genes were selected if their expression was correlated with epididymal fat weight above a very strict threshold ($r^2 > 0.7$). They hypothesize that these genes might be the bridge between acute fat-feeding and chronic obesity. Most of the genes used in the present computational model are from this selection or very close to this threshold because of their potential importance. The results by Scott and colleagues also suggest that metabolic disturbance leads to inflammation. Therefore, the current model serves also as a basis for future modeling of the inflammatory processes underlying T2D and obesity.

Our case study presents an executable model of the transcriptional and post-translational regulations of Mlxipl. The transcription factor Mlxipl is known to increase the transcription of genes involved in several metabolic pathways [6,7,15] and is significantly down-regulated in white adipose tissue during acute fat feeding and obesity [15]. Mlxipl can be linked to many aspects of obesity and T2D [6,7,18,15]. While the post-translational regulation of Mlxipl is fairly well understood, its regulation on the transcriptional and translational level is less clear [18]. Our computational model includes Mlxipl together with four interconnected metabolic pathways regulated by Mlxipl. As our data is taken from a tissue rather than from cells of the same type, we do not know exactly if the data relate to the major cellular constituents of white adipose tissue. But overwhelmingly the data refer to adipocyte functions rather than other cell types and are therefore highly likely to reflect the average fat cell. Hence, we build our computational model as a single cell model. The computational model is based on a metabolic network and a transcription interaction graph.

We analyze the model in conditions that represent the behavior of the network at 8 days and 15 weeks of fat feeding process. We use experimental data to calibrate our model by comparing the stable states of the model with the experimental data. This process has suggested modifications to the interaction graph and metabolic networks that need to be validated experimentally. Once a putative model reproduces the existing experimental data we formulate hypotheses as *in-silico* experiments and test the models as a way to direct actual laboratory experiments.

The computational cost of state-space analysis forced us to break down the network into four modules assuming that the components connecting the modules are constants in the later module. These components get all of their inputs in one module and have some or all of their outputs in the next module. This may cause some of the resulting component values to be too high or too low as it does not fully capture the dynamics of the whole system. However, all regulation loops in the full system are contained in one of the four modules, suggesting that most behaviors that do not exist in the partitioned model are transient. Furthermore, comparison with the experimental data revealed that the variables resulting from the network behave as expected, which emphasizes the validity of the whole model.

Our computational model and its analysis demonstrate the usefulness of Qualitative Networks to investigate cellular networks and their corresponding genomic data. With its individually adjustable target functions and its independence from quantitative data, Qualitative Networks can be used to model interesting biological interaction networks with available genomic data. This is especially emphasized by the Mlxipl case study. Although there is a good amount of knowledge on Mlxipl function, its regulation is not fully understood. The current case study provides insights into this regulatory process as well as highlights the importance of specific regulatory connections. The computational analysis reveals that the model is consistent with the experimental data. Our insights highlight the necessity of further regulators of Mlxipl in addition to the known ones. Furthermore, the analysis predicts various new modes of interactions between components in the network, and demonstrates that acetyl CoA and Mlxipl regulate the level of fatty acid production in a synergistic way. Future work of a more complex computational model of this system will further facilitate a better understanding of metabolic disturbance during obesity and T2D.

2 Methods

2.1 Qualitative Networks

In many cases, lack of quantitative data makes the construction of quantitative models impossible. Boolean Networks were used in recent years to construct abstracted computational models that are based mainly on the types of interactions between molecules. This is obviously a very abstract view, however, it has proven beneficial in numerous occasions [8,16,17]. In some cases, the restriction of Boolean Networks to the values of ON and OFF, seems too restricted and one would like to extend the modeling framework with a larger range of possible values.

Qualitative Networks are an extension of Boolean Networks; they use discrete variables ranging over finite domains instead of Boolean variables and allow functions to represent different types of interactions on top of activation and inhibition [12].

A *Qualitative Network* $Q(C, T, N)$ comprises a set of components, C, and a list of target functions, T. Each component $c_i \in C$ has a state which can take any integer value between 0 and N representing the qualitative level of the respective component. A target function $target_i \in T$ is a function $target_i : \{0, \ldots, N\}^C \rightarrow \{0, \ldots, N\}$ and it indicates in a given state of the system the specific value between 0 and N towards which each component c_i should move. For example, target functions can be as simple as the identity function, a Boolean function, the maximum or minimum, or a more complicated combination of other functions.

A *state* of the system is an assignment of a value between 0 and N to each of the substances. The system advances in discrete steps and, at each step, the level of a component can only increase or decrease by a single level. The consecutive level of each component c_i is calculated as follows:

$$c_i(t+1) = \begin{cases} c_i(t) - 1 & \text{if } target_i(S(t)) < c_i(t) \\ c_i(t) + 1 & \text{if } target_i(S(t)) > c_i(t) \\ c_i(t) & \text{if } target_i(S(t)) = c_i(t) \end{cases},$$

where $S(t)$ is the state of the network at time t. For example, consider the case of a network with three components c_1, c_2, and c_3. Let their values at time 1 are $c_1(1) = 1, c_2(1) = 2$ and $c_3(1) = 0$, and the target function $target_3(S(t)) := max(c_1(t), c_2(t))$ for the component c_3. Then $c_3(2)$ aspires to $target_3(S(1)) = max(1, 2) = 2$ and as $2 > 0$ the value of $c_3(2)$ becomes 1. The target functions are based on the mechanistic understanding of the biological interactions the model is built from.

As all variables range over finite domains, a *Qualitative Network* model has a finite number of states. As all variables are updated simultaneously (synchronously) an execution of such a model proceeds deterministically, i.e., regardless of the full history of an execution, every state proceeds to a fixed next state. It follows that every execution of a *Qualitative Network*[1] eventually ends in a cycle of states that are visited infinitely often. A state is defined as infinitely visited if there is an execution of the model in which the state appears infinitely often. In a biological system these states correspond to the stable states, hence states that are not infinitely visited often are considered as unstable. From these unstable states the system will always evolve towards a loop of infinitely visited states and remain there indefinitely. This suggests the following analysis technique: (a) compute the set of infinitely visited states (b) check properties of this set. For example, check whether in all infinitely visited states the value of a certain variable (substance) is above/below a given threshold. As this analysis aims

[1] As Boolean Networks are a special case, this applies also to Boolean Networks.

to validate the stability of the model, it is usually less important to identify one or few initial states. The analysis is usually applied on most (if not all) states of the system and corresponds to starting execution from an arbitrary state and letting the system stabilize [13].

We use the QNBuilder tool [14] to analyze *Qualitative Networks*. QNBuilder supports a simple and intuitive textual input format that defines the set of substances in the model. Some substances can be defined as constants and others are connected to target functions, that can be defined as look-up tables. Cells can be connected to grids and meshes in a simple way by relating to inputs from neighboring cells according to their locations in the grid. The QNBuilder uses the tool *CrocoPat* [1] for symbolic state analysis. CrocoPat computes a simple fixpoint of the states that can be reached in 1, 2, ... steps. As the number of states is finite, this computation stabilizes on the fixpoint of infinitely visited states. QNBuilder also includes requirements that are then checked on this set of infinitely visited states. In the case that a requirement does not hold, an infinitely visited state that does not conform to it can be extracted.

2.2 The Iterative Improvement Process [12]

When coming to check the accuracy of a putative biological model, two types of requirements have to be met. The first is a static requirement such as, a inhibits b or a activates b. Such requirements are integrated into the target functions and can be checked statically. The second is a dynamic requirement that relates to qualities of infinitely visited states. Such requirements are checked as explained above.

The main value of such models is that they can be contrasted with the experimental data and, hopefully, provide information that the biologist was not aware of initially. Accordingly, if a model fails to reproduce some of the experimental data, it means that our understanding of the process is incomplete, and the model should be refined through changes in the target functions, substances, interactions, or perhaps the range of possible values. Such changes can be justified by consulting the available literature or domain experts. When such information is not available, we may conclude that information is missing. We can try to modify the model so that all the experimental conditions are reproduced. Such modifications then need to be validated experimentally in order to verify that the model may indeed be correct.

Once a model reproduces the experimental data, we view it as a valid putative model and use it to further probe the system. We perform *in-silico* experiments by changing constants or requirements in the model. The results of such experiments can be used to highlight interesting avenues to explore experimentally.

3 Results

3.1 Case Study – Model of the Regulatory Network of Mlxipl

We demonstrate the usefulness of the Qualitative Networks through a case study and the insights that were obtained from its analysis. We construct a Qualitative

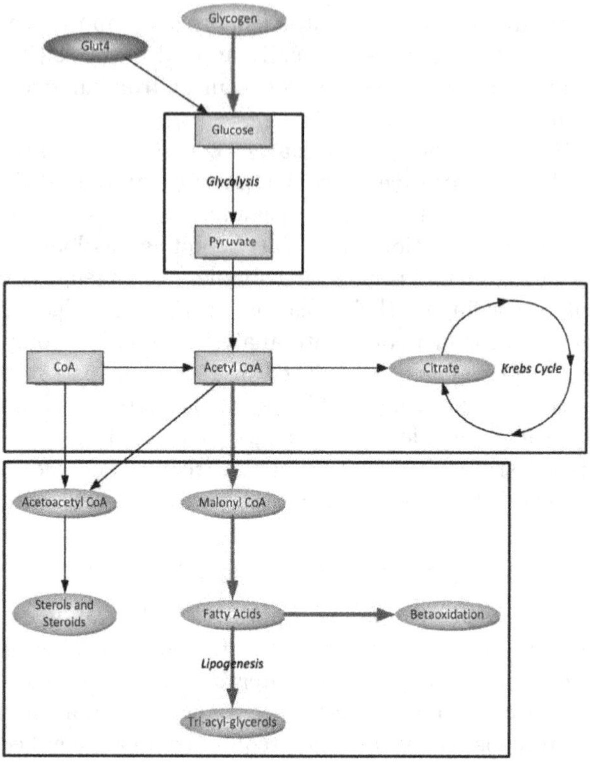

Fig. 1. Coarse representation of the metabolic network used for the computational model. Red arrows and red filling show the parts of the network that Mlxipl has influence on. The boxes indicate three of the separate models that are worked out in more detail.

Networks model of the regulatory network of Mlxipl, a general metabolic network that can be found in all mammalian cells (cf. Figure 1). There are four specific subnetworks making up this big metabolic network which were worked out in more detail as four separate models (due to computational constraints) according to the specific connections that are found in white adipose tissue (cf. Figures 2, 3, 4, and 5 and Figure 6 showing their connectivity). Figure 2 shows the transcription factor Mlxipl with its regulatory connections and some putative regulators of Mlxipl. Figures 3, 4 and 5 show biochemical reactions in the metabolic network underlying the present model including components regulated by Mlxipl.

We use granularity of three for all components in the network. The three possible levels correspond to: 0 representing down-regulation, 2 representing up-regulation, and 1 representing normal level of expression. The network includes two different kinds of interactions between components: regulatory connections on the gene or protein level (especially in the Mlxipl network), and biochemical reactions. These different interactions are represented by different target functions. For regulatory connections of the first kind (e.g., the control of PPARγ

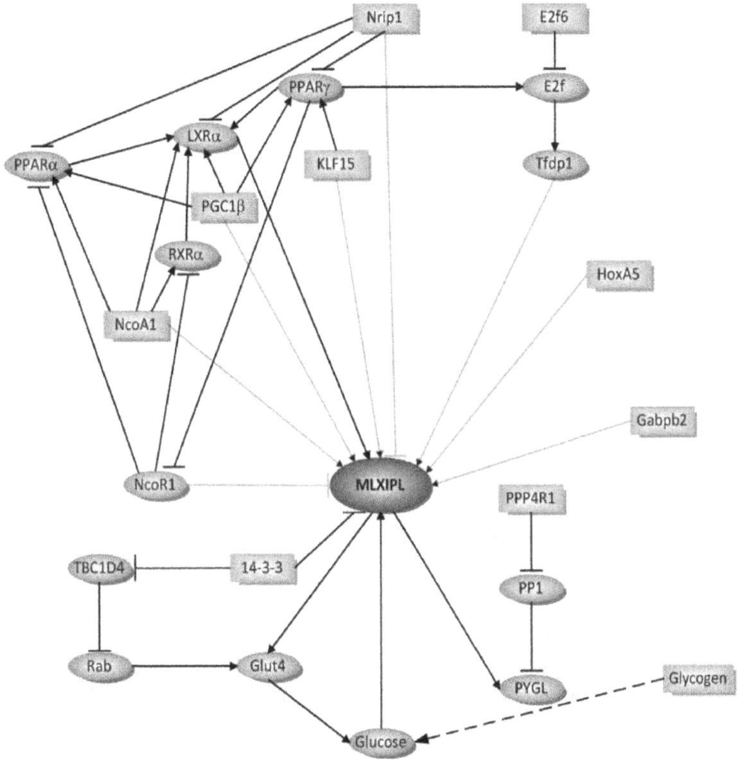

Fig. 2. Network of Mlxipl regulatory connections as used in the computational model. The gray arrows show putative inputs that are not verified experimentally. The green components are constant in the model, the blue ones are not constant and used in the requirement, the red one is key for consistency of the model. The dashed arrow indicates a biochemical reaction, the attached component is the enzyme of the reaction.

in Figure 2) target functions compare the amount of activating inputs with the amount of inhibiting inputs. For biochemical reaction of the second kind (e.g., the control of Fructose-6-P in Figure 3), the amount of the output depends on whichever input component has a lower level in this reaction, substrate or enzyme. The actual target functions are determined as explained below [12].

Consider a protein c_i of the network. An edge e_{ji} represents an interaction from c_j to c_i and has an associated weight α_{ji}. The edge is activating when $\alpha_{ji} > 0$ and inhibiting if $\alpha_{ji} < 0$. The target function of c_i includes the computation of the sum of activations and inhibitions that are scaled so that they always fall in the range 0 to N. Formally, the sum of activation on c_i is

$$act_i(s) = \frac{\sum\limits_{\alpha_{ji}>0} \alpha_{ji} c_j}{\sum\limits_{\alpha_{ji}>0} \alpha_{ji}},$$

and the sum of inhibition on c_i is

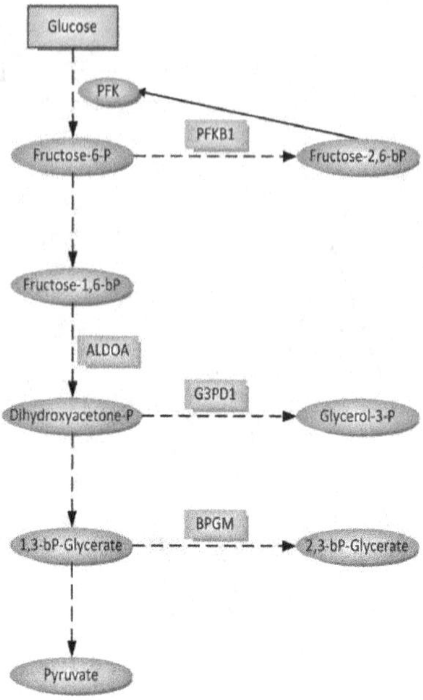

Fig. 3. Network representation of glycolysis used for the computational model. The purple component is a constant in this module, however it is an output in the Mlx-ipl module. Green components are constants in this module, the blue components are not constant and hence used in the requirement. Dashed arrows indicate biochemical reactions, the attached component is the enzyme of the reaction. If there is no component attached to the dashed arrow, the enzyme was not included in the model as its expression level did not significantly change from normal.

$$inh_i(s) = \frac{\sum\limits_{\alpha_{ji}<0} -\alpha_{ji}c_j}{\sum\limits_{\alpha_{ji}<0} \alpha_{ji}},$$

where s represents the state of the model. The target function for c_i is determined according to the difference between these two values. In the case that a component c_i has only inhibiting inputs, lack of inhibition is interpreted as activation. Formally,

$$target_i(s) = \begin{cases} \max(0, act_i(s) - inh_i(s)) & \text{If } \max(\alpha_{ji} > 0) \\ N - inh_i(s) & \text{If } \max(\alpha_{ji}) \leq 0 \end{cases}.$$

For example, the target function of PPARγ is the sum of its two activating components, PGC1β and KLF15 divided by 2, minus its inhibiting component, Nrip1.

If component c_i is controlled by a biochemical reaction, its target function is set to the minimum of the levels of its substrate and its enzyme. For example, Fructose-6-P tends towards the level of the level of the lowest between Glucose and PFK.

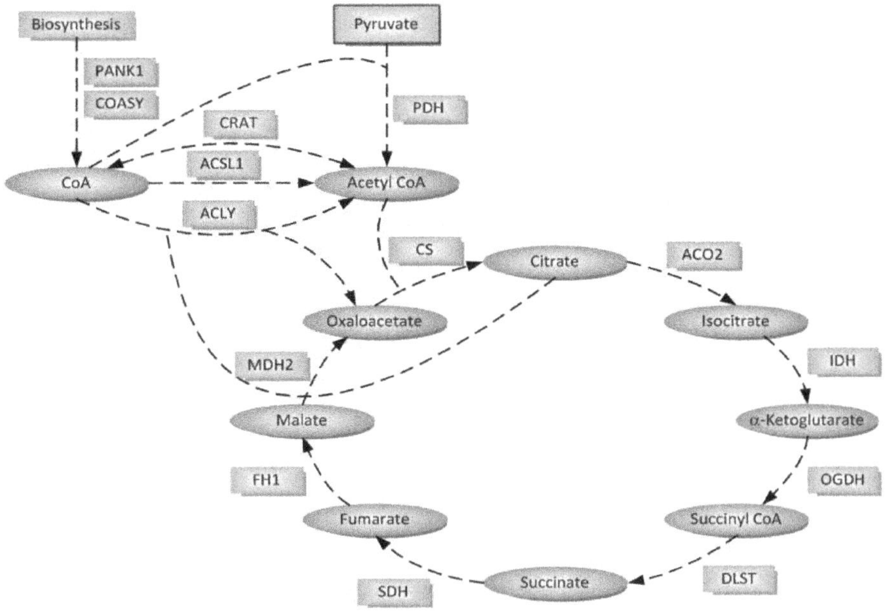

Fig. 4. Network surrounding acetyl CoA as used in the computational model. The purple component is constant in this module, however is an output in the glycolysis module. Green components are constants in this module, the blue components are not constant and used in the requirement. Dashed arrows indicate biochemical reactions, the attached component is the enzyme of the reaction.

For constant components, i.e. components that were not getting input from other components in the network, these values were extracted from the gene expression data (Table 1, data supplied by Scott and colleagues [15]). This data relates to two different time points after 8 days and 15 weeks of fat feeding process. These give rise to two versions of each model, differing in the values of the constant components, where there are differences in gene expression. For example, PP4R1 is constantly down-regulated in the 8 days model and constantly up-regulated in the 15 weeks model. We stress that these two version of the models use the same network structures and same target functions and differ only in these constant values. We used a threshold of 1 ± 0.1 to represent normal expression. A level above 1.1 is treated as up-regulated and a level below 0.9 is treated as down regulated[2]. In Table 1, green represents the value 0, white represents 1 and red represents 2.

[2] Setting the threshold for the different levels is an interesting biological question. From discussions with colleagues, it seems that a change of 10% in expression level is considered significant. It would be interesting to study models with a lower threshold for down- and up-regulation. Another interesting option is to allow more expression levels in the model and use extra levels for expressing down- and up-regulation in a more accurate way. Increasing the level of granularity of substances produces models that are too large to be analyzed. This is one of the topics of our future work.

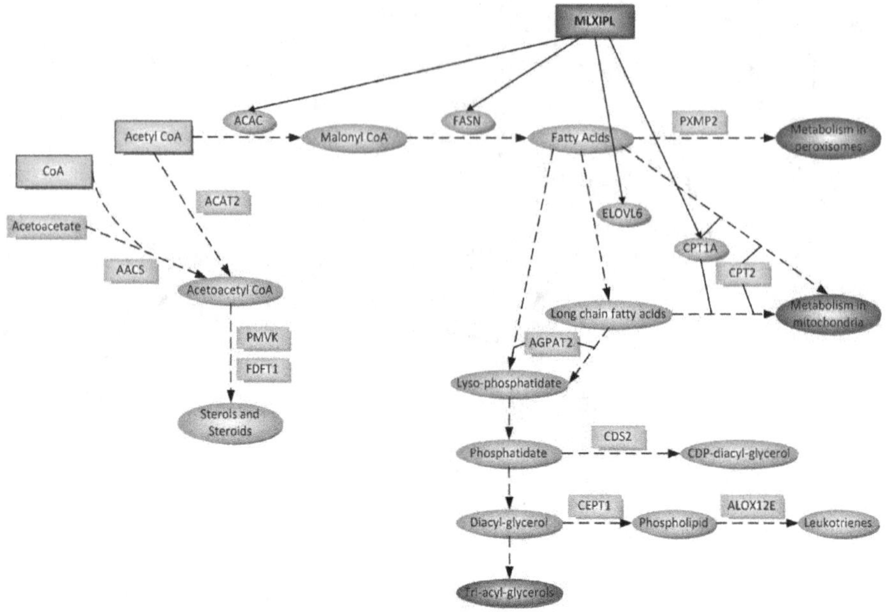

Fig. 5. Network representation of fatty acid metabolism used for the computational model with the exception of the components between Lyso-phosphatidate and Tri-acyl-glycerols. The purple components are constants in this module, however are outputs in the acetyl CoA module. Green components are constant in this module, the blue components are not constant and used in the requirement, and the red components are the key for consistency of the model. Dashed arrows indicate biochemical reactions, the attached component is the enzyme of the reaction.

The experimental data is used, in addition, to provide requirements on the network. For the present model, the requirements were derived from the gene expression data for eight days and 15 weeks of fat-feeding (cf. Table 1, [15]) for all non-constant components corresponding to the three levels of expression, as explained above. For example, in the model corresponding to 8 days, the stable states of the Mlxipl module (cf. Figure 2) have to satisfy the following requirement. The values of PPARγ, PPARα, LXRα, RXRα, TBC1D4, Glut4, PYGL, and Mlxipl must be 0 and the value of Ncor1 must be high. Similar requirements were derived for all modules and for both versions of the model, corresponding to 8 days and 15 weeks. The objective of the formal analysis is to verify that these requirements generally hold for the model. In the process of formal verification, all infinitely visited states were computed first and the requirements were then tested on *all* of these states.

3.2 New Biological Insights Revealed by Model Analysis

in-silico experiments were performed to test further assumptions. This was done in two ways, either by changing the values of constant components in the network

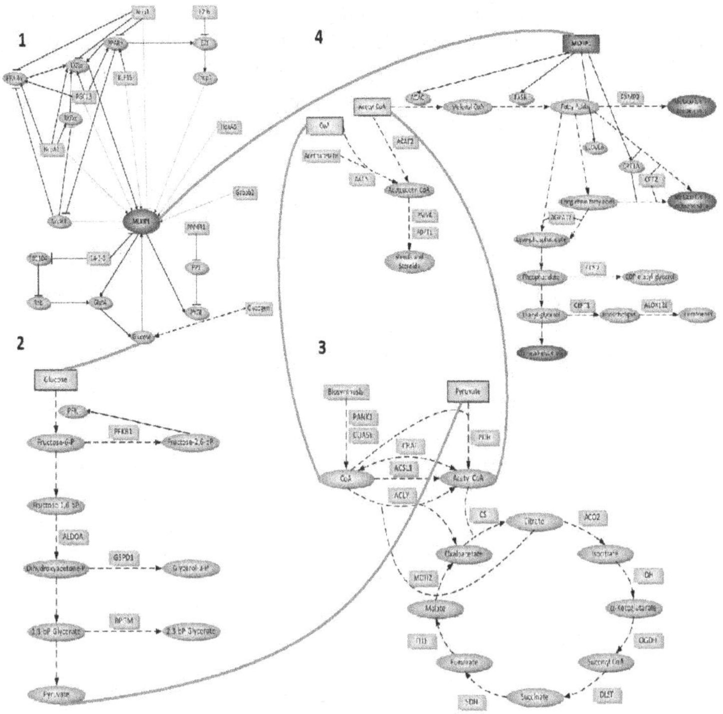

Fig. 6. Figure showing the four main parts of the metabolic network with their interactions as used in the computational model. (1) network of Mlxipl regulatory connections, (2) glycolysis, (3) network surrounding acetyl CoA and Krebs cycle, (4) fatty acid metabolism.

or by removing interactions (and components) from the network. While changing the values of the components, the values of other interesting components were monitored so that the effect of the changed components on the other components could be investigated. By removing interactions we tested if the model was still consistent with the data and hence if the existing connections are important or necessary to maintain the behavior of the system.

Ywhab is likely to be involved in Mlxipl down-regulation. It is not clear which one of the proteins of the 14-3-3 family is involved in the inhibition of Mlxipl, but Ywhab and Ywhag are the most likely candidates (cf. Table 1). During the consistency checking, we tried to find out which of these two proteins is most probable to inhibit Mlxipl in the present network. Considering all required levels of the other components in the Mlxipl network shows that, while both proteins are in accordance with the major requirement of Mlxipl being at a low level, the inclusion of Ywhag into the network was less consistent with the experimental data than including Ywhab. This indicates that Ywhab is the most likely

Table 1. Tables of gene expression data relative to gene expression with normal diet (represented by 1 in the model) sorted alphabetically and with regard to their model affiliation. Green indicates genes that are down-regulated (represented by 0 in the model), red shows genes that are up-regulated (represented by 2 in the model). Genes marked in yellow were not used as constants in the model.

MLXIPL Regulation

Gene	*8 days*	*15 weeks*
14-3-3		
Ywhab	0.86	0.60
Ywhag	0.93	0.92
E2f6	0.84	0.83
Gabpb2	0.84	0.82
Glut4	0.48	0.73
HoxA5	0.74	0.57
KLF15	0.63	0.55
LXRa	0.80	0.56
MLXIPL	0.45	0.69
NcoA1	0.79	0.73
NcoR1	1.15	0.96
Nrip1	1.33	1.30
PGC1b	0.58	0.84
PP1	0.83	1.08
PPARa	0.72	0.94
PPARg	0.76	0.71
PPP4R1	0.80	1.11
PYGL	0.68	0.78
RAB40B	0.43	0.51
RXRa	0.89	0.92
TBC1D4	0.65	0.83
TFDP1	0.80	0.92

Acetyl CoA Network and Krebs Cycle

Gene	*8 days*	*15 weeks*
AACS	0.56	0.59
ACAC	0.57	0.36
ACAT2	0.54	0.78
ACLY	0.71	1.31
ACO2	0.51	0.94
ACSL1	0.65	0.58
COASY	0.52	0.71
CRAT	0.70	0.50
CS	0.66	0.88
DLST	0.83	0.72
FASN	0.11	1.05
FDFT1	0.62	1.13
FH1	0.62	0.79
IDH	1.15	1.15
MDH2	0.68	1.03
OGDH	0.77	0.97
PANK1	0.37	0.86
PDH	0.58	0.82
PMVK	0.34	0.81
SDH	0.74	0.88

Glycolysis

Gene	*8 days*	*15 weeks*
ALDOA	0.58	0.90
BPGM	2.08	1.21
G3PD1	0.38	0.83
PFK	0.76	0.92
PFKFB1	0.63	0.69

Fatty Acid Metabolism

Gene	*8 days*	*15 weeks*
AGPAT2	0.52	0.82
ALOX12E	0.66	0.57
CDS2	0.72	1.08
CEPT1	0.78	0.80
CPT1A	1.50	1.42
CPT2	0.65	0.67
ELOVL6	0.31	1.73
PXMP2	0.64	0.63

inhibitor and, in addition, down-regulation of Mlxipl shows the consistency of the Mlxipl model with the experimental data. This consistency implies that new hypotheses can be tested using this model. With the model adjusted to our results regarding the proteins of the 14-3-3 family, we tested different hypotheses with respect to the number of regulatory inputs to Mlxipl. Our tests have shown that the three known regulators in the network are not sufficient to explain the behavior of Mlxipl. Further testing has shown that adding eight putative inputs to these three is, however, sufficient to have Mlxipl down-regulated. With regard to these putative inputs the model indicates that the three known inputs together with one arbitrary activating or inhibiting input of the eight putative ones are sufficient to explain Mlxipl down-regulation. However, we were not able to highlight specific inputs that are necessary for Mlxipl down-regulation. Nonetheless, the model has highlighted that all of the putative inputs are good candidates to be regulators of Mlxipl.

Table 2. Table showing the synergistic effect of acetyl CoA and Mlxipl on fatty acid production in the model. The numbers show the values of the components, green represents downregulation, white normal regulation and red upregulation.

Acetyl CoA	MLXIPL	Fatty Acids
0	0	0
0	1	0
0	2	0
1	0	0
1	1	1
1	2	1
2	0	0
2	1	1
2	2	2

Several genes must have additional inputs in the network. After checking all of the sub-network models for their consistency, we found that in the final model of lipid metabolism, betaoxidation of fatty acids in the mitochondria and peroxisomes, as well as fat production, were down-regulated as expected. This means that the model we derived from the complete network is consistent with the experimental data and allows new hypotheses about the network to be tested. In this process of state-space analysis, we were also able to highlight the genes TBC1D4, PP1 and CPT1A whose behavior was not explained by their inputs in the present network. Hence, we can assume that these genes must have additional regulatory inputs.

Also, the genes NcoR1, RXRα, PPARα and Tfdp1 from the Mlxipl module (cf. Figure 2), PFK from the glycolysis module (cf. Figure 3) and FASN and ELOVL6 in the acetyl CoA module (cf. Figure 4) have a different expression level at 8 days compared to 15 weeks (cf. Table 1). Comparing the models for these two time points showed that for these genes the changed level was not explained at 15 weeks. Further tests showed that the complete model was still consistent when applying these changed values to the 15 weeks model. Hence, we conclude that while these genes must have other regulatory inputs at both time points or only at 15 weeks, the rest of the network with all its connections represents the data at both time points.

Mlxipl and acetyl CoA have a synergistic effect on fatty acids production. To test the cooperative influence of acetyl CoA and Mlxipl on betaoxidation and fat production, we changed the value of one of these components while keeping the other at its normal level. In all cases the levels of betaoxidation and fat production remained low. Tests with both of the components being at an elevated level also did not change the levels of betaoxidation and fat production. This suggests that the other genes working on these pathways restrict the influence of acetyl CoA and Mlxipl on this part of fat metabolism. A similar series of tests were conducted to explore the effect of acetyl CoA and Mlxipl on the amount of fatty acids production. Our test have shown that neither acetyl CoA nor Mlxipl can be regarded as the main regulator of fatty acid production, but rather they seem to be acting in a synergistic way on the level of fatty acid production (cf. Table 2).

4 Discussion

We have presented the Qualitative Networks framework as an appropriate framework to modeling biological systems where little quantitative data is available. Qualitative Networks can be regarded as an abstraction of continuous models described by differential equations. Qualitative Networks do not require exact rates of reactions and are suitable for the kind of genomic data that is often available (as in our case study). They offer additional flexibility over Boolean Networks by allowing general target functions and multiple possible values of components. Overall, the Qualitative Networks framework is a very suitable approach for networks with no or only little quantitative data available and with complex biochemical interactions. Previously, Qualitative Networks were also used to construct a model of the crosstalk between the Wnt and Notch pathways in the Keratinocytes [12], and more recently to the modeling of cell fate specification during *C. elegans* vulval development [2].

Here we demonstrate the usefulness of the Qualitative Networks framework through a model of Mlxipl regulation. This model is based solely on gene expression data, and does not require exact reaction rates. The flexibility of target functions and multiple values enables to accurately represent enzymatic reactions and protein interactions in the same model. In accordance with several biological studies [6,15,18], the model reproduces the down-regulation of Mlxipl. Analysis of the model showed a down-regulation of the whole fat metabolism, which has also been demonstrated under similar experimental conditions [15]. This observation is especially important in terms of T2D and obesity as it is one of the major features of these diseases [15].

In addition, the model sheds light on aspects of the network that are not yet fully understood. For example, the model highlights the most likely candidate of the genes in the 14-3-3 family to inhibit Mlxipl [6,18]. Furthermore our model shows that Mlxipl must have other regulators in addition to well established ones and indicates which are the possible regulators. Additionally, analysis of the model suggests that Mlxipl and acetyl CoA have a synergistic effect on fatty acid production. Overall, this shows the applicability of the Qualitative Networks framework to the modeling of metabolic networks and its potential to highlight questions with implications for medical research.

4.1 Future Prospects

Several aspects of modeling were dictated by computational limitations that we hope to overcome in the future. Our main aim is to remove the partitioning of the network and analyze it as one model. This would prevent possible inconsistencies due to components being set to their stabilization values too early. Another goal is to increase the granularity of the target functions from 3 to a higher order, as having only three levels of expression cannot always capture the complexity of a system. As an intermediate goal, we could use different granularities for different components according to their roles and importance. As previously discussed, we would also like to consider different cut-offs for the expression level, for example 1 ± 0.5. In such a case, 1 would represent all genes with a medium level of expression and 0 and 2 would represent all genes with an expression deviating significantly from normal. The comparison of the results for different cut-offs could perhaps replace the usage of target functions with higher granularity.

We would also like to extend this model with additional biological data. For example, the control of the inputs that are currently modeled as constants. Considering the bigger picture, a major goal for the future would be to add the inflammatory pathways to this model, as this would help to get a deeper and more fundamental understanding of how these processes in adipose tissue contribute to T2D and obesity.

Acknowledgments

We would like to thanks Nir Piterman for critical reading of the manuscript and valuable comments and discussions. Part of this work was done while Antje Beyer was an intern at Microsoft Research, Cambridge, UK.

References

1. Beyer, D.: Relational Programming with CrocoPat. In: Proceedings of the 28th International Conference on Software Engineering (ICSE 2006), Shanghai, May 20-28, pp. 807–810. ACM Press, New York (2006), `http://www.cs.sfu.ca/~dbeyer/CrocoPat/`
2. Beyer, A., Fisher, J.: Unpublished results (2009)
3. von Dassow, G., Meir, E., Munro, E.M., Odell, G.M.: The segment polarity network is a robust developmental module. Nature 406, 188–192 (2000)
4. Fisher, J., Piterman, N., Hajnal, A., Henzinger, T.A.: Predictive modeling of signaling crosstalk during C. elegans vulval development. PLoS Comput. Biol. 3, e92 (2007)
5. Giurumescu, C.A., Sternberg, P.W., Asthagiri, A.R.: Intercellular coupling amplifies fate segregation during *Caenorhabditis elegans* vulval development. PNAS 103, 1331–1336 (2006)
6. Iizuka, Y., Horikawa, K.: ChREBP: a Glucose-activated Transcription Factor Involved in the Development of Metabolic Syndrome. Endocr. J. 55(4), 617–624 (2008)
7. Kooner, J.S., Chambers, J.C., Aquilar-Salinas, C.A., Hinds, D.A., Hyde, C.L., Warnes, G.R., Gómez Pérez, F.J., Frazer, K.A., Elliot, P., Scott, J., Milos, P.M., Cox, D.R., Thompson, J.F.: Genome-wide scan identifies variation in Mlxipl associated with plasma triglycerides. Nature Genetics 40, 149–151 (2008)
8. Li, F., Long, T., Lu, Y., Ouyang, Q., Tang, C.: The yeast cell-cycle network is robustly designed. Proc. Natl. Acad. Sci. USA 101, 4781–4786 (2004)

9. Lin, J., Yang, R., Tarr, P.T., Wu, P., Handschin, C., Li, S., Yang, W., Pei, L., Uldry, M., Tontonoz, P., Newgard, C.B., Spiegelman, B.M.: Hyperlipidemic Effects of Dietary Saturated Fats Mediated through PGC-1β Coactivation of SREBP. Cell 120, 261–273 (2005)
10. Mootha, V.K., Lindgren, C.M., Eriksson, K.-F., Subramanian, A., Sihag, S., Lehar, J., Puigserver, P., Carlsson, E., Ridderstråle, M., Laurila, E., Houstis, N., Daly, M.J., Patterson, N., Mesirov, J.P., Golub, T.R., Tamayo, P., Spiegelman, B.M., Lander, E.S., Hirschhorn, J.N., Altshuler, D., Groop, L.C.: PGC-1α-responsive genes involved in oxidative phosphorylation are coordinately downregulated in human diabetes. Nature Genetics 34, 267–273 (2003)
11. Neels, J.G., Olefsky, J.M.: Inflamed fat: what starts the fire? J. Clin. Invest. 116, 33–35 (2006)
12. Schaub, M.A., Henzinger, T.A., Fisher, J.: Qualitative networks: a symbolic approach to analyze biological signaling networks. BMC Systems Biology 1, 4 (2007), `http://www.biomedcentral.com/1752-0509/1/4/`
13. Schaub, M.A.: QNBuilder v0.1 User Manual, Draft (2008)
14. Schaub, M.A.: QNBuilder v0.1.c (2008), `http://cs.stanford.edu/people/mschaub/QNBuilderBeta/`
15. Scott, J.: Oxidative stress in adipose tissue is a unifying trigger for inflammation and insulin resistance (manuscript and personal communication)
16. Shmulevich, I., Zhang, W.: Binary analysis and optimization-based normalization of gene expression data. Bioinformatics 18, 555–565 (2002)
17. Shmulevich, I., Lahdesmaki, H., Dougherty, E.R., Astola, J., Zhang, W.: The role of certain Post classes in Boolean Network models of genetic networks. Proc. Natl. Acad. Sci. USA 100, 10734–10739 (2003)
18. Uyeda, K., Repa, J.J.: Carbohydrate hydrate response element binding protein, ChREBP, a transcription factor coupling hepatic glucose utilization and lipid synthesis. Cell Metabolism 4, 107–110 (2006)
19. Wellen, K.E., Fucho, R., Gregor, M.F., Furuhashi, M., Morgan, C., Lindstad, T., Vaillancourt, E., Gorgun, C.Z., Saatcioglu, F., Hotamisligil, G.S.: Coordinated Regulation of Nutrient and Inflammatory Responses by STAMP2 is Essential for Metabolic Homeostasis. Cell 129, 537–548 (2007)

Modelling Self-assembly in BlenX

Roberto Larcher[1,2], Corrado Priami[1,2], and Alessandro Romanel[1,2]

[1] CoSBi, 38123 Povo (TN), Italy
[2] DISI, Università di Trento, 38123 Povo (TN), Italy
{larcher,priami,romanel}@cosbi.eu

Abstract. The process through which disordered components spontaneously arrange themselves into patterns is called self-assembly. Molecular self-assembly describes the process by which molecules adopt a defined arrangement without external guidance (e.g. formation of membranes and protein complexes). These biological processes are essential to the functioning of cells. We investigate the usage of BlenX, a process calculi based programming language, for modelling molecular self-assembly of filaments, trees and rings. Moreover, we show how these structures can be used to model actin polymerization.

1 Introduction

The term self-assembly indicates a process in which disordered components form an organized structure or pattern only through local interactions among them, without an external coordination. Debates of how complex structures and functions can emerge from local interactions between simple components can be found in plenty of different fields (e.g. nanotechnologies [11], robotics [18], molecular biology [15,36], autonomous computation [24]). Here we concentrate on molecular self-assembly, the process through which molecules assemble in *complexes* without guidance from an outside source. We are particularly interested in inter-molecular self-assembly (i.e. the ability of proteins to form quaternary structures), a process which is crucial to the functioning of cells.

Conventional modelling approaches for molecular biology (e.g. ODEs) represent molecules and complexes by associating them with species identifiers or variables. This representation is extremely simple and effective. However there are modelling scenarios in which the explicit definition of all the possible complexes acting in a biological system and all the possible reactions in which they are involved is not always feasible or even possible. Examples are the modelling of processes in which small molecules combine to produce large complexes like chains or graphs of molecules (e.g. actin polymerization [2]) or in which the number of possible protein complexes and combinations of protein modifications tends to increase exponentially (e.g. multisite phosphorylation [2]).

Over the last decade, several computational approaches have been proposed to model and study complex interaction mechanisms in molecular biology (e.g. boolean networks [17], Petri nets [16], Bayesian networks [13], graphical gaussian models [25], process calculi [32], rule-based modelling [7]). Here we concentrate

C. Priami et al. (Eds.): Trans. on Comput. Syst. Biol. XII, LNBI 5945, pp. 163–198, 2010.

on the process calculi paradigm and in particular on BlenX [10], a programming language based on the Beta-binders process calculus [28]. The goal of this paper is to provide design patterns and examples of how to use BlenX for programming self-assembly processes. We consider a subset of BlenX focusing on the creation and modification of complexes; a theoretical study of the computational power of this subset is in [29]. The aim of this work is to show how the process calculi soul of BlenX allows us to model in an intuitive and effective way non-trivial structures (e.g. filaments, trees, symmetric rings). We also show how the expressivity of the language allows us to program and combine different growth and shape controls over these structures by means of the compositional nature of BlenX. We exemplify also how to use these structures to model actin polymerization processes, highlighting similarities and differences wrt other existing modelling techniques. All the complete programs presented in this paper can be downloaded at [1].

The rest of the paper is organized as follows. Sec. 2 presents informally the BlenX syntax and semantics with the help of an intuitive graphical notation. In Sec. 3 we first present examples of how to program the formation of polymers like filaments and trees and then we introduce a compositional programming pattern which allows us to implement shape controls over the formation of trees. In Sec. 4 we present a pattern for modelling the self-assembly of rings.

2 BlenX

BlenX is a stochastic programming language (i.e. quantitative information about speed and probability of actions is provided with systems specification) explicitly designed to model interactions of (biological) entities. A BlenX program is made up of an optional *declaration* file for the user-defined constants and functions, an *interfaces definition* file for the quantitative information about the system, and a *program* file for the program structure.

BlenX represents a protein as a computational entity, a *box*, composed by a set of *interfaces* and an *internal program*. Interfaces (e.g., protein domains) are associated with a structure and are the places where a protein can interact with other proteins; the internal program, instead, codifies for the mechanism that transforms an interaction into a protein conformational change, which can result in the modifications of other interface structures. This is inspired by the available knowledge of protein structures and functions (see for example [35]).

Two boxes belong to the same species when they are structurally equivalent [33]; we refer to classes of structurally equivalent boxes as box species or simply species. A BlenX box is defined using the keyword *bproc*. For instance the following piece of code defines a box called *B*:

```
let B : bproc = #(x,C),#(y,D),#(z,E) [ P ];
```

`#(x,C)`, `#(y,D)` and `#(z,E)` are box interfaces defined by a *subject* (e.g. `x`) and a structure (e.g. `C`). The subjects and structures of a box are all different and an interface has an associated state that can be either bound or free (see below).

Subjects can be referred by the internal program `P`, which is described by a process built on top of the following primitives:

```
x?(y) | x!(y) | ch(x,A) | ch(r,x,A)
```

The input action `x?(y)` intuitively means that a process is willing to receive on the channel `x` a name that will replace the target variable `y`. The output action `x!(y)` sends a name `y` along a channel `x`. When the name sent and the target variable are irrelevant (i.e. we only want to represent pure synchronization and not exchange of information) we will freely omit the object `y` of the two primitives simply writing `x?()` and `x!()`. The primitive `ch(x,A)` changes the structure of an interface with subject `x` into `A`. The primitive `ch(r,x,A)` is the same as the previous one, but allows to associate a specific stochastic rate `r` to the action. BlenX allows the user to define a set of global stochastic rates at the beginning of the program file, e.g., a rate for all the change actions without specific rates, rates for global names (see [10] for details).

The basic actions can be combined by sequential composition (.), parallel composition (|), choice (+), guarded replication (`rep`) and `if` statement, in order to construct complex protocols that abstract protein conformational changes. An example of internal program is:

```
if (x,A) and ( not (x,bound) or (y,B) ) then
   x?(y).y!() + y!().x?()
endif
| rep x?().ch(r,y,C).nil
```

`nil` indicates the deadlock process, i.e. a process that does nothing. The condition of an `if` statement is a boolean expression where the atoms refer to interfaces via their unique subjects and check whether they have a given structure (e.g `A`) or whether they are in a given state (e.g. `not bound`). The choice operator + composes processes that can be alternatively executed, i.e. the execution of one of the processes prevents the execution of the others. The parallel operator | composes processes that execute concurrently and allows processes to communicate or synchronize when they perform complementary input/output actions on the same channel. The `rep` operator permits the definition of repetitive (recursive) and possibly infinite behaviours (see [34]).

The internal program of a box can be specified either directly in the box definition or by using identifiers of processes defined through the keyword `pproc`:

```
let P : pproc = x?().y!(a) + y!();
```

The graphical notation of boxes we use throughout this paper is depicted as follows:

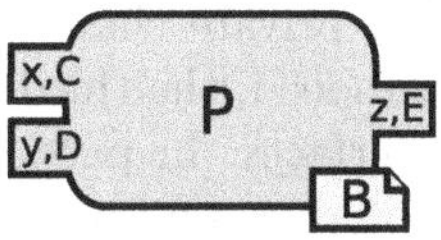

The small squares on the border of the box are the interfaces; `x`, `y` and `z` here are the interface subjects; `C`, `D` and `E` are the interface structures; `P` is the internal program and `B` is a label identifying a box species or a set of box species derived from the evolution of a certain box. All these elements are omitted when not necessary.

Two boxes can interact by composing or communicating only if they possess *compatible* interfaces. In our graphical notation, complexes are depicted as graph of boxes, where bonds are represented as lines connecting the interfaces of the proteins composing the complex. As an example, the picture

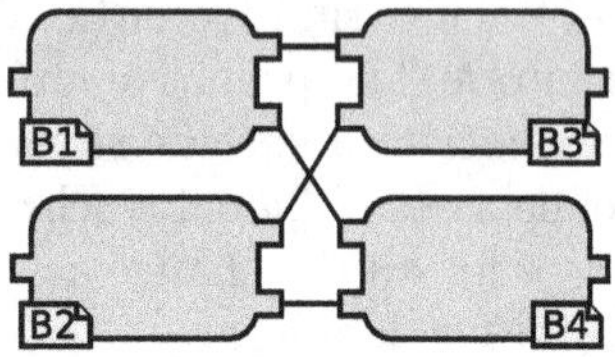

represents a complex made up of four proteins. In BlenX, the ability of boxes to form and break down complexes is given by the compatibility of their interfaces structure.

BlenX defines compatibility in the interfaces file by first listing the finite set of considered structures and then by associating a triple of real values to compatible structure pairs:

```
{C,D,E}%%{(C,D,1.2,3.5,2),(C,E,2.1,0,inf)}
```

The intuition is that we can use `C`, `D` and `E` interface structures and that `C` and `D` are compatible as well as `C` and `E`; all the other pairs not listed above are not compatible. The meaning of the real values associated to structure pairs is the key for introducing the interaction mechanism of *protein complexes* formation, i.e., proteins linked by non-covalent bonds [2].

According to the previous interface compatibility example, two boxes possessing interfaces with structure `C` and `D`, respectively, can bind, with rate 1.2, to form a complex:

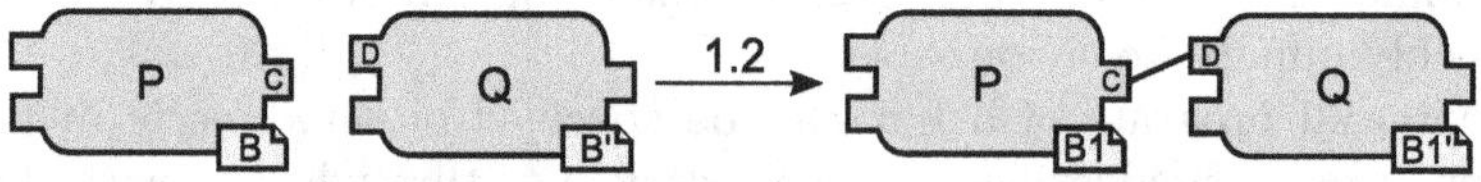

and can unbind with rate 3.5:

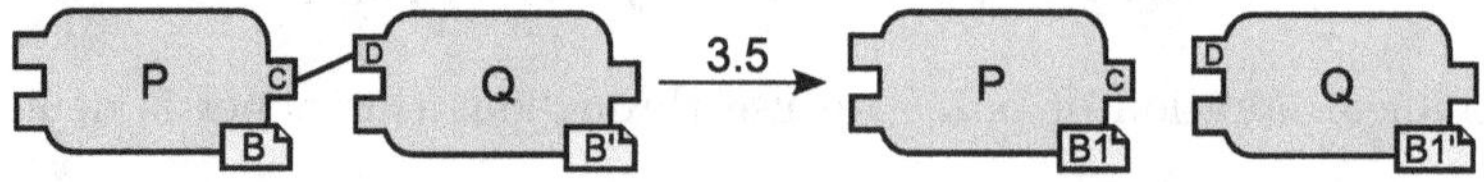

The binding represents a kind of private channel through which only the two bound boxes can communicate. Indeed, the third value 2 enables the two boxes to communicate over bound interfaces. In particular, if the first box does an output `x!(z)` and the second box does an input `y?(k)`, then the name `z` flows from the first box to the second one, generating a replacement[1] of all the bound occurrences of the target variable k in Q′ with the name z.

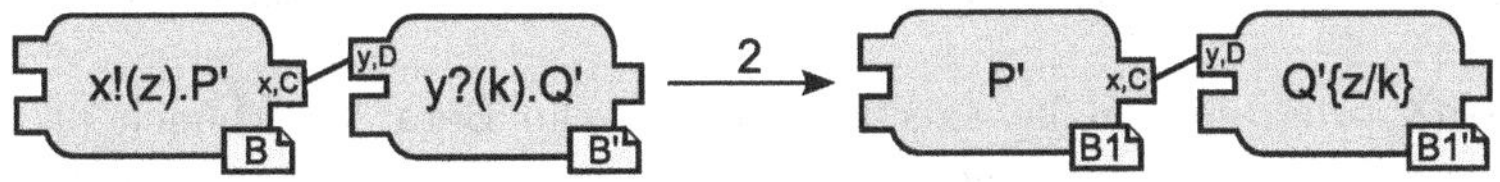

Note that we use compatibility of interfaces to realise the exact complementarity of channels used in process calculi. BlenX permits communications between different subjects of interfaces if their structures are compatible.

Each box can also evolve independently from the other boxes by synchronizing internally an input and an output on the same channel, or by executing a change action:

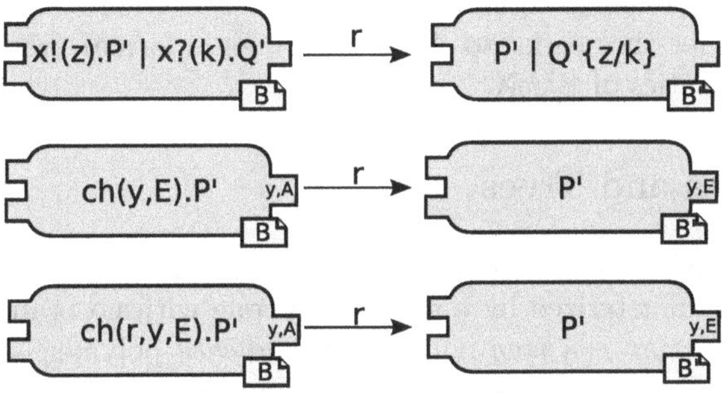

In the first example x is a global name and its rate is specified at the beginning of the program as previously mentioned. Global rates are also used to obtain the rate r associated to the execution of the first change action.

The `rep` operator executes indefinitely many times internal actions, producing as many copies in parallel as needed of the `rep` argument. For example, the process `rep x?().P` produces as many copies of P as needed by synchronizing the input with `x!()` outputs:

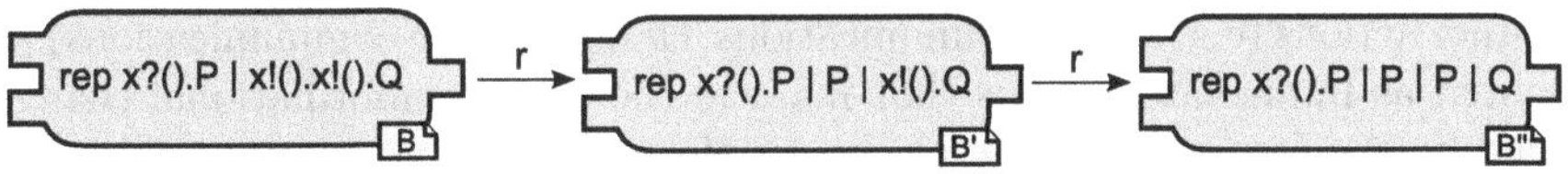

Actions guarded by `if` statements can be executed only if conditions evaluates to true wrt the actualn interfaces; once these actions are executed, the continuation process of the `if` can proceed and the condition is deleted:

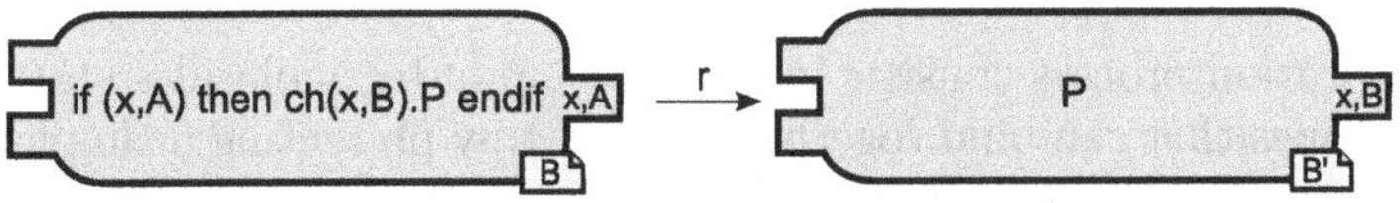

In addition BlenX supports immediate actions (using the keyword `inf` as rate value), i.e., actions that have precedence wrt actions associated with a finite

[1] We rely here on the notion of substitution typical of process calculi [22].

rate. Immediate actions embed in BlenX a two level priority mechanism, where immediate actions have high priority and actions with stochastic rates have low priority. This is a key mechanism for all the modelling patterns we present in this paper.

A BlenX program can be executed within the Beta Workbench [9], where an implementation of an efficient variant of the Gillespie algorithm [14] allows using the quantitative information of BlenX to run stochastic simulations. In the program file we can specify the length of the simulation (e.g. number of steps, limit time) and the initial number of boxes in the system:

```
run 10 B1 || 35 B2 || 2400 B3
```

Here, we presented only the subset of BlenX needed in this work and refer the reader to [10] for a complete and exhaustive presentation of the full language. Moreover, we refer the reader to Appendix A for a formal presentation of the operational semantics of BlenX.

3 Filaments and Trees

An important feature of polymers is their structure. The simplest structure is a *linear* chain, characterized by a single backbone with no branches. A related unbranching structure is a *ring* polymer. A *branched* polymer is composed of a main chain with one or more side chains. Examples of polymers are actin, DNA and microtubules.

We exploit here BlenX programming to model linear and branched polymerization processes, namely the self-assembly of *filaments* and *trees*. Ring polymers are delayed to the next section. For the sake of simplicity, the models presented in this section can only grow, i.e., we do not consider reversible processes in which boxes can detach from a filament. Although this is a simplification the programs presented give the flavor of the BlenX potential. In all programs the only reactions associated with finite rates are the bindings between boxes. All the other actions (e.g. intra-communications, changes, inter-communications) are executed as immediate actions that manage the set of conformational changes caused by the formation of complexes. Conformational changes are assumed to be atomic with the complexation because they are so fast that the time they take to execute can be ignored.

3.1 Filaments

A polymerization process consists in the creation of big molecules starting from small molecules that can bind together. We start by presenting a simple scenario in which a box `M`, representing a monomer, can bind to boxes of its own species and generate filaments of boxes. For simplicity, and for a fine control over the creation of the filaments, we introduce also a *seed* box `S`. Its role is recruiting the first monomer and starting the formation of a filament. The graphical intuition of the structures of boxes `S` and `M` is given in the following picture:

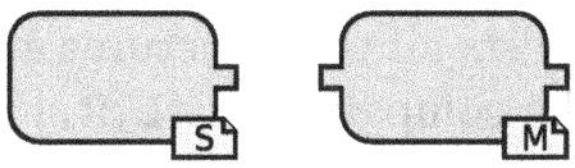

Box S has only one interface, used to bind to a free monomer, while M is equipped with two interfaces, the *left* one used to bind to the last box of a growing filament (can be both a box S or M) and the *right* one used to bind to a free monomer. The dynamics of the model is depicted in Fig. 1 (arrows decorated with a star represents more then one step).

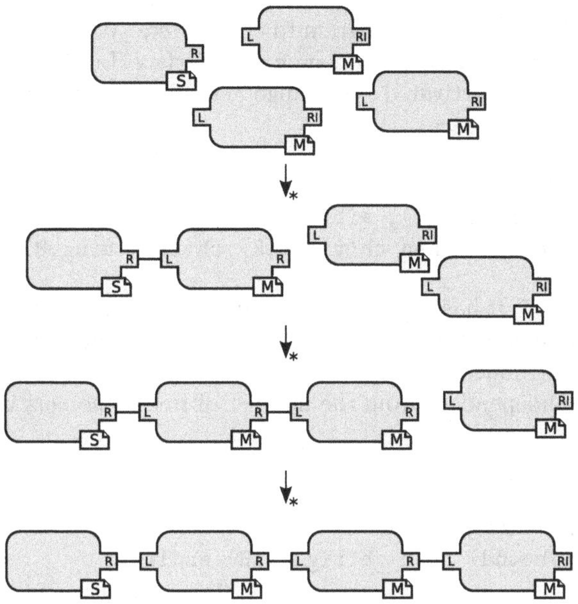

Fig. 1. Example of filament formation

Notice that a filament can grow only on the right side. The seed S starts the creation of the filament, while the growing process involves only the binding of M monomers. A box M can accept the binding of another monomer on its right interface only if it is already part of a growing filament, i.e., its left interface is bound to a filament.

We start with the definition of the two boxes, S and M:

```
let S : bproc =
 #(right,R)
 [ nil ];
```

```
let M : bproc =
 #(left,L), #(right,RI)
  [
    if (left,bound) then
      ch(right,R)
    endif
  ];
```

The interfaces definition file lists all the structures appearing over the interfaces of our boxes and specifies the compatibility `(L,R,1,0,0)`. Due to the compatibility between `L` and `R`, a monomer `M` can bind to the seed `S`. After the binding, the internal program of `M` recognizes that something is bound to the **`left`** interface and changes the structure of the right interface from `RI` (that stands for `R` inactive) into `R`. In this way, the **`right`** interface of the bound `M` gets the capability to bind to the left interface of another free monomer (see Figure 1).

3.2 Trees

Here we upgrade the previous program to create trees. We modify the monomer box `M` adding a third interface, the **`branching`** interface. Like the **`right`** interface, the **`branching`** one is activated via change actions only when `M` has something bound on its **`left`** interface.

```
let M : bproc =
   #(left,L), #(right,RI), #(branching,BI)
   [ if (left,bound) then ch(right,R).ch(branching,B) endif ];
```

We add also a box `T`. It has two interfaces and its role is to bind to the **`branching`** interface of a monomer in a filament and to start the formation of a branch. This species is introduced because we want the rate of branches formation over filaments to be independent from the number of free monomers in the model. `T` is defined as:

```
let T : bproc =
   #(left,TL), #(right,TRI)
   [ if (left,bound) then ch(right,TR) endif ];
```

Also for `T`, the internal program is instructed for changing the structure of the **`right`** interface once the box gets bound on the **`left`** interface. For trees, we specify two additional affinities: `(B,TL,1,0,0)` enables the complexation of a box `M` that is part of a filament with a free box `T`; `(L,TR,1,0,0)` enables the complexation of a box `T` that is bound to a monomer with a free monomer `M`. A possible run of this program with one box `S`, three boxes `M` and one box `T` is depicted in Fig. 2.

3.3 Introducing Controls

We introduce programs to build trees with constraints over the depth of the branches and then by imposing a minimum distance among branches. Finally we combine the two models to obtain trees with both the characteristics. In order to make the merge procedure of models more general we introduce a BlenX design pattern.

For each box we partition the interfaces in *interaction interfaces* and *state interfaces*, denoted by $\mathcal{II}$ and $\mathcal{SI}$, respectively. Interaction interfaces are used

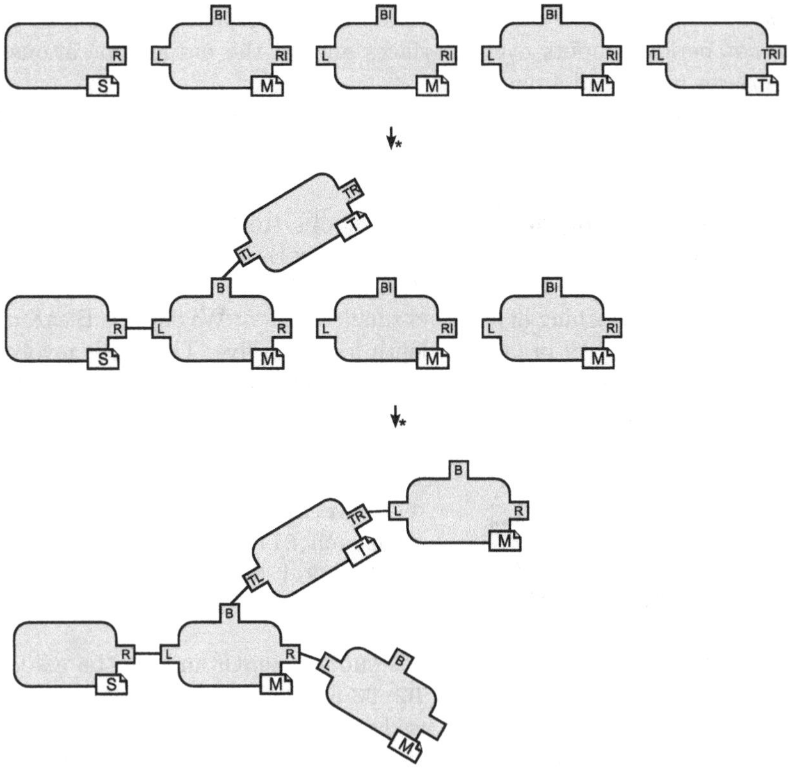

Fig. 2. Generation of a branching tree

by a box to interact with other boxes, while the state interfaces are used to store information regarding the box state. Note that storing information in the interfaces simplifies the internal coding; indeed all the parallel processes can access this information easily by managing and checking interface structures through change actions and if statements. The internal program of each box is composed by three parallel processes. We call them *interaction modifier process* (IMP), *state modifier process* (SMP) and *messages receiver process* (MRP).

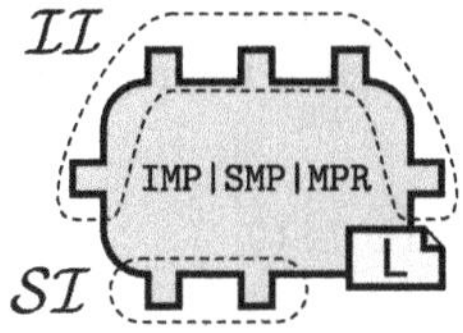

Processes IMP, SMP and MRP are defined following a list of criteria that restrict their behaviour and the possible actions they can perform:

- IMP can modify and check the state of the interfaces belonging to $\mathcal{II}$ and $\mathcal{SI}$. Moreover, it cannot perform inputs and outputs;

- SMP can modify and check the state of the interfaces belonging to $\mathcal{SI}$. It cannot perform inputs over interfaces and all the output operations over interfaces must send a name;
- MRP receives incoming messages from $\mathcal{II}$ interfaces and executes outputs over each received names.

Controlling branching depth: each node of a tree can be associated with a branching depth level. A node has branching depth level n if there are exactly $n-1$ branches from the node to the root. A tree has branching depth less than n if all its nodes have branching depth level smaller than n. We define a BlenX model that generates trees with branching depth less than five. The boxes involved in this model are monomers, branches and seeds. The BlenX graphical and textual representation of monomers is:

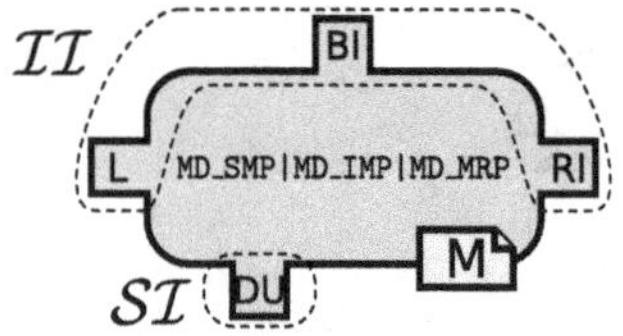


```
let M : bproc =
 #(right,RI), #(left,L),
 #(branch,BI), #(depth,DU)
  [ MD_SMP | MD_IMP | MD_MRP ];
```

The only interface belonging to $\mathcal{SI}$ has subject `depth` and can be associated with five different structures: `DU`, `D1`, `D2`, `D3` and `D4`. Structure `DU` is used when the box is free, while the others are used when the box is bound and represent the node branching depth level. We refer to this interface as *depth* interface.

The *interaction modifier process* `MD_IMP` changes the structure of the right interface when the box binds on the left interface and changes the structure of the branching interface depending on the structure of the depth interface. In particular, the structure has to be changed in order to let the box branch only if the strucure exposed on the depth interface is one among `D1`, `D2` and `D3`. Hence, `MD_IMP` is defined as follows:

```
let MD_IMP : pproc =
   if (left, bound) then ch(right,R) endif
   | if not ((depth,D4) or (depth,DU)) then ch(branch,B) endif;
```

We construct the *messages receiver process* `MD_MRP` mechanically, composing in parallel similar processes, one for each interface belonging to $\mathcal{II}$:

```
let MD_MRP : pproc =
   rep right?(channel).channel!()
   | rep left?(channel).channel!()
   | rep branch?(channel).channel!();
```

The last process composing the internal program is the *state modifier process* `MD_SMP`. Its role is to modify the structure associated with the depth interface in order to make it represent the number of branches from the root to the current box. This task is performed through an exchange of messages among the *state modifier* processes of different boxes. Thus MD_SMP is composed by two parallel processes, one receiving and using messages for updating the state interfaces, and another sending messages to bound boxes and trigger their update:

```
let MD_SMP : pproc =
   MD_r_SMP | MD_s_SMP
```

MD_r_SMP defines a sequential process composed by an input on a channel called depth_msg which guards a sum of inputs over specified channels whose firing enables proper modifications of the depth interface:

```
let MD_r_SMP : pproc =
   depth_msg?().(
      d1?().ch(depth,D1)
      + d2?().ch(depth,D2)
      + d3?().ch(depth,D3)
      + d4?().ch(depth,D4)
   );
```

The interacting partner of MD_r_SMP is:

```
let MD_s_SMP : pproc =
     right!(depth_msg).(
      if (depth,D1) then right!(d1) endif
      + if (depth,D2) then right!(d2) endif
      + if (depth,D3) then right!(d3) endif
      + if (depth,D4) then right!(d4) endif
     )
   | branch!(depth_msg).(
      if (depth,D1) then branch!(d1) endif
      + if (depth,D2) then branch!(d2) endif
      + if (depth,D3) then branch!(d3) endif
     );
```

This process is composed by two parallel processes. The output on right can be consumed only when the box binds to another box on the corresponding right interface. When consumed, the output sends the name depth_msg to MD_MRP of the bound box, and it is propagated inside the box by an output on depth_msg without object. This output synchronizes with the MD_r_SMP process of that box and enables a sum of input processes on channels d1, d2, d3 and d4. The choice operator is resolved according to the next message sent by MD_s_SMP and only one input is fired. The initial output of MD_s_SMP enables a sum of mutually exclusive if, each one sending through the right interface a different channel depending on the depth interface structure. MD_MRP sends a message on this channel that synchronizes with one of the inputs of MD_s_SMP causing the properly update of the depth interface. Notice that this communication protocol propagates the depth information to the newly attached box and is executed atomically with a sequence of immediate actions.

The MD_s_SMP process performs a sequence of immediate actions on the branch interface, similar to the ones described for the right interface, to propagate the depth information also along the branches. In this case the depth D4 case is not considered because a branch cannot grow from a monomer with branching depth equal to four.

Now we proceed with the description of the branching box T:

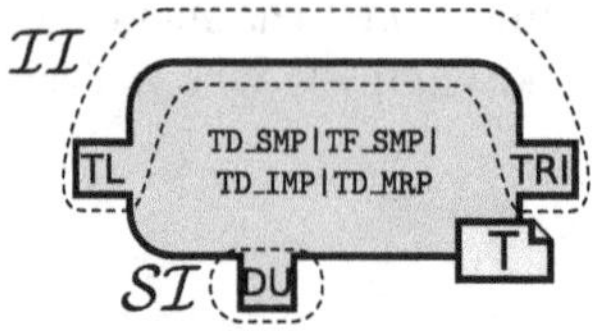


```
let T : bproc =
  #(right,TRI), #(left,TL), #(depth,DU)
   [ TD_SMP | TD_IMP | TD_MRP ];
```

Also in this case the interfaces of the box are the same we introduced in the previous section. The interaction modifier process TD_IMP has only to change the right interface structure when the box binds with another box on the left:

```
let TD_IMP : pproc =
   if (left, bound) then ch(right,TR) endif;
```

The message receiver process TD_MRP is defined mechanically following the pattern previously described.

The state modifier process TD_SMP propagates the depth information to the new filament nodes. In this case, given that it is a branch node, we increase the depth level and we do not consider the depth level one (D1), because the branch node contributes to its depth level and so it has at least depth level two:

```
let TD_SMP : pproc =
   depth_msg?().(
      d1?().ch(depth,D2)
      + d2?().ch(depth,D3)
      + d3?().ch(depth,D4)
   )
   | right_site!(depth_msg).(
      if (depth,D2) then right!(d2) endif
      + if (depth,D3) then right!(d3) endif
      + if (depth,D4) then right!(d4) endif
   );
```

The seed box S does not have any state interface and its BlenX graphical and textual description is:

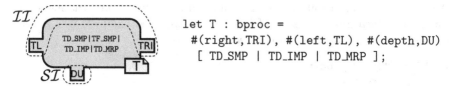

```
let S : bproc =
  #(right,R)
  [ SD_SMP | SD_IMP | SD_MRP ];
```

The interaction capabilities of this box never change and therefore the interaction modifier process is the empty process nil. The state modifier process SD_SMP has

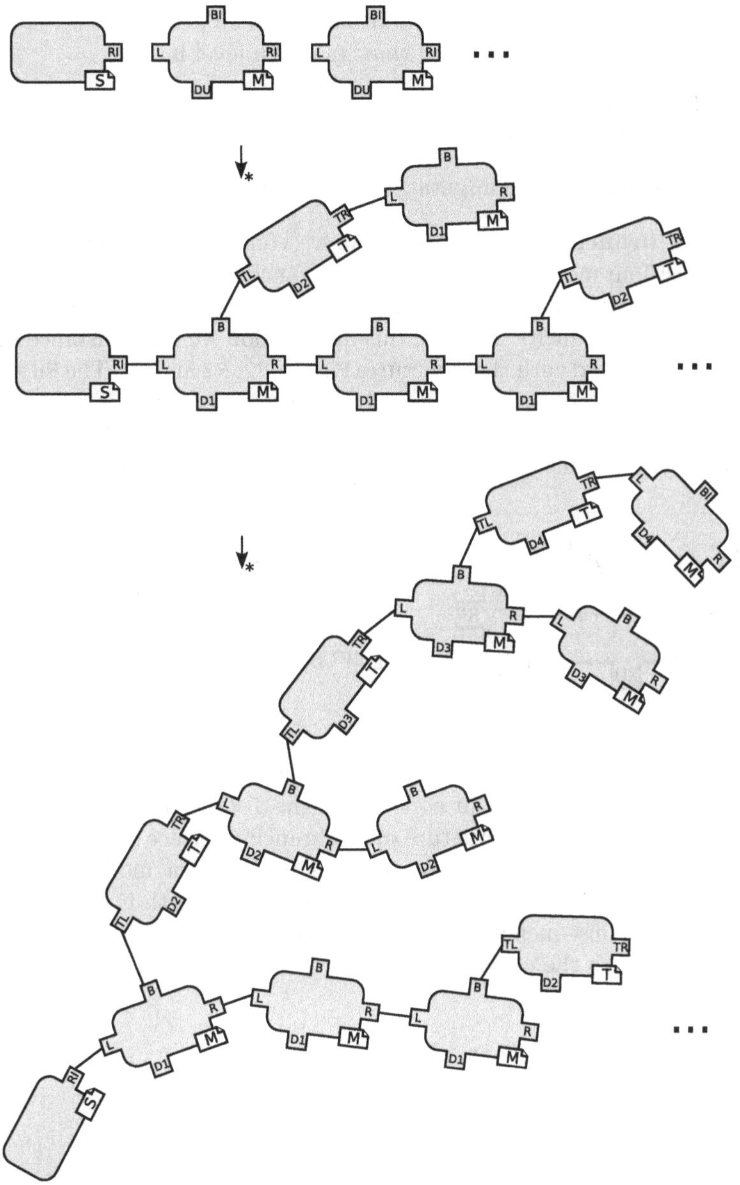

Fig. 3. Example of a tree generated with the branching depth control program. Notice how at level four the monomer branching interfaces are inactive. The dots indicate the presence of other boxes in the system.

only to propagate the initial depth information when a filament is started, i.e., the first box it binds to is informed that its depth level is one:

```
let SD_SMP : pproc =
   right!(depth_msg).right!(d1);
```

Fig. 3 shows an example of computation of the branching depth control program.

Controlling branching distance: now we consider the generation of trees having at least four monomers between their branches. We store in each monomer belonging to a filament its distance from the nearest branch (the seed node is considered to be a branch). To store this information we use a distance interface that can be associated with the structures FU, F1, F2, F3 and F4. The FU structure is used when the monomer box is free while the others when it is bound to other boxes. The number indicates the distance from the nearest branch with the exception of F4 that means the distance is greater or equal to four. The BlenX code of M is:

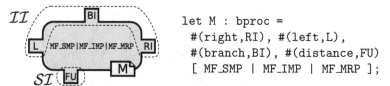

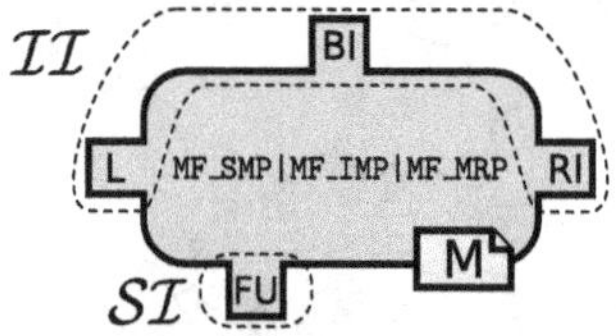


```
let M : bproc =
  #(right,RI), #(left,L),
  #(branch,BI), #(distance,FU)
  [ MF_SMP | MF_IMP | MF_MRP ];
```

The interaction modifier process MF_IMP permits the polymerization on the right interface when the box binds on the left and permits branching formation only if the nearest branch is at least four nodes far. If the distance interface has structure F4, the process changes the structure of the branch interface into B, allowing the branch formation. Then it enables an if that checks for modifications of the distance interface, that can happen when branching growth is in the proximity of the box. As a consequence the MF_SMP process changes the distance interface structure from F4 to the correct one.

```
let MF_IMP : pproc =
     if (left, bound) then ch(right,R) endif
   | if (distance,F4) then ch(branch,B).
          if not (distance,F4) then ch(branch,BI) endif
     endif;
```

We proceed with the BlenX code description of the state modifier process MS_SMP:

```
let MF_SMP : pproc =
   ( if (distance,F1) then right!(distance_msg).right!(f2) endif
   + if (distance,F2) then right!(distance_msg).right!(f3) endif
   + if (distance,F3) or (distance,F4) then
        right!(distance_msg).right!(f4)
     endif)
```

```
|rep distance_msg?().(
     if(distance,F1) then
        f1?() + f2?() + f3?() + f4?()
     endif
   + if(distance,F2) then
        f1?().ch(distance,F1).propagate!(f2)
      + f2?() + f3?() + f4?()
     endif
   + if(distance,F3) then
        f1?().ch(distance,F1).propagate!(f2)
      + f2?().ch(distance,F2).propagate!(f3)
      + f3?() + f4?()
     endif
   + if(distance,F4) or (distance,FU) then
        f1?().ch(distance,F1).propagate!(f2)
      + f2?().ch(distance,F2).propagate!(f3)
      + f3?().ch(distance,F3)
      + f4?().ch(distance,F4)
     endif
 )
 | rep propagate?(distance).(
      ( if (left,bound) then
           left!(distance_msg).
           left!(distance)
        endif
      + if not (left,bound) then kill!() endif)
      |
      ( if (right,bound) then
           right!(distance_msg).
           right!(distance)
        endif
      + if not (right,bound) then kill!() endif)
   )
 | rep kill?().nil;
```

The first parallel component propagates the distance information to the box that binds to the right interface. If its distance is n, the distance of the newly attached box will be $n+1$. In order to communicate this information, the process sends first a message containing the channel `distance_msg` to make the receiving box aware that a message regarding its distance from a branch is incoming. After that, depending on the structure of its distance interface, it can send three different channels: `f2`, `f3` and `f4`. They represent respectively distance two, three and distance greater or equal to four. In this case the process cannot send distance one (represented by the channel `f1`) because a monomer is at least at distance one from a branch and so the distance of a neighbor is at least two. The channel `f1` is sent by branch and seed nodes when they bind to a monomer.

The second parallel component manages the incoming distance messages and updates the distance interface, if required. After an input operation on `distance_msg`, four `if` processes made up of alternatives selected according to the distance interface become active. These four processes wait for an output action over one of the `fn` channels (where $\mathtt{n} \in \{1,2,3,4\}$). If an inter-communication happens then the distance interface is updated only if the new distance is smaller than the current one. If the interface is updated, the distance, increased by one, is propagated by the third parallel component that is activated through an output operation over the `propagate` channel. Notice that the second parallel component is under a replication operator because it is executed each time another process in parallel performs an output over the `distance_msg`. The process performing the propagation ignores the origin of the received information (left or right interface). Thus, it sends the distance message both over the right and the left interfaces with two processes in parallel. These processes only differs for the interface they operate over. They consist of two processes composed by means of a choice operator. The first process is triggered if the interface is bound; in this case it communicates the received distance over it. The other is executed if the interface is not bound and it triggers an output operation over a channel called `kill`. Its role consists in avoiding pending communications in the case the interface is not bound.

Now we introduce the definition of the branching box `T`. In this case the state of the branches boxes does not depend on the bound boxes and hence they do not have any state interface:

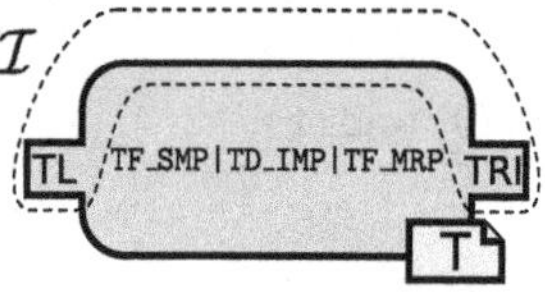


```
let T : bproc =
 #(right,TRI), #(left,TL)
 [ TF_SMP | TD_IMP | TF_MRP ];
```

Note that interface modifier process `TD_IMP` is exactly the same of the previous model. The state modifier process `TF_SMP`, even without state interfaces has to cooperate with the state modifier processes of the other boxes in order to send them distance information. In particular, when something binds to the box, `TF_SMP` has to communicate to the new neighbour that its new distance is one. The communication of the distance is done as previously:

```
let TF_SMP : pproc =
   left!(distance_msg).left!(f1)
   | right!(distance_msg).right!(f1);
```

The seed box `S` has no state information to store either. So its structure is the following:

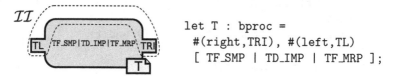

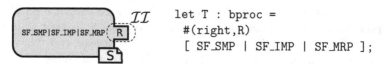

```
let T : bproc =
 #(right,R)
 [ SF_SMP | SF_IMP | SF_MRP ];
```

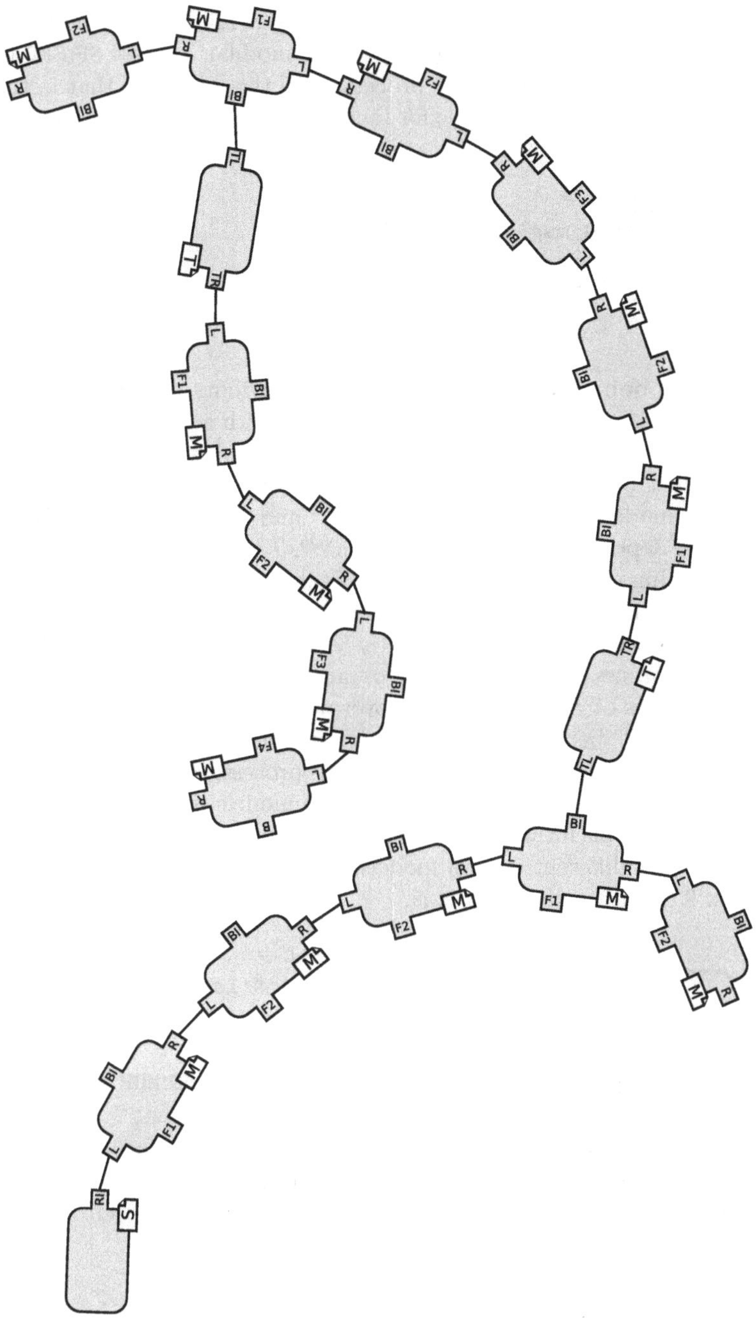

Fig. 4. Example of a tree generated by the branching distance control program

The interaction modifier process SF_IMP is empty (nil) because the interaction capabilities of the box never change. The state modifier process SF_SMP has the same role of the branch box TF_SMP process, with the difference that in this case there is only one interface available for binding. Thus, the information regarding the distance is sent only over it:

```
let SF_SMP : pproc =
   right!(distance_msg).right!(f1);
```

Fig. 4 shows an example of a tree guarded by the branching distance control program.

Merging the behaviours: we merge the behaviours of the two models to generate trees with depth level less than five and with a branches distance of at least four monomers. We create a new seed box S, a new branching box T and a new monomer box M by starting from their corresponding definitions in the two previous models. We denote with S^d, T^d and M^d the boxes definition for the branching depth control model and with S^b, T^b and M^b the boxes definition for the branching distance control model. Each box B (with B ranged over by S, T and M) has the same interaction interfaces of B^d and B^b and a set of state interfaces which is the union of the state interfaces of B^d and B^b. The state modifier process of each box B is obtained composing in parallel the state modifier processes of B^d and B^b. Moreover, the message receiver process is the same of B^d and B^b because the two boxes have the same interaction interfaces. Finally we define a new interaction modifier process. Thanks to the adopted design pattern, the process of merging the two models is easy. We only need to define a new IMP. This intuitive procedure is enough general to allow the merging of more than two different control models.

The box M of this merged model is:

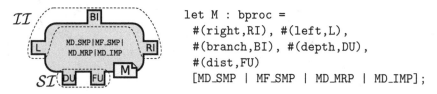

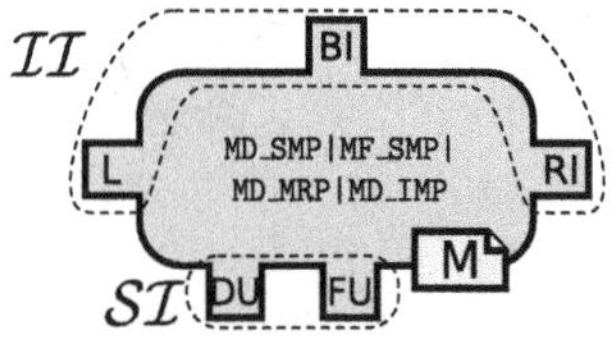


```
let M : bproc =
  #(right,RI), #(left,L),
  #(branch,BI), #(depth,DU),
  #(dist,FU)
  [MD_SMP | MF_SMP | MD_MRP | MD_IMP];
```

We define an IMP that allows the binding on the right interface only if the box is bound on the left interface and that allows the growth of a branch only if the box has level less then five and it is far at least three monomers from a branch:

```
let M_IMP : pproc =
   if (left, bound) then ch(right,R) endif
   |
   if (distance,F4) and not (depth,D4) and not (depth,DU) then
      ch(branch,B).
      if not (distance,F4) then ch(branch,BI) endif
   endif;
```

The definition of the merged branch box T is easier because the interaction modifier processes of the branch boxes of the models we are merging are the same. Thus we can use one of them:

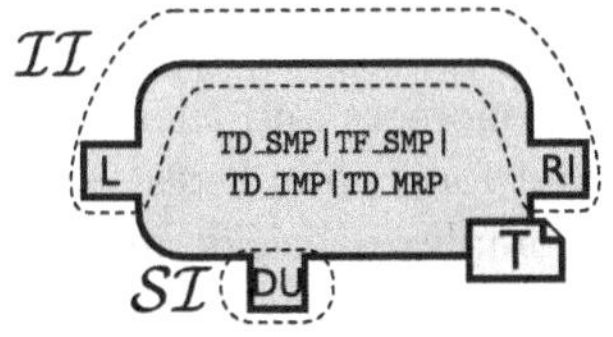


```
let T : bproc =
 #(right,TRI), #(left,TL),
 #(depth,DU)
 [ TD_SMP | TF_SMP | TD_IMP | TD_MRP];
```

Similarly, we can build the merged seed box S of this model as follows:

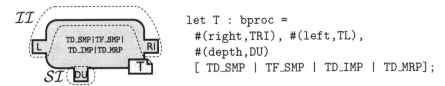

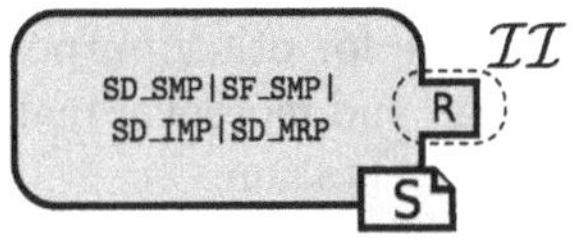


```
let S : bproc =
 #(right,R)
 [ SD_SMP | SF_SMP | SD_IMP | SD_MRP];
```

Note that in these models we used structure interfaces as counters. Since structure interfaces are not numbers, no arithmetic on them is possible (e.g. we cannot increase or decrease the values represented by interface structures with plus or minus operators). Hence we have to represent each possible number with a different interface structure and implement addition and subtraction operations with processes that changes interface structures in the appropriate way (see for example the previous definition of TD_SMP process). It is clear that when dealing with wide ranges of numbers our approach results in very verbose code. Anyway, since all these operations can be implemented following the same template, a possible solution is to implement them by adding to the language a set of macros automatically expandable in BlenX source code at compile time. Another solution to this problem can be found in the BlenX extension presented in [31], where interface structures are interpreted as concrete values and can be manipulated through mathematical operations.

3.4 Modelling Actin Polymerization and Related Works

Actin is a small globular protein present in almost all the known eukaryotes. It is one of the most conserved protein among species because it is involved in fundamental cell processes, e.g. muscle contraction, cell motility, cell division and cytokinesis, vesicle and organelle movement, cell signaling, and the establishment and maintenance of cell junctions and cell shape. All these tasks are performed due to the ability of actin to polymerize in long filaments [20].

Because of their importance in cell life, actin filaments are widely studied and modelling techniques are often employed. ODE-based approaches define all the complexes involved in the modeled system as variables or identifiers already present in the initial state of a model. This is a limitation because these approaches introduce strong approximations [26,5,23] given that actin filaments are complexes of molecules that can assume a finite but really huge number of arrangements. ODE-based techniques can provide useful information and abstractions, but their application to represent the low-level dynamics underlying the formation and evolution of actin filaments is usually quite difficult and sometimes even impossible.

Better results can be obtained by adopting methods that introduce species for representing the possible conformations of the actin monomers and then introduce polymerization reactions for composing them and form complexes. These kind of models, requiring only the definition of the complexes bricks, avoid to list all the possible arrangements that actin filaments can assume. Preliminary attempts to use this approaches are in [12,21] where the authors define ad hoc programs implementing the Gillespie algorithm to perform simulations. The limitation in this case is that there is no distinction between the model and the simulator. This makes it impossible to reuse the work done for other purposes (for example for performing simulations of different models) and moreover makes it difficult to check for the correctness of the model implementation.

A distinction among the model and the simulator is in [4] where the authors define a model of actin polymerization in stochastic π-calculus [30] and simulate it through the SPiM simulator [27]. This approach has numerous advantages. The process calculi paradigm [32] allows indeed applying the metaphor of autonomous entities that are interacting with each other in order to build complex structures. Moreover, models can be written and modified without acting on the simulator implementation and the correctness of the simulator has to be verified only once against the semantics of the considered process calculus and not against each written model. The concept of complex is handled in [4], but complexes generated during simulation are not explicitly listed in the output of the simulation. Actin structures can be extracted from the output using the geometric plotting, but since the information used for the plotting relies on the coordinates associated to processes, there are cases in which the interpretation is ambiguous. In these cases it is not possible, for example, to retrieve the time-course of specific actin structures.

As the previous approach, also ours has a clear distinction between the simulator and the model definition and is based on the process calculi metaphor. Here we do not aim presenting a detailed model. Our intention is to demonstrate that the self-assembling structures presented in Sec. 3 can be used to model real biological examples. Thus we start looking at the similarity between the first model we presented and the actin filament polymerization process. In the model we have `M` monomers that can bind together forming a structure that grows in a long filament. Furthermore we showed how to extend the first model in order to create filaments that can branch and form tree structures. Actin filaments have a similar behaviour. They interact with a multitude of molecules, in particular they can bind with the Arp2/3 complex, a seven subunit protein. This complex has the capability to bind existing actin filaments and become a nucleation site for actin monomers. This leads to the formation of tree-like structures that are important for processes like cell locomotion, phagocytosis and intracellular motility of lipid vesicles. Also the introduction of controls over the development of the structure can find a parallelism in the biological world. In fact when Arp2/3 complex binds to a filament, because of structural constraints, makes the complexation of other branching proteins around its proximity impossible. The intuitive model described here can be improved in order to mimic more realistically the behaviour of the actin molecules. At the url [1] it is possible to download some BlenX models that behaves as the models presented in [4].

The expressivity and the ductility of this approach is paid in terms of simulation performance. For example stochastic simulation approaches are significantly slower then deterministic one (such as ODEs); moreover, in the case of ad hoc programs defined for describing models it is possible to introduce specific optimization and tricks that are not applicable to the general case. Hence we have to deal with a trade off between expressivity and performance.

Rule-based approaches (e.g. [7]) seems also to be candidate for modelling actin polymerization processes, since in general can be used to model self-assembling systems [6]. In [19], for example, a rule-based model of genetic regulation (with growing DNA and RNA strands and moving ribosomes) that avoids the combinatorial state-explosion is presented.

4 Rings

Here we investigate how BlenX can be used to model the formation of ring polymers. In this section we do not make use of the previously defined design pattern because we want to underline the flexibility of the language in allowing different programming styles. Here we introduce also *reversibility*, an important characteristic of many biological processes omitted in the previous modelling scenarios.

We start our ring modelling by pointing out an example that shows which are the difficulties we have to face. Consider a scenario in which we have a population of boxes `A` equipped with two distinct interfaces with structures `L` and `R`. If we specify a binding affinity for `L` and `R` structures, initialize the system with an initial amount of boxes `A` and run stochastic simulations, we can observe in the output of the simulations the formation of various kinds of complexes as depicted in Fig.5. The figure reveals the difficulties we have in controlling the formation of non-trivial complexes by only acting on the specification of interface capabilities (similar aspects and problems have been discussed in [6] for the κ-calculus). In this section we define a model that generates rings of fixed size avoiding the creation of long chains. We start without introducing reversibility and then we extend the model in order to introduce this capability.

Assume to have an initial population of boxes `A`, each containing the same internal program. Each box has an interface with structure `L` (left) and an interface with structure `R` (right), which are compatible, i.e. `(R,L,1,0,inf)`. Moreover, each box has an interface used to store the information regarding the length of the chain to which the box belongs; initially, the structure of that interface is `S1`:

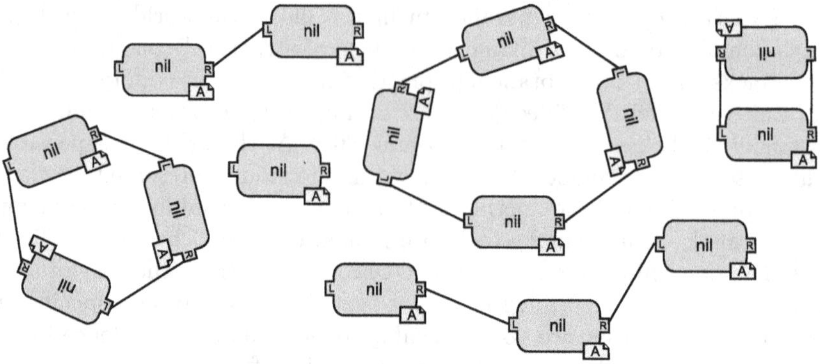

Fig. 5. Examples of ring complexes

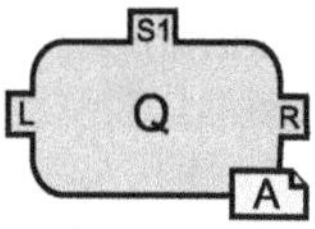

Boxes A can bind together, forming chains that grows till they reach a size in which they close, forming a ring. We consider the ring being the stable form. The size of the ring can be controlled at programming level. As a characteristic of self-assembly, we want rings to be only the result of local interaction between boxes, avoiding any global coordination.

Each box can be in one of these states: not bound to any other box; bound either on the left or on the right; bound both on the left and on the right. Hence, the internal program implements a state machine with the above three states; moving from one state to another is controlled by the structure of the box interfaces. After any binding operation, the involved boxes start a protocol made up of a sequence of immediate actions (high priority), through which they exchange information (by using communication) used to change the internal program state and their interface structures. The definition of A is:

```
let A : bproc =
    #(left,L), #(right,R), #(num,S1)
    [
        b0!()
        | rep b0?().Bound0
        | rep b1?().Bound1
        | rep b2?().Bound2
    ];
```

Processes Bound0, Bound1 and Bound2 implement the three different states of our state machine. In order to implement a recursive behaviour, we use the replication operator. For example, entering the state Bound1 is implemented by performing an internal communication on the channel b1.

The process Bound0 is defined by composing in parallel a sum of sequential processes performing inputs on the left and right interfaces and a sum of sequential processes performing outputs of the name one on the left and right interfaces:

```
let Bound0 : pproc =
  left?(val).ch(right,RF).Set
  + right?(val).ch(left,LF).Set
  | left!(one)
  + right!(one);
```

```
let Set : pproc =
  val!() |
  one?().ch(num,S2).b1!()
  + ...
  + n-1?().ch(num,SN).b1!();
```

After consuming the left input, the structure of the free right interface is changed from R to RF (where RF stands for R final). Similarly, after consuming the right input, the structure of the free left interface is changed from L to LF. Since RF is compatible only with L, and LF is compatible only with R, this means that we allow the chains to grow only by adding single boxes and not chains of boxes. Moreover, the non compatibility of RF and LF avoids a chain to close in a ring. Both inputs enable the process Set that changes the num interface according to the information received from the input. Notice that the value N reflects the length of the self-assembly rings we are programming (e.g. is equal to 3 in case of triangles) and that we cannot use it as a variable in the current version of BlenX. In order to obtain a runnable code we have to substitute it with a name respecting the actual size of rings we intend to create.

To better understand this mechanism, consider what happens when two initial boxes bind together to form a chain of length two:

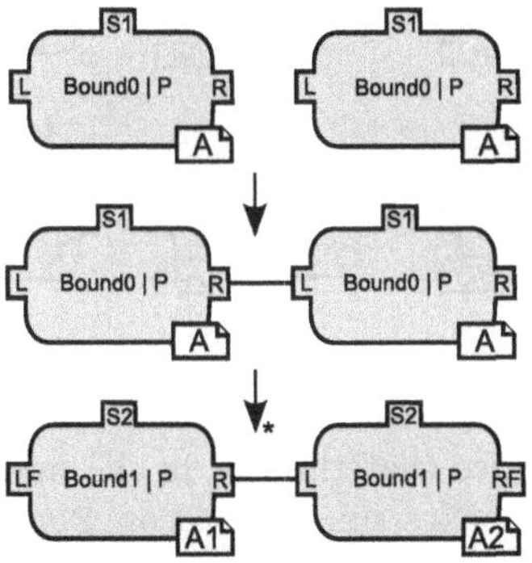

After binding together, the two boxes start a sequence of immediate actions that lead to the last complex depicted in the previous figure; each box has recorded in the structure of its interfaces that belongs to a chain of length 2 and the left and right limits of the chain have interface with structures LF and RF, respectively. Moreover, the two boxes are now in the state encoded by the process Bound1 that embodies different controls composed with the choice operator. The first control process manages the flow of information incoming from the right interface, when its structure is RF:

```
if (right,RF) then
    right?().ch(right,R).left!().b2!().SIC_right
endif
```

The `if` condition is satisfied when a free box binds on the right. After consuming the right input, the process changes the right interface again into the structure `R`, propagates a signal on the left, activates the process `Bound2` and finally starts the process `SIC_right` that sends on the right the actual size of the chain and increments its own size:

```
let SIC_right : pproc =
    if (num,S2) then x!(two).ch(num,S3) endif
    + ...
    + if (num,SN-1) then x!(n-1).ch(num,SN) endif;
```

where `N` is the size of the rings we want to create. Notice that also in this case we use names to represent interface structures, and hence with `n-1` we mean the name representing the structure denoting size `N-1` (e.g. if `N` is 4 then `n-1` is `three` and `SN-1` is `S3`). To better understand this mechanism, consider the following graphical representation:

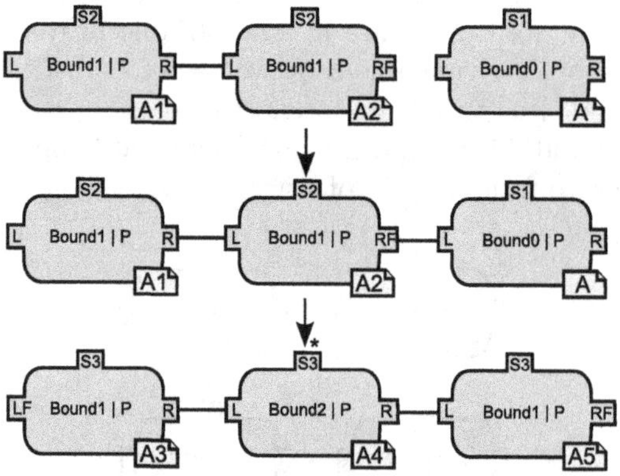

A single box binds on the right side of a chain of length two. After the binding creation, a sequence of immediate actions leads to the creation of the last complex depicted in the figure. Notice that the box in the middle now runs the process `Bound2`, that the external boxes run the process `Bound1` and that all the three boxes record the information on the length of the chain. This is obtained by creating a flow of immediate communications from `A2` to `A` passing through `A1`. In particular, using the second control process embodied in the process `Bound1`, `A1` changes its `num` interface after receiving a signal from the right and restarting `Bound1`:

```
if (right,R) then right?().b1!().IncreaseCounter endif
```

The process `IncreaseCounter` changes the `num` interface with a structure that represents the actual size incremented by one:

```
let IncreaseCounter : pproc =
    if (num,S2) then ch(num,S3) endif
    + ...
    + if (num,SN-1) then ch(num,SN) endif;
```

Specular control processes for the left interface are embodied in Bound1. Moreover, other two control processes are part of Bound1 to control the growing of the chain on both sides and to close the chain in a ring when it reaches the appropriate length. The process controlling the right size has the form:

```
if (right,RF) and (num,SN) then ch(right,RC).
    if (right,bound) then ch(right,R).b2!() endif
endif
```

If the right interface has structure RF and the num interface has structure N, then the right structure is changed into RC and the process is blocked till the right interface is bound. Since the process controlling the left side is specular, when the chain reaches a length of N the right and left side interfaces assume structures RC and LC, respectively, and remain blocked till they bind together. However, since the compatibility of these two structures is defined as (RC,LC,inf,0,0)) they immediately close and the two blocked processes can continue their execution, leading to a complex made up of structurally equivalent boxes and hence representing a completely symmetric ring. As an example, if N is equal to 3, when a chain of size 3 is formed it immediately closes in a symmetric triangle:

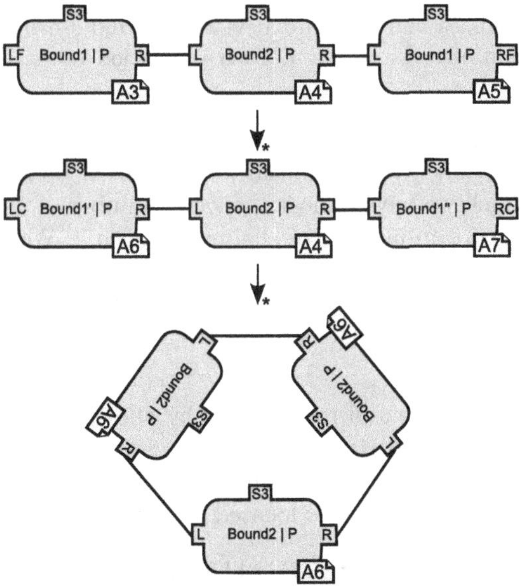

Note that although in the triangle formation the state Bound2 is not important, it becomes essential in the formation of rings of size greater then three. Indeed, the process Bound2 has the structure

```
let Bound2 : pproc =
    right?().left!().b2!() + left?().right!().b2!();
```

and is in charge to propagate signals from the left to the right and viceversa. Note that in our actual implementation each signal received by a box in state Bound2 cause only the change of the num interface with a structure corresponding to the increment of the current one.

Consider the previous program with N equal to 4. By running 100 different stochastic simulations with an initial population of 1000 boxes A we observe the average dynamics and the standard deviation reported in Fig. 6.

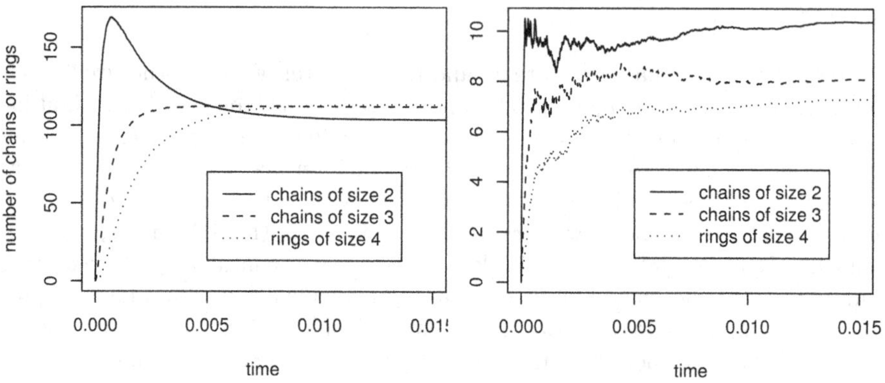

Fig. 6. Average dynamics and standard deviation of 100 stochastic simulations regarding the formation of rings of size 4. Each simulation takes in average around 2 seconds.

The results show that given an initial population of boxes, we reach a final configuration in which we have almost the same number of chains of size 3 and rings of size 4, and certain number of chains of length 2. We would like instead the system to generate as many rings as possible, because we assume rings be the only stable complex. In order to achieve this behaviour, we have to encode reversibility in our program, i.e., chains of size 2 and 3 can always disassembly.

Adding reversibility: We add reversibility by allowing only the unbinding of boxes located at the left and right sides of a chain. In order to implement this protocol we have first to modify the definition of the process Bound1, because it is the state characterizing the boxes located at the sides of a chain.

```
if (left,LF) then ch(r,right,RU).b1!() endif
+ if (right,RF) then ch(r,left,LU).b1!() endif
+ if (not (left,bound)) and (not (right,bound)) then
    ch(left,L).ch(right,R).ch(num,S1).b0!()
  endif
```

where `r` is a rate describing the speed of the reversible reaction. Each chain side can decide to unbind by changing the structure of the bound interface (left interface for the right side and viceversa). After executing the change action the box is immediately detached from the chain, because pairs `(LU,R,0,inf,0)` and `(RU,L,0,inf,0)` are defined in the compatibilities specification. After the unbinding, for both sides, the condition of the third `if` becomes true and hence the interfaces and the internal programs of the unbound boxes return back in the initial configuration (recall the recursive behaviour of box A):

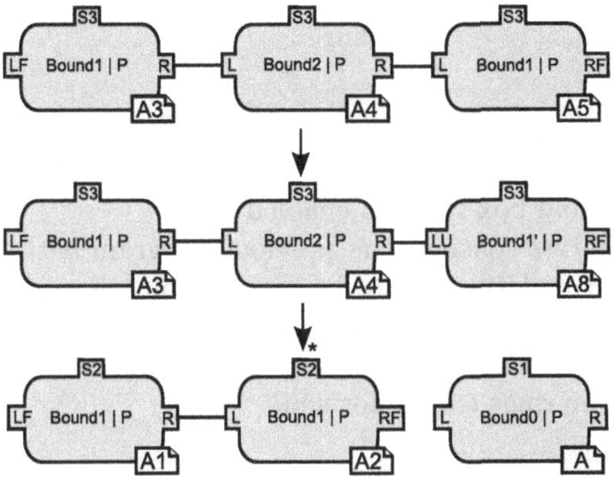

The last complex of the picture shows how the remaining chain also returns back in a consistent configuration, where the boxes have the correct internal state and their interfaces `num` have the correct structures. In order to obtain this result, we have to modify the processes `Bound1` and `Bound2` as well. In particular, `Bound2` becomes a sum of processes, where one process controls the binding status of the right interface and recognizes when there is an unbinding:

```
if not(right,bound) then
    ch(right,RF).left!(dec).b1!().DecreaseCounter
endif
```

and another process does the same for the left interface (the `DecreaseCounter` process is very similar to the `IncreaseCounter` one but decrements the structure of the `num` interfaces). The process propagates on the left a name `dec`. This means that wrt the program without reversibility here we propagate in the complex not only a signal, but names `inc` (increasing of one in the length) and `dec` (decreasing of one in the length). All the processes in `Bound1` and `Bound2` in charge of capturing and processing the communications are adapted accordingly. Indeed, in the `Bound2` sum, we have a process in charge to process and propagate an output from the right:

```
right?(act).left!(act).(
    act!().b2!()
    | inc?().IncreaseCounter + dec?().DecreaseCounter
)
```

and another for processing and propagating an output from the left. Notice that, depending on the received name, the increasing or decreasing process is enabled and process Bound2 is enabled again. Similarly, Bound1 has an alternative in charge to process and propagate an output from the right

```
if (right,R) then
   right?(act).(
       act!().b1!()
       | inc?().IncreaseCounter + dec?().DecreaseCounter
   )
endif
```

and another from the left interface. Also in this case, depending on the received name, the increasing or decreasing process is enabled and the process Bound1, representing the actual box state, is enabled again.

Also in this case we consider the previous program with N equal to 4 and we run stochastic simulations with an initial population of 1000 boxes A. We can observe the average dynamics and standard deviation reported in Fig. 7. Note that by adding reversibility we have that the number of rings continue to increase, till no more rings can be formed.

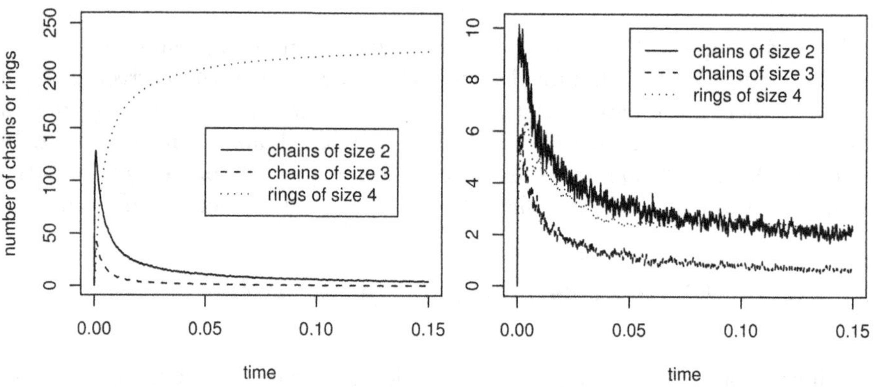

Fig. 7. Average dynamics and standard deviation of 100 stochastic simulations regarding the formation of rings of size 4 with reversibility. Each simulation takes in average around 10 seconds.

5 Conclusion

We investigated the usage of BlenX to program molecular self-assembly. We started by introducing programs that describe the formation of protein filaments and protein trees. Then, we described how BlenX can be used to introduce controls over the growth structure of trees. We developed a design pattern and used it to program two different controls over the formation of trees having branching depth less than five and branching distance equal to or greater than four.

Then, we showed how the proposed design pattern allows us to compose the two control programs exploiting the compositional nature of BlenX. In the second part of the paper we investigated also how to program the formation of ring populations of fixed size, with and without reversibility in filament formations. We used a different programming style wrt the initial programs, just to highlight the flexibility of the language.

This work shows the potential of BlenX in modelling these kind of scenarios. We plan to use this work as a starting point for modelling and studying real biological phenomena (e.g. actin polymerization, microtubules formation). Moreover, tools for making easier the visualization and interpretation of the output of the simulations that generate large number of complexes [8] are under development.

Acknowledgments

We would like to thank Alida Palmisano and Michele Forlin for their precious help, comments and suggestions.

References

1. `http://www.cosbi.eu/downloads/attachment/Models.tar.gz`
2. Alberts, B., Johnson, A., Lewis, J., Raff, M., Roberts, K., Walter, P.: Molecular biology of the cell (IV ed.). Garland science (2002)
3. Cardelli, L.: On process rate semantics. Theor. Comput. Sci. 391(3), 190–215 (2008)
4. Cardelli, L., Caron, E., Gardner, P., Kahramanoğulları, O., Phillips, A.: A Process Model of Actin Polymerisation. Electronic Notes in Theoretical Computer Science 229(1), 127–144 (2009)
5. Carlsson, A.E., Wear, M.A., Cooper, J.A.: End versus side branching by Arp2/3 complex. Biophysical journal 86(2), 1074–1081 (2004)
6. Curien, P., Danos, V., Krivine, J., Zhang, M.: Computational self-assembly. Theor. Comput. Sci. 404(1-2), 61–75 (2008)
7. Danos, V., Feret, J., Fontana, W., Harmer, R., Krivine, J.: Rule-based modelling of cellular signalling. In: CONCUR, pp. 17–41 (2007)
8. Dematté, L., Larcher, R.: Custom visualization of biological structures: an application to BlenX complexes. Technical Report TR-16-2009 CoSBi (2009)
9. Dematté, L., Priami, C., Romanel, A.: The Beta Workbench: a computational tool to study the dynamics of biological systems. Brief. Bioinform. 9(5), 437–449 (2008)
10. Dematté, L., Priami, C., Romanel, A.: The BlenX Language: A Tutorial. In: Bernardo, M., Degano, P., Zavattaro, G. (eds.) SFM 2008. LNCS, vol. 5016, pp. 313–365. Springer, Heidelberg (2008)
11. Drexler, E.K.: Nanosystems: molecular machinery, manufacturing, and computation. Wiley, Chichester (1992)
12. Fass, J., Pak, C., Bamburg, J., Mogilner, A.: Stochastic simulation of actin dynamics reveals the role of annealing and fragmentation. Journal of theoretical biology 252(1), 173–183 (2008)
13. Friedman, N., Linial, M., Nachman, I., Pe'er, D.: Using Bayesian networks to analyze expression data. Journal of Computational Biology 7(3), 601–620 (2000)

14. Gillespie, D.T.: Exact stochastic simulation of coupled chemical reactions. Journal of Physical Chemistry 81(25), 2340–2361 (1977)
15. Hasty, J., McMillen, D., Collins, J.J.: Engineered gene circuits. Nature 420(6912), 224–230 (2002)
16. Heiner, M., Gilbert, D., Donaldson, R.: Petri nets for systems and synthetic biology. In: Bernardo, M., Degano, P., Zavattaro, G. (eds.) SFM 2008. LNCS, vol. 5016, pp. 215–264. Springer, Heidelberg (2008)
17. Kauffman, S.A.: Metabolic stability and epigenesis in randomly constructed genetic nets. Journal of theoretical biology 22(3), 437–467 (1969)
18. Klavins, E.: Automatic synthesis of controllers for assembly and formation forming. In: International Conference on Robotics and Automation (2002)
19. Kuttler, C., Lhoussaine, C., Nebut, M.: Rule-based modeling of transcriptional attenuation at the tryptophan operon. In: Formal Methods in Molecular Biology (2009)
20. Lodish, H.F.: Molecular cell biology. W.H. Freeman, New York (2003)
21. Matzavinos, A., Othmer, H.G.: A stochastic analysis of actin polymerization in the presence of twinfilin and gelsolin. Journal of theoretical biology 249(4), 723–736 (2007)
22. Milner, R.: Communicating and mobile systems: the π-calculus. Cambridge Universtity Press, Cambridge (1999)
23. Mogilner, A., Edelstein-Keshet, L.: Regulation of actin dynamics in rapidly moving cells: a quantitative analysis. Biophysical journal 83(3), 1237–1258 (2002)
24. Nagpal, R.: Programmable self-assembly using biologically-inspired multiagent control. In: AAMAS 2002: Proceedings of the first international joint conference on Autonomous agents and multiagent systems, pp. 418–425 (2002)
25. Neumann, A.: Graphical gaussian shape models and their application to image segmentation. IEEE Transactions on Pattern Analysis and Machine Intelligence 25(3), 316–329 (2003)
26. Pantaloni, D., Boujemaa, R., Didry, D., Gounon, P., Carlier, M.F.: The Arp2/3 complex branches filament barbed ends: functional antagonism with capping proteins. Nature cell biology 2(7), 385–391 (2000)
27. Phillips, A., Cardelli, L.: Efficient, Correct Simulation of Biological Processes in the Stochastic Pi-calculus. In: Calder, M., Gilmore, S. (eds.) CMSB 2007. LNCS (LNBI), vol. 4695, pp. 184–199. Springer, Heidelberg (2007)
28. Prandi, D., Priami, C., Quaglia, P.: Communicating by compatibility. JLAP 75, 167 (2008)
29. Prandi, D., Zunino, R.: Computing by Complexes. Technical Report TR-11-2009 CoSBi (2009)
30. Priami, C.: Stochastic π-calculus. The Computer Journal 38(6), 578–589 (1995)
31. Priami, C., Quaglia, P., Romanel, A.: BlenX Static and Dynamic Semantics. In: Bravetti, M., Zavattaro, G. (eds.) CONCUR 2009. LNCS, vol. 5710, pp. 37–52. Springer, Heidelberg (2009)
32. Priami, C., Regev, A., Shapiro, E.Y., Silverman, W.: Application of a stochastic name-passing calculus to representation and simulation of molecular processes. Inf. Process. Lett. 80(1), 25–31 (2001)
33. Romanel, A., Priami, C.: On the decidability and complexity of the structural congruence for beta-binders. Theor. Comput. Sci. 404(1-2), 156–169 (2008)
34. Sangiorgi, D., Walker, D.: The π-calculus: a Theory of Mobile Processes. Cambridge University Press, Cambridge (2003)
35. Stock, A.M., Goudreau, P.N., Robinson, V.L.: Two-component signal transduction. Annu. Rev. Biochem. 69, 183–215 (2000)
36. Yin, P., Choi, H.M.T., Calvert, C.R., Pierce, N.A.: Programming biomolecular self-assembly pathways. Nature 451(318-322) (2008)

A Operational Semantics

In the paper BlenX has been presented using its *concrete syntax*. Its operational semantics is instead presented by relying on an *abstract syntax*.

Let $\mathcal{N}$ be a countably infinite set of *names* (ranged over by x, y, n, x_1, x', $\cdots$) and let $\mathcal{T}$ be finite set of *sorts* (ranged over by Δ, Γ, Δ', Δ_0, $\cdots$) such that $\mathcal{T} \cap \mathcal{N} = \emptyset$. Moreover, let $\delta : \mathcal{N} \to \mathbb{R}$ a function that associates stochastic rates to names and $\alpha : \mathcal{T}^2 \to \mathbb{R}^3$ a symmetric function, called *affinity*, that associates triples of stochastic rates to sort pairs. The abstract syntax of BlenX is given in the following:

$$
\begin{array}{llll}
B ::= \mathsf{Nil} \;\big|\; I[P]_n \;\big|\; B \parallel B & & M ::= \mathsf{nil} \;\big|\; *\,\pi.\,P \;\big|\; \pi.\,P \;\big|\; M + M \;\big|\; \langle C \rangle M \\
I \;::= S(x,r,\Gamma) \;\big|\; S(x,r,\Gamma)\; I & & \pi \;\;::= x?y \;\big|\; x!y \;\big|\; \mathsf{ch}(x,\Gamma,r) \\
S \;::= \oplus \;\big|\; \otimes & & C \;::= (x,\Gamma) \;\big|\; (x,S) \;\big|\; op_u\; C \;\big|\; C\; op_b\; C \\
P \;::= M \;\big|\; P|P & &
\end{array}
$$

A BlenX system, written (B, ξ), is a pair made up of a bio-process B and an environment ξ. We denote with $\mathcal{S}$ the set of all possible systems (ranged over by S, S', S_1, $\cdots$).

Bio-processes are generated by the non-terminal symbol B of the grammar. A bio-process can be either empty (Nil), a *box* ($I[P]_n$) or the parallel composition of bio-processes ($B||B$). In the definition of box $I[P]_n$, I represents its interaction capabilities, P its internal engine, and n is used as an *identifier* to address the box at hand.

I is a non-empty string of *interfaces* of the form $S(x, r, \Delta)$, where S denotes the state of the interface, which can be either *free* ($\oplus$) or *bound* ($\otimes$), the name x is the *subject* of the interface, and Δ is a sort representing the structure of the interface. We sometimes use *active* as synonymous of free, and *complexed* as synonymous of bound. The subject x of the interface $S(x, r, \Delta)$ of a box $S(x, r, \Delta)\, I[P]_n$ is a binding for the free occurrences of x in P. We write $I = I_1 I_2$ or $I = I_1 I_2 I_3$ to mean that I is given by the juxtaposition of other interfaces. The metavariables I^*, I_1^*, $\cdots$ stay for either an interface or the empty string, and the above notation for juxtaposition is extended to these metavariables in the natural way. Moreover, we use the functions $\mathsf{sub}(I)$ and $\mathsf{sorts}(I)$ to extract from I the set of its subjects and the set of its sorts, respectively. We say that a box $I[P]_n$ is well-formed if all the subjects and sorts of the interfaces composing I are distinct. Moreover, we define an equivalence relation $\doteq$ on boxes that allows reasoning up-to renaming of interface subjects permitting a more natural definition of the operational semantics.

Definition 1. *We denote with* $\doteq$ *the smallest equivalence relation on boxes that satisfies the axiom:* $I_1^*\, S(x, r, \Delta)\, I_2^*[P]_z \doteq I_1^*\, S(y, r, \Delta)\, I_2^*[P\{y/x\}]_z$ *provided* $y \notin fn(P) \cup sub(I_1^*\, S(x, r, \Delta)\, I_2^*)$.

Definition 2. *Boxes related by means of $\doteq$ are identified.*

The non-terminal symbol P generates *processes* (ranged over by P, P', P_1, $\cdots$). A process can be either the parallel composition of two processes ($P|P$), or the replication of an action-guarded process ($*\pi.\,P$), or the empty process (nil), or an action-guarded process ($\pi.\,P$), or the non-deterministic choice of action-guarded or replicated action-guarded processes ($M + M$). Replications, deadlock processes, action-guarded processes and non-deterministic choices can be guarded by a condition C ($\langle C \rangle M$). The condition C is a boolean expression that controls the execution of the process M. Unary and binary operators used in conditions are such that $op_u \in \{\neg\}$ and $op_b \in \{\wedge, \vee\}$, while atoms refer to interfaces via their unique subjects and check whether they have a given sort (e.g Δ) or whether they are in a given state (e.g. $\otimes$). Given an interface I, if C evaluates to *true* on I (denoted with $[\![C]\!]_I = tt$), then M can fire. Considering that the only binder in the definition of processes is the name y in $x?y$, function $\mathsf{fn}(P)$ returns the free names of P and is defined in the usual way.

Boxes can be bound the one with the other through their interfaces to form complexes which are best thought of as graphs with boxes as nodes. Intuitively, the *environment* component ξ of the BlenX system (B, ξ) is used to record these bindings. In detail, ξ is a set of pair sets of the shape $\{\Delta_1 n_1, \Delta_2 n_2\}$ meaning that the two boxes addressed by n_1 and n_2 are bound together through the interfaces with sorts Δ_1 and Δ_2, respectively.

Below, the typical post-fixed notation $\{a/b\}$ is used to mean the substitution of the entity b with the entity a, with the usual conventions about renaming and α-conversion when the substitution is applied to processes. This notation is naturally extended to other domains, e.g. to sets. For example, $\xi\{\Delta m/\Gamma n\}$ denotes the substitution of the occurrences of Γn with Δm in ξ.

Definition 3. *A BlenX system (B, ξ) is well-formed iff (1) all the boxes composing B are well-formed, (2) all the boxes identifiers are distinct, (3) each element Δx in the environment ξ appears only once in ξ and (4) Δx appears in the environment ξ iff a box in B with identifier x and bound interface with sort Δ exists.*

We denote with $\overline{\mathcal{S}} \subset \mathcal{S}$ the set of well-formed systems. Given a well-formed BlenX system, its dynamics is formally specified by the operational semantics reported in Tab. 2. Let $\Theta = \{|_0, |_1, +_0, +_1\}^*$ the set of *addresses* with metavariable ϑ. The reduction relation $\xrightarrow{\omega}_r$ is defined using the auxiliary relation $\xrightarrow{\overline{\omega}}_r$ where $r \in \mathbb{R}_{\geq 0} \cup \{\infty\}$ and where $\omega \in \Omega$ and $\overline{\omega} \in \overline{\Omega}$ are *labels*. Labels in $\overline{\Omega}$ and Ω, defined in Tab.1, guarantee to identify reductions univocally permitting hence a correct generation of the underlying Continuous Time Markov Chain (CTMC).

The reduction relation $\xrightarrow{\overline{\omega}}_r$ describes the behaviour of single boxes, while reduction relation $\xrightarrow{\omega}_r$ describes the behaviour of systems.

Theorem 1. *Reduction relation $\xrightarrow{\omega}_r$ preserves well-formedness of BlenX systems.*

Table 1. Definition of labels

$\overline{\Omega} = \{\vartheta x?y \mid \vartheta \in \Theta \wedge x, y \in \mathcal{N}\} \cup$ $\{\vartheta x!y \mid \vartheta \in \Theta \wedge x, y \in \mathcal{N}\} \cup$ $\{\vartheta(\Delta, \Gamma) \mid \vartheta \in \Theta \wedge \Delta, \Gamma \in \mathcal{T}\} \cup$ $\{\vartheta(\vartheta_1, \vartheta_2) \mid \vartheta, \vartheta_1, \vartheta_2 \in \Theta\}$	$\Omega = \{n\vartheta \mid \vartheta \in \Theta\} \cup \{n\vartheta(\vartheta_1, \vartheta_2) \mid \vartheta, \vartheta_1, \vartheta_2 \in \Theta\} \cup$ $\{n\Delta \mid \Delta \in \mathcal{T}\} \cup \{n\underline{\Delta} \mid \Delta \in \mathcal{T}\} \cup$ $\{n\Delta\vartheta!y \mid \vartheta \in \Theta \wedge y, n \in \mathcal{N} \wedge \Delta \in \mathcal{T}\} \cup$ $\{n\Delta\vartheta?y \mid \vartheta \in \Theta \wedge y, n \in \mathcal{N} \wedge \Delta \in \mathcal{T}\} \cup$ $\{(n_1\Delta_1, n_2\Delta_2) \mid n_1, n_2 \in \mathcal{N} \wedge \Delta_1, \Delta_2 \in \mathcal{T}\} \cup$ $\{(n_1\underline{\Delta_1}, n_2\underline{\Delta_2}) \mid n_1, n_2 \in \mathcal{N} \wedge \Delta_1, \Delta_2 \in \mathcal{T}\} \cup$ $\{(n_1\vartheta_1, n_2\vartheta_2) \mid \vartheta_1, \vartheta_2 \in \Theta \wedge n_1, n_2 \in \mathcal{N}\}$

Rules in Tab. 2 describe formally the intuitive semantics previously given in Sec. 2. The correspondence between concrete and abstract syntax is straightforward.

Rules (r1) and (r2) describe the execution of output and input prefixes, respectively. Rule (r3) describes how an interface sort can be changed through the corresponding action. If Γ does not clash with the sorts of the other interfaces in I, then the sort associated to x is turned into Γ; the label records information about the sorts involved in the change operation. Rule (r4) describes how guarded replication is unfolded, while (r5) describes *intra-communications*, i.e. communications between processes within the same box. Rule (r5) stores on the label information about the positions of input and output prefixes. Rule (r6) describes the behaviour of the conditional operator: if the process guarded by the condition is not in deadlock and the condition is true in I, then the process can proceed and the condition is deleted. Rules (r7-10) collect processes contexts and update labels in the appropriate way.

Rules (r11-14) describe how to lift from $\xrightarrow{\overline{\omega}}_r$ to $\xrightarrow{\omega}_r$ relations, i.e. from box to system level. Outputs are lifted only if the communication object y does not clash with the subjects of the interfaces in I. Since interface subjects are bindings for processes and can be substituted at hand using $\doteq$ relation, the side condition prevents from the extrusion of private information, hence preserving a consistent system configuration. Also outputs are lifted only if y does not clash with the subjects of the interfaces in I, preventing from name captures (i.e. no receiving name has to be captured by interface subjects). Note that since boxes are identified up to $\doteq$ relation, then there always exists a box configuration satisfying the condition. A change action is lifted by applying a substitution on the environment. In particular, since the interface could be involved in a binding, the environment is consistently updated by possibly refreshing the previous value of the interface with Γ. Here notice that the requirement about the freshness of Γ in rule (r3) guarantees the freshness of Γn in the updated environment. Note that for all these rules labels are updated in the proper way, by preserving and adding only relevant information.

Boxes can interact the one with the other in various ways: they can bind together, unbind, or communicate when bound. These interactions are based on the existence of the affinity function α which returns a triple of real values representing the binding, unbinding and inter-communication stochastic rates of the two argument values. We use $\alpha_b(\Delta,\Gamma)$, $\alpha_u(\Delta,\Gamma)$ and $\alpha_c(\Delta,\Gamma)$ to mean, respectively, the first, the second, and the third projection of $\alpha(\Delta,\Gamma)$. Rules (r15), (r16), (r20) and (r21) describe the dynamics of binding and unbinding, respectively. In both cases the modification of the binding state of the relevant interfaces is reflected in the interface markers, which are changed either from $\oplus$ to $\otimes$ or the other way round. Also, the association $\{\Delta_1 n_1, \Delta_2 n_2\}$ recording the actual binding is either added to the environment or removed from it. The third kind of interaction between boxes, called *inter-communication*, is ruled by (r19). This involves an input and an output action firable in two distinct boxes over interfaces with associated values Δ_1 and Δ_2. Information passes from the box containing the sending action to the box enclosing the receiving process. Here notice that inter-communication depends on the affinity of Δ_1 and Δ_2 rather than on the fact that input and output actions occur over exactly the same name. Inter-communication is enabled only under the proviso that the two interfaces are already bound together. Rules (r17) and (r18) collects bio-processes contexts.

Labels of binding, unbinding and inter-communication actions record boxes identifiers and, respectively, the sorts of the interfaces involved in the bindings or unbindings and the ϑ addresses of the inputs and outputs involved in the communications.

Given a well-formed BlenX system S, the following definitions (similarly to [3]) describe how to extract its Labeled Stochastic Transition Graph and its Continuous Time Markov Chain.

Definition 4. *Let $S \in \overline{\mathcal{S}}$. The Labeled Stochastic Transition Graph (LSTG) of S, referred as Φ^S, is obtained with the following procedure:*

$$\begin{aligned}
\Phi^S &= \textstyle\bigcup_n \Phi_n \ \textit{where} \\
\Phi_0 &= \{S \xrightarrow{\omega}_r S' \mid r > 0 \wedge \nexists \omega' \ s.t.\ S \xrightarrow{\omega'}_\infty S''\} \cup \{S \xrightarrow{\omega}_\infty S'\} \\
\Phi_{n+1} &= \{S \xrightarrow{\omega}_r S' \mid r > 0 \wedge \nexists \omega' \ s.t.\ S \xrightarrow{\omega'}_\infty S'' \wedge S \ \textit{is a state of}\ \Phi_n\} \cup \\
&\quad\ \{S \xrightarrow{\omega}_\infty S' \mid S \ \textit{is a state of}\ \Phi_n\}
\end{aligned}$$

Definition 5. *Let $S \in \overline{\mathcal{S}}$. The Continuous Time Markov Chain (CTMC) of S, referred as $|\Phi^S|$, is obtained from Φ^S in the following way:*

$$\begin{aligned}
|\Phi^S| = \{\ & S \xrightarrow{r} S' \mid S \neq S' \wedge \exists S \xrightarrow{\omega}_{s \in \mathbb{R}} S' \in \Phi^S \wedge r = \textstyle\sum r_i \ s.t.\ S \xrightarrow{\omega_i}_{r_i} S' \in \Phi^S\} \cup \\
\{\ & S \xrightarrow{\infty(n/m)} S' \mid n = \#\{\omega \mid S \xrightarrow{\omega}_\infty S' \in \Phi^S\} \wedge m = \#\{\omega \mid S \xrightarrow{\omega}_\infty S'' \in \Phi^S\}\}
\end{aligned}$$

Note that our CTMCs contain immediate transitions with associated probabilities. Given these probability, standard algorithms can be used to eliminate immediate transitions from the CTMC, preserving the probabilities of transitions and sojourn times.

It is important to underline that although the presented semantics is based on an individual level view, the BlenX stochastic simulator (the kernel of the

Beta Workbench [9]) implements an abstract machine which is based on a population level view and which dynamics is equivalent wrt the semantics given here. Following results presented in [33], it is indeed possible to collect and count in an efficient way structurally congruent boxes and isomorphic complexes. Note that whereas in the individual view identifiers are important to distinguish single boxes, in the population view they can be ignored. However, the individual level abstraction presented in this appendix is general enough to explain the dynamics of BlenX systems and to be used as a basis for other different abstraction views and implementation strategies, like the population level view which is used only as a mean to implement an efficient variant of the Gillespie's algorithm.

Since the Beta Workbench supports the population level view, the simulation results presented in the paper have to be interpreted following this abstraction view, in which BlenX system states are multisets of structurally congruent boxes and isomorphic complexes and time-courses refer to the cardinality of the classes of boxes and complexes in time. Note that this information can be extracted also from the states of the previous given CTMC definition by ignoring boxes identifiers.

Table 2. BlenX operational semantics

$$\text{(r1)}\ I[x!y.\,P]_n \xrightarrow{x!y}_0 I[P]_n \qquad \text{(r2)}\ I[x?y.\,P]_n \xrightarrow{x?z}_0 I[P\{z/y\}]_n$$

$$\text{(r3)}\ \frac{I = I_1^*\, S(x,r,\Delta)\, I_2^* \wedge \Gamma \notin \mathsf{sorts}(I)}{I[\mathsf{ch}(x,\Gamma,r).\,P]_n \xrightarrow{(\Delta,\Gamma)}_r I_1^*\, S(x,r,\Gamma)\, I_2^*[P]_n} \qquad \text{(r4)}\ \frac{I[\pi.\,P]_n \xrightarrow{\varpi}_r I'[P']_n}{I[*\pi.\,P]_n \xrightarrow{\varpi}_r I'[*\pi.\,P \mid P']_n}$$

$$\text{(r5)}\ \frac{\begin{array}{c} I[P_1]_n \xrightarrow{\vartheta_1 x!z}_{r_1} I[P_1']_n \wedge I[P_2]_n \xrightarrow{\vartheta_2 x?z}_{r_2} I'[P_2']_n \wedge \\ ((S(x,r,\Delta) \in I) \vee (x \notin \mathsf{sub}(I) \wedge \delta(x) = r)) \end{array}}{I[P_1 \mid P_2]_n \xrightarrow{(\vartheta_1,\vartheta_2)}_r I'[P_1' \mid P_2']_n} \qquad \text{(r6)}\ \frac{I[M]_n \xrightarrow{\varpi}_r I'[M']_n \wedge [\![C]\!]_I = tt}{I[\langle C\rangle M]_n \xrightarrow{\varpi}_r I'[M']_n}$$

$$\text{(r7)}\ \frac{I[M]_n \xrightarrow{\varpi}_r I[M']_n}{I[M+N]_n \xrightarrow{+_0\varpi}_r I[M']_n} \qquad \text{(r8)}\ \frac{I[N]_n \xrightarrow{\varpi}_r I[N']_n}{I[M+N]_n \xrightarrow{+_1\varpi}_r I[N']_n}$$

$$\text{(r9)}\ \frac{I[P]_n \xrightarrow{\varpi}_r I[P']_n}{I[P \mid Q]_n \xrightarrow{|_0\varpi}_r I[P' \mid Q]_n} \qquad \text{(r10)}\ \frac{I[Q]_n \xrightarrow{\varpi}_r I[Q']_n}{I[P \mid Q]_n \xrightarrow{|_1\varpi}_r I[P \mid Q']_n}$$

$$\text{(r11)}\ \frac{I_1^* \otimes(x,r,\Delta)\, I_2^*[P]_n \xrightarrow{\vartheta x!y}_r I[P']_n \wedge y \notin \mathsf{sub}(I)}{(I_1^* \otimes(x,r,\Delta)\, I_2^*[P]_n, \xi) \xrightarrow{n\Delta\vartheta!y}_r (I[P']_n, \xi)} \qquad \text{(r12)}\ \frac{I[P]_n \xrightarrow{\vartheta(\vartheta_1,\vartheta_2)}_r I[P']_n}{(I[P]_n, \xi) \xrightarrow{n\vartheta(\vartheta_1,\vartheta_2)}_r (I[P']_n, \xi)}$$

$$\text{(r13)}\ \frac{I_1^* \otimes(x,r,\Delta)\, I_2^*[P]_n \xrightarrow{\vartheta x?y}_r I[P']_n \wedge y \notin \mathsf{sub}(I)}{(I_1^* \otimes(x,r,\Delta)\, I_2^*[P]_n, \xi) \xrightarrow{n\Delta\vartheta?y}_r (I[P']_n, \xi)} \qquad \text{(r14)}\ \frac{I[P]_n \xrightarrow{\vartheta(\Delta,\Gamma)}_r I'[P']_n}{(I[P]_n, \xi) \xrightarrow{n\vartheta}_r (I'[P']_n, \xi\{\Gamma n/\Delta n\})}$$

$$\text{(r15)}\ (I_1^* \oplus(x,r,\Delta)\, I_2^*[P]_n, \xi) \xrightarrow{n\underline{\Delta}}_0 (I_1^* \otimes(x,r,\Delta)\, I_2^*[P]_n, \xi)$$

$$\text{(r16)}\ (I_1^* \otimes(x,r,\Delta)\, I_2^*[P]_n, \xi) \xrightarrow{n\Delta}_0 (I_1^* \oplus(x,r,\Delta)\, I_2^*[P]_n, \xi)$$

$$\text{(r17)}\ \frac{(B_1, \xi) \xrightarrow{\omega}_r (B_1', \xi')}{(B_1 \parallel B_2, \xi) \xrightarrow{\omega}_r (B_1' \parallel B_2, \xi')} \qquad \text{(r18)}\ \frac{(B_2, \xi) \xrightarrow{\omega}_r (B_2', \xi')}{(B_1 \parallel B_2, \xi) \xrightarrow{\omega}_r (B_1 \parallel B_2', \xi')}$$

$$\text{(r19)}\ \frac{(B_1, \xi) \xrightarrow{n_1\Delta_1\vartheta_1!z}_{r_1} (B_1', \xi) \wedge (B_2, \xi) \xrightarrow{n_2\Delta_2\vartheta_2?z}_{r_2} (B_2', \xi) \wedge \{\Delta_1 n_1, \Delta_2 n_2\} \in \xi}{(B_1 \parallel B_2, \xi) \xrightarrow{(n_1\vartheta_1, n_2\vartheta_2)}_{\alpha_c(\Delta_1,\Delta_2)} (B_1' \parallel B_2', \xi)}$$

$$\text{(r20)}\ \frac{(B_1, \xi) \xrightarrow{n_1\underline{\underline{\Delta_1}}}_{r_1} (B_1', \xi) \wedge (B_2, \xi) \xrightarrow{n_2\underline{\underline{\Delta_2}}}_{r_2} (B_2', \xi)}{(B_1 \parallel B_2, \xi) \xrightarrow{(n_1\underline{\underline{\Delta_1}}, n_2\underline{\underline{\Delta_2}})}_{\alpha_b(\Delta_1,\Delta_2)} (B_1' \parallel B_2', \xi \cup \{\{\Delta_1 n_1, \Delta_2 n_2\}\})}$$

$$\text{(r21)}\ \frac{(B_1, \xi) \xrightarrow{n_1\Delta_1}_{r_1} (B_1', \xi) \wedge (B_2, \xi) \xrightarrow{n_2\Delta_2}_{r_2} (B_2', \xi)}{(B_1 \parallel B_2, \xi \cup \{\{\Delta_1 n_1, \Delta_2 n_2\}\}) \xrightarrow{(n_1\Delta_1, n_2\Delta_2)}_{\alpha_u(\Delta_1,\Delta_2)} (B_1' \parallel B_2', \xi)}$$

Rule-Based Modeling of Transcriptional Attenuation at the Tryptophan Operon

Céline Kuttler[1], Cédric Lhoussaine[1], and Mirabelle Nebut[1,2]

[1] University of Lille
[2] BioComputing group, LIFL & IRI (CNRS UMR 8022 & USR 3078)

Abstract. Transcriptional attenuation at E.coli's tryptophan operon is a prime example of RNA-mediated gene regulation. In this paper, we present a discrete stochastic model of the fine-grained control of attenuation, based on chemical reactions. Stochastic simulation of our model confirms results that were previously obtained by master or differential equations. Our approach is easier to understand than master equations, although mathematically well founded. It is compact due to rule schemas that define finite sets of chemical reactions. Object-centered languages based on the π-calculus would yield less intelligible models. Such languages are confined to binary interactions, whereas our model heavily relies on reaction rules with more than two reactants, in order to concisely capture the control of attenuation.

1 Introduction

Transcriptional attenuation is a control mechanism deployed by bio-synthetic operons across bacterial species [14,15,38]. Operons are sequences of jointly transcribed genes, bio-synthetic operons encode enzymes for the synthesis of amino acids. Attenuation prematurely interrupts an ongoing round of the operon's transcription, in situations where the environment already contains a high concentration of the corresponding amino acid. Summarizing, it works as follows. First, the amino acid concentration determines the speed at which a ribosome translates the nascent messenger RNA (mRNA). Second, the ribosome's position controls how the mRNA folds into a two-dimensional structure. Finally, the mRNA structure sets the end point of transcription.

Although attenuation has been investigated within bacterial systems since the 1970s [19,36], it attracted significantly less interest than the control of transcription initiation, that is mediated by DNA binding proteins. This changed in the 2000s after the discovery of regulatory mechanisms in higher organisms that exploit RNA properties [4]. Quantitative investigations of RNA-mediated regulation gained momentum for therapeutic approaches and synthetic biology [3].

E.coli's tryptophan (*trp*) operon is the best understood bio-synthetic operon. It allows the bacterium to synthesize the amino acid tryptophan upon need. Tryptophan regulation in *E.coli* relies on two further mechanisms beyond transcriptional attenuation, that are not considered in this paper.

C. Priami et al. (Eds.): Trans. on Comput. Syst. Biol. XII, LNBI 5945, pp. 199–228, 2010.

Santillan and Zeron (2004) [30] modeled all three levels of *trp* regulation in *E. coli* through delay differential equations (DDE), without investigating attenuation in detail. DDEs are usually directly derived from informal biochemical reactions. The main drawback of such deterministic models is that they only provide observations of average behavior. In particular, they do not account for possible stochastic noise from which multi-modal states may arise. In that case, the average behavior does not correspond to any of the actual states. Since regulatory systems involve few biological entities, a criterion known to increase stochastic effects, one may wonder if the deterministic assumption is appropriate regarding *E.coli*'s *trp* operon. This calls for stochastic modeling. The first stochastic treatment of attenuation at the *trp* operon indeed dates back to 1977 [34].

Elf and Ehrenberg (2005) [11] analyze the sensitivity of attenuation through probability functions and, more generally, discrete master equations. This approach benefits from a rich probability theory that gives valuable insights and measurement capabilities. However, apart from rare exceptions, master equations can only be evaluated numerically, and not solved symbolically. Each biological system requires an ad-hoc master equation or probability function that is usually hard to design from the mechanistic intuition of the system.

Discrete event models for stochastic simulation are commonly described by chemical reactions. These can be studied within formal rule-based modeling languages [6,7,8,21], where molecular systems are understood as multisets of molecules, that are rewritten by chemical reactions. Reaction speeds are derived from rate constants and cardinalities of sets. The stochastic semantics of chemical reactions is given in terms of continuous time Markov chains (CTMCs). The algorithm of Gillespie (1976) [12] allows direct stochastic simulation, starting from a given multiset of molecules and a set of chemical reactions. Rule-based models are intuitive in the sense that they describe molecular interactions and are simpler to modify and extend than models based on classical mathematical functions.

Certain authors [5,26] support the idea that *binary reactions* are sufficient to represent biochemical knowledge. They do so to advocate formal object-centric representations that are confined to binary interactions, namely recent languages based on the stochastic π-calculus [10,17,22,25,28]. However, rewriting n-ary to binary reactions is tedious and requires sufficient expressiveness of formal languages. Sequences of reactions need to be executed within atomic transactions, so that no other interactions intervene.

Contribution. In this work, we present the first formal rule-based stochastic model of transcriptional attenuation at *E. coli*'s tryptophan operon. We cover a similar extent of biological knowledge as Elf and Ehrenberg (2005) [11], but take a different methodological approach in the tradition of stochastic models of gene expression [1]. Our representation is based on chemical reactions. We use 71 reactions to faithfully cover the *trp* operon's control by attenuation, summarizing the rich narrative account in the biological literature [13,20,33]. We obtain a concise description by two ingredients, 13 *rules schemas* (introduced in

Section 3) from which we generate our 71 chemical reactions, and *n-ary chemical reactions.*

By means of *rule schemas*, we represent finite sets of chemical reactions in a compact manner, which differ only in the choice of certain molecule parameters that are quantified over e.g. folding or binding state, or location. This idea is well known from logic programming [23], unification grammars [31] or term rewriting [2].

N-ary reactions are indispensable to intelligible representations of the *trp* attenuator, as our work indicates. We hypothesize the same holds for many other biological cases. By n-ary reactions, we refer to rules with three or more inputs, as opposed to binary reactions. They allow to incorporate *global control* into models, of which our work distinguishes three categories.

First are *conditions for rule application.* Here, one among a rule's multiple inputs is neither consumed nor modified by the rule's application. However if this molecule wasn't available, the rule could not be applied. For instance, we use this mechanism to model that transcription only aborts if the nascent transcript has folded into the termination hairpin, i.e. the corresponding rule checks this later's presence.

In the second category, an actor undergoes a state change as a *side effect* of the reaction between two others. For instance, after translation has progressed beyond a certain threshold, a state change hinders the corresponding mRNA components to form hairpins.

The third category allows to *switch between abstraction levels*: upon application of a rule, one actor is replaced by the enumeration of its individual constituents, or vice versa. We use this mechanism for dedicated control segments of mRNA that can interact as a whole with other segments, or be processed stepwise. Depending on the circumstances our model opts for their representation either as a whole, or as the sequence of their constituents.

Paper outline. We review the biological background in Sect. 2, introduce our rule-based language in Sect. 3, and review related languages in Sect. 3.4. As a first example, we model the multi-step race between transcription and translation by four rule schemas in Sect. 4.1. We quantitatively investigate the hypersensitivity of this *basic model* by simulation with the Kappa Factory[1] [8]. The *concurrent elongation model* of Sect. 4.2 extends, it explicitly renders the simultaneity of transcription and translation of the same mRNA. We use it to investigate the quantitative impact on the multi-step race when only part of the mRNA is present at the beginning of the simulation. In Sect. 5, we present our model of the detailed attenuation mechanism at *E.coli*'s *trp* operon. We qualitatively reproduce and confirm results of Elf and Ehrenberg [11] regarding the probability of uninterrupted transcription into the full operon as a function of the rate of *trp*-codon translation, and discuss the quantitative differences between our results and theirs.

[1] This preliminary implementation of the κ-calculus was available to us as beta-testers, the web-based tool *Cellucidate* (`http://cellucidate.com`) is its successor.

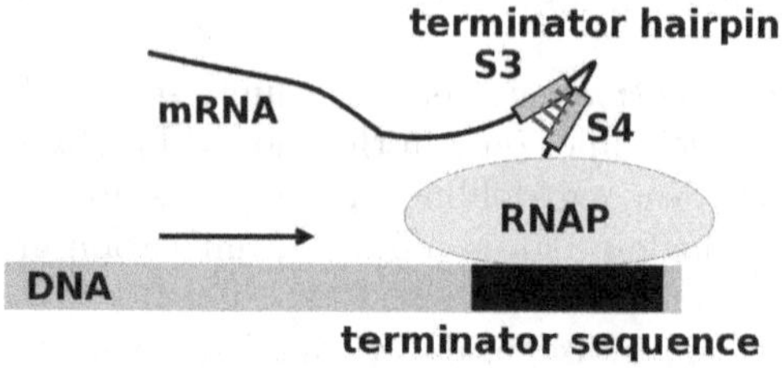

Fig. 1. Transcription terminates if the most recent portion of mRNA is folded into a hairpin, when RNAP reaches a terminator DNA sequence

2 Transcriptional Attenuation

Transcriptional attenuation prematurely interrupts a gene's ongoing transcription, or that of an *operon*, when the cell does not actually need the corresponding proteins. *E.coli*'s *trp* operon encodes enzymes for the biosynthesis of the amino acid tryptophan (Trp). Their production is attenuated if the bacterium's environment provides sufficient amounts of tryptophan to feed on.

In this section, we first review the principles of gene expression in bacteria. Then, we introduce ribosome-mediated transcriptional attenuation, a regulatory mechanism used across bacterial species, and how specifically it functions at *E.coli*'s tryptophan (*trp*) operon [13,36,37]. We omit certain aspects documented in the biological literature that do not enter our formal model presented in this paper, such as the role of transfer RNA in translation, or the redundancy of the genetic code.

Transcription copies information content from a DNA sequence into an mRNA sequence. It is carried out by an enzyme called RNA polymerase. RNAP initiates its work by binding to a distinguished short DNA sequence which indicates the beginning of a gene (or operon), from where RNAP starts assembling an mRNA molecule. In the following *elongation phase* RNAP advances stepwise over DNA, extending the growing mRNA nucleotide by nucleotide, progressing at an average rate of 50 nucleotides per second. Transcription *terminates* when RNAP encounters a terminator sequence on DNA, if an additional condition is then fulfilled. Figure 1 illustrates this additional condition, that depends on a property of the mRNA being transcribed. The linear mRNA sequence can fold into stable secondary structures, which due to their shape are called *hairpins*. In order for transcription to terminate, while the RNAP encounters a terminator sequence on DNA, the most recent portion of the transcript must be folded into a hairpin.

Translation reads out an mRNA molecule into the corresponding sequence of amino acids. It initiates with the binding of a *ribosome* to the free end of an mRNA, the other end of which is still being elongated by an RNAP. The ribosome advances along the mRNA towards the RNAP in steps of *codons*, which

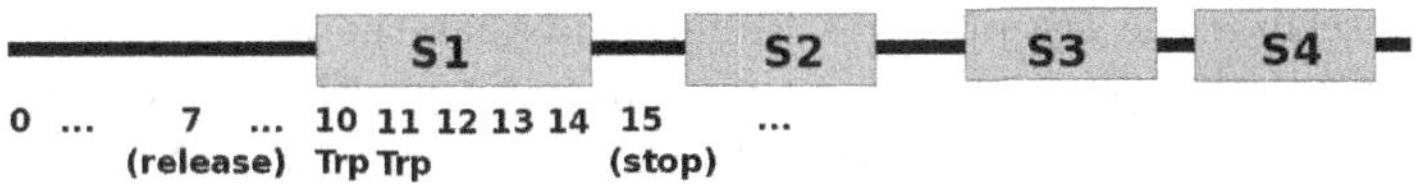

Fig. 2. Leader region of the *trp* mRNA. Adjacent pairs of the four segments S_1 to S_4 fold into alternative hairpins. The *anti-terminator* hairpin $S_2 \cdot S_3$ promotes transcription of the operon, whereas the *terminator* $S_3 \cdot S_4$ aborts it.

are words of three mRNA nucleotides. For each codon, the ribosome adds the corresponding amino acid to the growing sequence. These amino acid sequences later fold into three dimensional structures known as proteins. While the average rate of translation is 15 codons per second, each step of the ribosome is actually limited by the abundance of the currently required amino acid. The ribosome slows down on codons for which the corresponding amino acid is in short supply.

Transcriptional attenuation subtly couples the termination of an ongoing round of transcription to the translation efficiency of the first part of the nascent mRNA, where this latter is limited by a critical amino acid (tryptophan, for the *trp* operon). The so-called *leader* sequence consists of the operon's first few dozen nucleotides. Attenuation boils down to a *race* between the RNAP transcribing the leader DNA, and the ribosome translating the leader mRNA. In a nutshell, the attenuation race is as follows. If the amino acid of interest is *abundant*, the ribosome advances at its maximal speed, and the terminator hairpin forms. Transcription then aborts. Conversely if the critical amino acid is *rare*, the ribosome stalls early within the leader. The stalled ribosome inhibits the terminator hairpin, hence the RNAP wins the race, and transcription continues into the operon's protein coding regions.

Trp leader architecture. The leader's architecture is fundamental to attenuation at *E.coli*'s tryptophan operon. We distinguish four segments S_1 to S_4 within the leader mRNA, see Fig. 2. Each pair of adjacent segments folds into a hairpin, if neither of the required segments is masked by a ribosome. Three different secondary structures can occur within the leader mRNA of the *trp* operon. They are named by their respective roles in attenuation. The pairing of S_1 with S_2 represents the *pause* hairpin, that between S_2 and S_3 is the *anti-terminator* hairpin, and the *terminator* shown in Fig. 1 is the pairing of S_3 with S_4.

Hairpin co-occurrence and mutual exclusion. Each leader segment can only participate in one hairpin at the same time. Most importantly, the anti-terminator prohibits the terminator hairpin by sequestering S_3. Because both require S_2, the pause hairpin excludes the anti-terminator. On the other hand, the pause hairpin ($S_1 \cdot S_2$) and the terminator ($S_3 \cdot S_4$) can co-occur, since they do not compete for a shared segment. With respect to our model, we need to mention that hairpin formation is fater than any other reaction in the system. It is also important to bear in mind that the segments become available one by one while

RNAP transcribes the *trp* operon's leader, and that the leader transcript progressively forms hairpins whenever the ribosome's position allows. Hence, the leader mRNA never indeed remains unfolded, as simplifyingly shown in Fig. 2.

The role of hairpins. The impact of hairpins on transcription is significant, they determine *pausing* of the RNAP and *termination* of transcription. As opposed to this, mRNA hairpins do not impair translation. A translating ribosome disrupts hairpins along its way, without significantly slowing down. We now detail on the pause and terminator hairpins at *E.coli*'s *trp* operon.

Pause hairpin. After RNAP has transcribed the segments S_1 and S_2, it remains stalled on a strong DNA *pause site*, while the mRNA rapidly folds into the pause hairpin. This combination resembles the conditions for transcription termination, however, it is reversible: RNAP resumes transcriptions after a ribosome has arrived along the transcript and disrupted the pause hairpin. Let us consider the details of this *initial configuration for the attenuation race* in Fig. 3 (left). RNAP is stalled on the pause site on DNA, more precisely on the nucleotide that we refer to as DNA_0. It has so far transcribed the leader up to and including its segments S_1 and S_2, that have folded into the pause hairpin. The approaching ribosome disrupts the pause hairpin with its step onto the 7th codon of the leader mRNA, which is the first step from the initial conformation in Fig. 3. The attenuation race now starts.

Terminator hairpin. The DNA leader of the *trp* operon contains a *terminator sequence*, just after the portion that encodes S_4, see Fig. 1. When RNAP arrives here, the terminator mRNA hairpin can form. The combination of terminator mRNA hairpin and terminator DNA sequence aborts transcription. However if the anti-terminator is already present when S_4 is completed - it can appear as soon as S_2 and S_3 have been transcribed - the terminator is prevented. In this case RNAP continues unhindered through the terminator sequence, and reaches the enzyme-coding region of the operon (the *structural genes*).

Trp codons within the leader mRNA. We have not yet mentioned the codons 10 and 11 of the leader mRNA (see Fig. 2), the translation of which each requires one tryptophan molecule. These two *control* codons determine the outcome of attenuation race. They act as sensors for the tryptophan concentration, and determine the speed of the ribosome's forward movement.

If tryptophan is in *rare*, the ribosome stalls on the control codons, hence its footprint does not advance far enough to mask the second segment. Soon later, the anti-terminator hairpin forms between S_2 and S_3, and transcription continues into the structural genes. This is depicted as *read-through configuration* in Fig. 3.

Conversely if tryptophan is *abundant*, the ribosome efficiently translates through the control codons. From the time point the ribosome has reached the 13th codon, and until it dissociates from the stop codon 15, the ribosome's footprint masks S_2, which prevents the anti-terminator [29]. The ribosome's unbinding delay from the stop codon is generally one second, which is a considerably long time scale, compared to all other reactions in the system. While S_2 remains

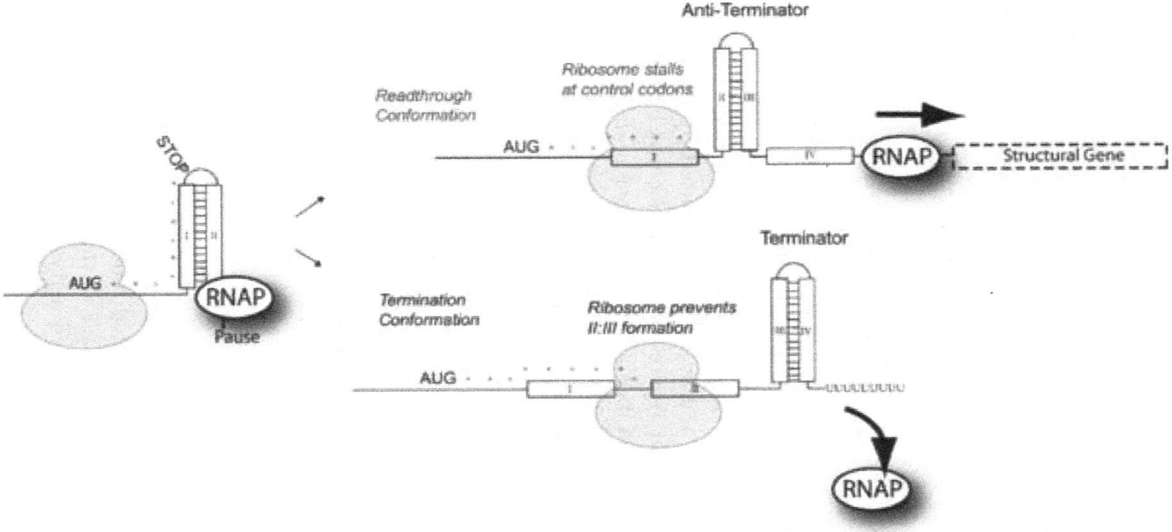

Fig. 3. Starting point and possible outcomes of the attenuation race at *E.coli*'s *trp* operon. *Initial conformation (left)*: RNAP is paused by the pause hairpin, awaiting to be released by the ribosome's next step. *Readthrough conformation*: when tryptophan supply is low, the ribosome stalls on the control codons, the anti-terminator hairpin forms and transcription continues into the operon. *Termination conformation*: when tryptophan supply is high, the ribosome rapidly translates over the control codons. Before it unbinds from the mRNA, the terminator hairpin forms and transcription aborts. Figure reproduced with permission from Elf and Ehrenberg (2005) [11].

blocked by the ribosome, RNAP continues transcription, it completes S_3 when reaching the 36th DNA nucleotide, and S_4 at the 47th DNA nucleotide. The terminator hairpin then forms and transcription aborts – this is the *termination configuration.*

Basal read-through due to premature ribosome release. A third possible outcome of the race is not covered by Fig. 3. When tryptophan supply is high, the ribosome occasionally dissociates from the stop codon sooner than expected. In that case S_3 can already have been transcribed, but S_4 not yet. Hence S_1, S_2 and S_3 are available at the same time. With equal probability, either the pause hairpin or the anti-terminator forms, and in case of the latter, transcription continues. This *basal read-through* of the operon has been experimentally observed for $10 - 15$ % of initiated transcripts when tryptophan is abundant [18].

3 Rule Schemas for Chemical Reactions

In this section we first provide a formal and minimal rule-based language tailored to our needs (Sect. 3.1). We define chemical reactions, that operate on multisets of complex molecules with attributes such as RNAP · DNA(23). Herein, the infix operator · indicates a complex between RNAP that is bound within a DNA sequence, more precisely at the position 23 stated by the attribute value of the DNA nucleotide. Other attributes of molecules could be the compartment of a molecule, or information on its states, for instance folding or binding state.

In Sect. 3.2, we present a language of *rules schemas*, that allows to define finite sets of chemical reactions in a compact manner. Rule schemas are like chemical reactions, except that attribute values are now extended to expressions with variables. All *variables are universally quantified over finite sets*, such that a rule schema defines a finite set of reactions. An example of a complex molecule is the term RNAP · DNA($x + 1$) where x is a variable with values in $\{0, \ldots, 50\}$. We introduce our language's stochastic semantics in Sect. 3.3.

As discussed in Sect. 3.4, more general ruled-based languages might have been used for our modeling study. The language in this paper is not intended as a contribution on its own, for the sake of its simplicity we however chose it for our presentation. Indeed, we relied on the software tool for another rule-based language to implement our models of Sect. 4 and 5.

3.1 Chemical Reactions

In order to define the syntax of attributed molecules, we fix a possibly infinite set of attribute values $\mathcal{C}$ and a finite set $\mathcal{N}$ of molecule names. We assume that each molecule name $N \in \mathcal{N}$ has a fixed arity $ar(N) \geq 0$, which specifies its number of attributes.

A *molecule* M, defined in Table 1, is a complex of attributed molecules. We write $M_1 \cdot M_2$ for the complex of M_1 and M_2. For instance, if RNAP, DNA $\in \mathcal{N}$ and $47 \in \mathcal{C}$ then RNAP · DNA(47) is a molecule complex consisting of an RNAP that is bound to the DNA nucleotide at position 47. A *chemical solution* S is a multiset of molecules.

Table 1. Chemical reactions where $N \in \mathcal{N}$, $c_1, \ldots, c_n \in \mathcal{C}$, $ar(N) = n$ and $k \in \mathbb{R}^+ \cup \{\infty\}$

Molecules	$M \in Mol$	::=	$N(c_1, \ldots, c_n) \mid M_1 \cdot M_2$
Solutions	$S \in Sol$	::=	$M \mid S_1 \,,\, S_2$
Reactions	$S_1 \rightarrow_k S_2$		

Table 2. Rule schemas where $x \in \mathcal{V}$, $c \in \mathcal{C}$, $f \in \mathcal{F}$, $e_1, \ldots, e_n \in Exp$, $N \in \mathcal{N}$, $ar(N) = n$, $D_1, \ldots, D_n \subseteq \mathcal{C}$ are finite sets, and $k \in \mathbb{R}^+ \cup \{\infty\}$

Expressions	$e \in Exp$	::=	$x \mid c \mid f(e_1, \ldots, e_n)$
Schematic molecules	$M \in SMol$	::=	$N(e_1, \ldots, e_n) \mid M_1 \cdot M_2$
Schematic solution	$S \in SSol$	::=	$M \mid S_1 \,,\, S_2$
Rule schema	$\forall x_1 \in D_1 \ldots \forall x_n \in D_n.\ S_1 \rightarrow_k S_2$ where $\mathcal{V}(S_1) \cup \mathcal{V}(S_2) \subseteq \{x_1, \ldots, x_n\}$		

A *chemical reaction* is a rule that rewrites a solution S_1 into a solution S_2, it is assigned a possibly infinite stochastic rate constant $k \in \mathbb{R}^+ \cup \{\infty\}$. For instance, the following reaction states that an RNAP bound to the DNA nucleotide at position 23 may advance to the DNA nucleotide at position 24. The speed of this reaction is 50 sec^{-1}:

$$\text{RNAP} \cdot \text{DNA}(23) \,,\, \text{DNA}(24) \rightarrow_{50} \text{DNA}(23) \,,\, \text{RNAP} \cdot \text{DNA}(24)$$

In order to represent transcription, one would need many similar rules for the many other DNA nucleotides with different positions. This motivates the introduction of rule schemas, that allow to define such sets of chemical reactions in a compact manner.

3.2 Rule Schemas

In order to define rule schemas for chemical reactions, we introduce *variables* x for attribute values and *expressions* such as $x + 1$, in order to compute corresponding attribute values. By *universal quantification over a finite set*, we generalize the above chemical reaction to the following rule schema:

$$\forall x \in \{1, \ldots, 49\}.$$
$$\text{RNAP} \cdot \text{DNA}(x) \,,\, \text{DNA}(x+1) \rightarrow_{50} \text{DNA}(x) \,,\, \text{RNAP} \cdot \text{DNA}(x+1) \qquad (0)$$

We thus need a set $\mathcal{V}$ of variables that are ranged over by x, and a finite set $\mathcal{F}$ of function symbols $f \in \mathcal{F}$ with arities $ar(f) \geq 0$. Furthermore, we assume an interpretation $[\![f]\!] : \mathcal{C}^{ar(f)} \rightarrow \mathcal{C}$ for every $f \in \mathcal{F}$. An *expression* e with values in $\mathcal{C}$ is a term with the abstract syntax given in Table 2. In our modeling case studies, we will assume that symbol $+ \in \mathcal{F}$ of arity two is interpreted as addition on natural numbers. We freely use infix syntax as usual, i.e. we write $e_1 + e_2$ instead of $+(e_1, e_2)$. Given a variable assignment $\alpha : \mathcal{V} \to \mathcal{C}$, every expression $e \in \mathit{Exp}$ denotes an element $[\![e]\!]_\alpha \in \mathcal{C}$ that we define as follows:

$$[\![c]\!]_\alpha = c \qquad [\![x]\!]_\alpha = \alpha(x) \qquad [\![f(e_1, \ldots, e_n)]\!]_\alpha = [\![f]\!]([\![e_1]\!]_\alpha, \ldots, [\![e_n]\!]_\alpha)$$

A *schematic molecule* M is like a molecule, except we now allow for expressions in attribute positions rather than attribute values only. A *schematic solution* $S \in \mathit{SSol}$ is a multiset of schematic molecules. As usual, we write $\mathcal{V}(S)$ for the set of variables that occur in molecules of S. A *rule schema* specifies the domains of variables occurring in the schematic solutions of the rule by universal quantification over finite sets.

For every variable assignment $\alpha : \mathcal{V} \to \mathcal{C}$ that maps variables to values in their domain, we can instantiate the rule schema to finitely many reactions. A schematic molecule M is mapped to a molecule $[\![M]\!]_\alpha \in \mathit{Mol}$. Similarly, schematic solutions $S \in \mathit{SSol}$ get instantiated to solutions $[\![S]\!]_\alpha \in \mathit{Sol}$:

$$\begin{aligned} [\![N(e_1, \ldots, e_n)]\!]_\alpha &= N([\![e_1]\!]_\alpha, \ldots, [\![e_n]\!]_\alpha) \\ [\![M_1 \cdot M_2]\!]_\alpha &= [\![M_1]\!]_\alpha \cdot [\![M_2]\!]_\alpha \\ [\![S_1 \,,\, S_2]\!]_\alpha &= [\![S_1]\!]_\alpha \,,\, [\![S_2]\!]_\alpha \end{aligned}$$

A rule schema is instantiated to a set of chemical reactions, by enumerating the chemical reactions for all variable assignments licensed by the quantifiers:

$$\begin{aligned} &[\![\forall x_1 \in D_1 \ldots \forall x_n \in D_n .\, S_1 \to_k S_2]\!] = \\ &\qquad \{[\![S_1]\!]_\alpha \to_k [\![S_2]\!]_\alpha \mid \alpha : \mathcal{V} \to \mathcal{C},\ \alpha(x_1) \in D_1, \ldots, \alpha(x_n) \in D_n\} \end{aligned}$$

3.3 Stochastic Semantics and Simulation

For the sake of completeness, we recall the stochastic semantics of chemical reactions and how to use them for the stochastic semantics with Gillespie's algorithm. This underlines that our biological modeling case studies are indeed expressed in a formal modeling language.

The semantics of a set of chemical reactions is a continuous time Markov chain (CTMC). Note that, for modeling convenience, we allow infinite rate constant ∞. Chemical reactions with infinite rates always have the highest priority and are executed immediately, that is without time delay. Such *extended* CTMCs with infinite rate constants can actually be converted to regular CTMCs by elimination of immediate transitions, while preserving sojourn time (i.e. how long the Markov chain stays in a given state) and probability transitions (that is, given a current state, the probability to make a transition to another given state)[2].

The states of the extended CTMCs are congruence classes $[S]_\equiv$ of chemical solutions S with respect to the least congruence relation $\equiv$ that makes complexation and summation associative and commutative:

$$\begin{array}{ll} M_1 \cdot M_2 \equiv M_2 \cdot M_1 & (M_1 \cdot M_2) \cdot M_3 \equiv M_1 \cdot (M_2 \cdot M_3) \\ S_1 \,,\, S_2 \equiv S_2 \,,\, S_1 & (S_1 \,,\, S_2) \,,\, S_3 \equiv S_1 \,,\, (S_2 \,,\, S_3) \end{array}$$

[2] For such an elimination procedure, see [22] and references therein.

Table 3. Stochastic semantics of chemical reactions with finite and infinite rate constants

$$\frac{L \subseteq \{1,\ldots,n\} \qquad \oplus_{i\in L} M_i \equiv S \qquad S \rightarrow_k S'}{\oplus_{i=1}^{n} M_i \xrightarrow[L]{k} S' \;,\; \oplus_{i\notin L} M_i}$$

$$\frac{r = \sum_{\{(L,k) \mid S \xrightarrow[L]{k} S_1 \equiv S'\}} k \qquad \neg\exists L \exists S''. S \xrightarrow[L]{\infty} S''}{S \xrightarrow{r} S'}$$

$$\frac{n = \sharp\{L \mid S \rightarrow S_1 \equiv S'\} \qquad m = \sharp\{L \mid S \rightarrow S_2\}}{S \xrightarrow{\infty(n/m)} S'}$$

In Table 3, we introduce transitions $S \xrightarrow[L]{k} S'$ stating that S can be reduced to S' by applying a chemical reaction with rate constant $k \in \mathbb{R}^+ \cup \{\infty\}$ to the subset of molecules in S with positions in L. Positions are the indices in multisets such as M_1 , ... , M_n that we also write as $\oplus_{i=1}^{n} M_i$. We next introduce two transitions

- $S \xrightarrow{r} S'$, where $r \in \mathbb{R}^+$ sums up all rate constants of chemical reactions reducing S to S', as many times as they apply for some index set L, provided that no immediate reaction can occur,
- $S \xrightarrow{\infty(r)} S'$ where the corresponding probability is $r = n/m$. The number of occurrences of immediate reactions leading from S to a solution congruent to S' is n, and the number of all occurrences of immediate reactions starting from S is m.

Such transitions are invariant under structural congruence, i.e. for all $S_1 \equiv S_1'$ and $S_2 \equiv S_2'$ it holds that $S_1 \xrightarrow{r} S_2$ (resp. $S_1 \xrightarrow{\infty(r)} S_2$) if and only if $S_1' \xrightarrow{r} S_2'$ (resp. $S_1' \xrightarrow{\infty(r)} S_2'$). We can thus define $[S]_\equiv \xrightarrow{r} [S']_\equiv$ by $S \xrightarrow{r} S'$ and $[S]_\equiv \xrightarrow{\infty(r)} [S']_\equiv$ by $S \xrightarrow{\infty(r)} S'$ as the transitions of the extended CTMC.

Gillespie's algorithm for stochastic simulation takes as input a finite set of chemical reactions and a chemical solution S. If reactions with infinite rate constants are applicable, it computes n and m as defined above for each immediately reachable solution S', and returns such an S' with probability n/m jointly with a null time delay. Otherwise, it computes the overall rate of all possible transitions $R = \sum_{\{r \mid S \xrightarrow{r} S_1\}} r$, returns with probability r/R a solution S_1 with transition $S \xrightarrow{r} S_1$ jointly with a time delay drawn randomly from the exponential distribution with rate r.

3.4 Language Design Choices and Related Rule-Based Languages

Models with rule schemas are more compact than if only simple reactions were used, thus easier to read. Attributed molecules and expressions that manipulate them were introduced, in the context of biological modeling languages, in [17]. Note that even if rule schemas could be defined solely by means of variables, function symbols allow a better control and precision of the collection of reactions that are generated. For example, without function symbols, we would need to resort to *name sharing* to represent DNA sequences[3]. Each DNA nucleotide would bear two parameters, one referring to its predecessor, the other to its successor. Given link names $\{\ell_0, \ldots, \ell_{50}\}$, our previous rule (0) on page 207 reads as

$$\forall x, y, z \in \{\ell_0, \ldots, \ell_{50}\}.$$
$$\text{RNAP} \cdot \text{DNA}(x, y)\ ,\ \text{DNA}(y, z) \rightarrow_{50} \text{DNA}(x, y)\ ,\ \text{RNAP} \cdot \text{DNA}(y, z)$$

Then starting from a DNA sequence $\text{DNA}(\ell_0, \ell_1), \text{DNA}(\ell_1, \ell_2), \ldots, \text{DNA}(\ell_{49}, \ell_{50})$, this rule schema instantiates into more ground rules than needed. For example, the rule $\text{RNAP} \cdot \text{DNA}(\ell_1, \ell_{10})\ ,\ \text{DNA}(\ell_{10}, \ell_{45}) \rightarrow_{50} \text{DNA}(\ell_1, \ell_{10})\ ,\ \text{RNAP} \cdot \text{DNA}(\ell_{10}, \ell_{45})$ is a meaningless instance of the above schema. Indeed, it is never applicable if the above DNA sequence is not modified as it is expected in a correct model.

All models written with our rule schemas can be compiled, by instantiation, to finite collections of simple and formally well-defined chemical reactions. Although reactions do not define a Turing-complete language [5], their expressiveness is sufficient for our purposes. Furthermore, such collections of reactions are supported by standard tools for stochastic simulation such as Dizzy [27] or the rule-based language BioCham [6].

Alternative rule-based languages with higher expressiveness are Turing complete, e.g. the graph rewriting language Kappa [9], BioNetGen [16], and bigraphs [21]. Their pattern based graph rewriting rules resemble schemas, but their semantics is not based on instantiations to ground rules. They rather directly apply to arbitrary subgraphs satisfying the pattern. In contrast to our approach, such patterns may describe infinitely many reactions. Furthermore, stochastic simulation is possible without inferring all those reactions on before hand. This generation process is uncritical in the present paper, since the overall number of reactions remains small, but is the bottleneck in other applications, where it grows exponentially [8,35]. Another promising language is LBS [24]. Its general purpose semantics allows for translations to different concrete semantics such as ODEs and CTMCs. LBS also features compact description of reactions with yet another approach by means of parameterized modules, species expressions and "non-deterministic" species. These formal rule-based languages were designed and used so far to tackle protein-protein interactions that occur in cellular signaling such as metabolic pathways. In contrast to this, our rule-based model deals with a fine-grained mechanism of gene regulation.

[3] This is actually how we implemented our model of Sect. 4 in the Kappa Factory.

Table 4. Rules for n transcription steps, in race with m translation steps

$\forall i \in \{0,\ldots,\text{n-1}\}.\textsc{Rnap} \cdot \textsc{Dna}_i \ , \ \textsc{Dna}_{i+1} \rightarrow_{e_1} \textsc{Dna}_i \ , \ \textsc{Rnap} \cdot \textsc{Dna}_{i+1}$	(1)
$\textsc{Rnap} \cdot \textsc{Dna}_n \rightarrow_{\infty} \textsc{Rnap} \ , \ \textsc{Dna}_n$	(2)
$\forall i \in \{0,\ldots,\text{m-1}\}.$ $\text{Ribosome} \cdot \text{m}\textsc{Rna}_i \ , \ \text{m}\textsc{Rna}_{i+1} \rightarrow_{e_2} \text{m}\textsc{Rna}_i \ , \ \text{Ribosome} \cdot \text{m}\textsc{Rna}_{i+1}$	(3)
$\text{Ribosome} \cdot \text{m}\textsc{Rna}_m \rightarrow_{\infty} \text{Ribosome} \ , \ \text{m}\textsc{Rna}_m$	(4)

4 Hyper-Sensitivity of Multi-step Races

In this section we illustrate rule schemas for chemical reactions with a simple yet interesting example, borrowed from Elf and Ehrenberg (2005) [11]. Abstracting away from its detailed control by mRNA hairpins, transcriptional attenuation boils down to a plain race between the two competing multi-step processes of transcription and translation. As intuition easily confirms, the probability that transcription wins the race decreases as the ribosome speeds up, and vice versa. We present two rule-based models for this multi-step race.

The *basic model* of Sect. 4.1 investigates the hyper-sensitivity of attenuation depending on the respective number of transcription versus translation steps (n vs m). Using Elf and Ehrenberg's rate constants for transcription and translation, we reproduce the results of Fig. 2 in [11].

In Sect. 4.2 we enrich our basic model by what we call *concurrent elongation.* An additional parameter m_0 denotes the number of codons contained in the initial solution, the remaining codons are dynamically spawn by the RNAP at simulation time. We show the impact of this additional level of concurrency, with respect to the attenuation race, through simulation.

4.1 Basic Model of Transcription and Translation

Elf and Ehrenberg demonstrated that the relative change in the probability that transcription wins the race can be much sharper, than the relative change in the ribosome's speed. As our work confirms, this *hyper-sensitivity* of attenuation is determined by the number of transcription steps (n) versus translation steps (m). We give a basic rule-based model that allows to reproduce the results of Elf and Ehrenberg. As we believe, our framework is easier to understand and less prone to error than the master equation approach, while compact and mathematically well-founded.

Model. The following initial solution describes the starting point of the multi-step race, where the RNAP and ribosome are bound to the first positions of DNA and mRNA respectively:

$$\textsc{Rnap} \cdot \textsc{Dna}_0 \ , \ \oplus_{i=1}^{n}\textsc{Dna}_i \ , \ \text{Ribosome} \cdot \text{m}\textsc{Rna}_0 \ , \ \oplus_{i=1}^{m}\text{m}\textsc{Rna}_i$$

We use the following notational conventions. Molecule names are $\mathcal{N} = \{$RNAP, mRNA, DNA, Ribosome$\}$, attribute values $\mathcal{C} = \mathbb{N}_0$, function symbols $\mathcal{F} = \{+\}$, variables $\mathcal{V} = \{i\}$, and value parameters $n, m \in \mathcal{C}$. Because our model's attributed molecules bear only few arguments, for the sake of presentation we slightly differ from the formal syntax introduced in Sect. 3. We write attributes as indices for molecule names, instead of parenthesizing them, e.g. DNA_i instead of DNA(i). Moreover we write $\oplus_{i=1}^{n}\text{DNA}_i$ instead of DNA(1), ..., DNA(n). Finally, we emphasize that each mRNA_i denotes one *codon*, which biologically speaking is a sequence of three individual mRNA nucleotides, that the ribosome reads out in one step.

Our model's rule schemas are listed in Table 4. Rule 1 for the n steps of the transcribing RNAP from one DNA nucleotide to the next remains as in Sect. 3.2, where it was the running example. The translation rule 3 is analogous and reflects the ribosome's m steps over codons. The remaining rules 2 and 4 model the dissociation from the respectively last positions of DNA and mRNA. Note that, bearing rate ∞, dissociation occurs without advance of the simulation clock. Hence it does not quantitatively affect our simulation results compared to the model of Elf and Ehrenberg, that does not include dissociation. In order to incorporate the control conditions at the *trp* operon, we will refine the dissociation rules in Sect. 5.

Simulation. The plot reporting our simulation results in Fig. 4 is organized as follows. The y-axis gives the probability that RNAP wins the race, on a scale between zero and one. It corresponds to the proportion of simulations in which RNAP dissociates before the ribosome does. The x-axis reports the translation rate on a logarithmic scale, that we vary from 0.01 to 100 codons per second in our simulations with the Kappa Factory [8].

The three models that each contribute one curve in the plot only differ in their numbers of translation (m) versus transcription (n) steps. We combined ($m = 1$ vs $n = 1$), ($m = 1$ vs $n = 50$), and ($m = 10$ vs $n = 50$).

Let us compare the sensitivity of these three models. When $m = 1$ and $n = 1$, the probability curve decreases gently, already showing some non-linearity. Increasing the number of transcription steps to $n = 50$ steepens the curve, i.e. increases the sensitivity. The transition becomes even sharper when the number of translation steps reaches higher values ($m = 10$, $n = 50$). Such values hold for systems where, unlike at *E.coli*'s *trp* operon, attenuation is the sole control mechanism [19]. It is worthwhile pointing out that in this model each of the m translation steps potentially slows down the ribosome's advance, while at *E.coli*'s *trp* operon, only 2 in 9 steps do.

Our rate constants are calculated as in [11], to ensure the outcomes of the three races are comparable. We keep the total time to perform the series of n transcription steps constant, such that $1/e_1' = 1$ sec. Thus, the rate constant for one individual transcription step out of n is $e_1 = n \cdot e_1'$. For one translation step the rate constant is $e_2 = m \cdot e_2'$. Hereby e_2' is the average rate for m translation steps, which varies logarithmically between 0.01 and 100.

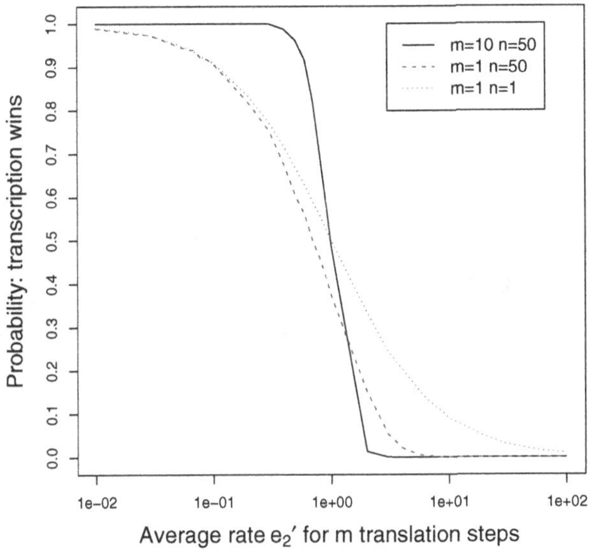

Fig. 4. Probability that transcription wins in the basic model, as a function of the average translation rate e_2', for different numbers of translation (m) versus transcription (n) steps

As the next section will show, our model can smoothly be extended by additional concurrent issues, that are more difficult to handle within the master equation approach.

4.2 Concurrent Elongation of mRNA

In the basic model, the multi-step race was represented by a ribosome and an RNAP advancing along two independent strands of mRNA and DNA. Here we add what we call *concurrent elongation* to the multi-step race. The idea is to reflect that RNAP still elongates a transcript when the ribosome starts translating its older end. Translation can now become limited by the slower transcription: the ribosome can only translate those codons that have previously been produced by the RNAP. Our simulation results demonstrate that the outcome of the race depends on the length of the initially available mRNA.

Model. Compared to the basic model, we now explicitly elongate the previously available mRNA in each transcription step. As Fig. 5 illustrates, a portion of the mRNA is available to the ribosome from the beginning of the race. In the basic model the parameters n and m denoted the respective lengths of the DNA and mRNA sequences for the attenuation race. Here the transcript dynamically grows from an initial length (for which we introduce the new parameter m_0) to its final length m.

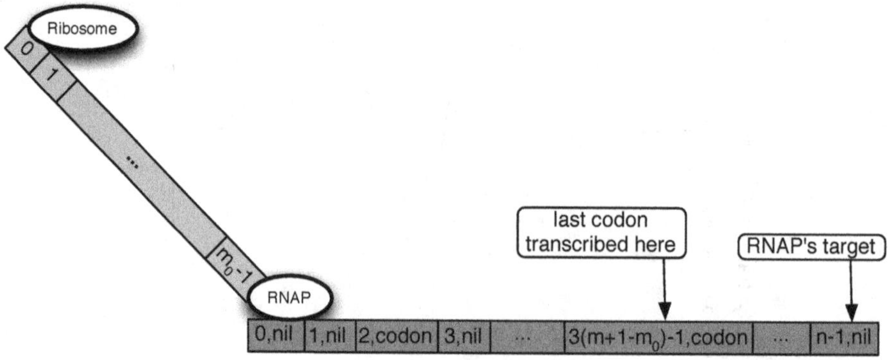

Fig. 5. General initial solution for the *concurrent elongation* model, containing an mRNA of length m_0 and a DNA of length n. The DNA is composed such that upon simulation, every three steps of the RNAP one new codon is spawn; the final solution contains m codons.

We use two more function symbols for integer arithmetics than previously, $\mathcal{F} = \{+, -, /\}$, attribute values $\mathcal{C} = \mathbb{N}_0 \cup \{\text{codon}, \text{nil}\}$, the previous molecule names $\mathcal{N} = \{\text{RNAP}, \text{mRNA}, \text{DNA}, \text{Ribosome}\}$, and variables $\mathcal{V} = \{i, x\}$. DNA molecules now come with a second attribute with values in $\{\text{codon}, \text{nil}\}$, noted as an upper index and with the following meaning. When RNAP leaves the ith nucleotide $\text{DNA}_i^{\text{codon}}$, it produces a new codon. As opposed to this, $\text{DNA}_i^{\text{nil}}$ indicates that no new codon is spawn when RNAP passes from the ith nucleotide to the next.

The choice of an appropriate *initial solution* is crucial to the proper functioning of this model, because we want the polymerase to spawn one new codon every three DNA nucleotides. Assuming that RNAP is initially bound to DNA_0 the solution must be such that, for $i \textit{ mod } 3 = 2$, nucleotides are of the form $\text{DNA}_i^{\text{codon}}$, and otherwise $\text{DNA}_i^{\text{nil}}$. Correspondingly the first two nucleotides must be $\text{DNA}_0^{\text{nil}}$ and $\text{DNA}_1^{\text{nil}}$, followed by the nucleotide $\text{DNA}_2^{\text{codon}}$, and so forth respecting the pattern nil, nil, codon. If only one codon is part of the initial solution ($m_0 = 1$) we obtain:

$$\begin{array}{l} \text{Ribosome} \cdot \text{mRNA}_0 \ , \\ \text{RNAP} \cdot \text{DNA}_0^{\text{nil}} \ , \ \text{DNA}_1^{\text{nil}} \ , \ \text{DNA}_2^{\text{codon}} \ , \\ \text{DNA}_3^{\text{nil}} \ , \ \text{DNA}_4^{\text{nil}} \ , \ \text{DNA}_5^{\text{codon}} \ , \ \ldots \ , \ \text{DNA}_n^{\text{nil}} \end{array}$$

For the sake of simplicity we do not show the DNA position corresponding to mRNA_m. When $m_0 = m + 1$, the initial solution reduces to that of Sect. 4.1:

$$\begin{array}{l} \text{Ribosome} \cdot \text{mRNA}_0 \ , \ \oplus_{i=1}^{m} \text{mRNA}_i \ , \\ \text{RNAP} \cdot \text{DNA}_0^{\text{nil}} \ , \ \text{DNA}_1^{\text{nil}} \ , \ \text{DNA}_2^{\text{codon}} \ , \\ \text{DNA}_3^{\text{nil}} \ , \ \text{DNA}_4^{\text{nil}} \ , \ \text{DNA}_5^{\text{codon}} \ , \ \ldots \ , \ \text{DNA}_n^{\text{nil}} \end{array}$$

Figure 5 illustrates the general case. In addition to the constraint on DNA nucleotide alternation, we assume that the initial solution contains m_0 codons, that

Table 5. Rules for *concurrent elongation* (n steps), where the transcribing RNAP adds one new codon to the solution every three DNA nucleotides

$$\forall i \in \{0,\ldots,n-1\}.\forall x \in \{\text{codon},\text{nil}\}.$$
$$\text{RNAP}\cdot\text{DNA}_i^{\text{nil}}\ ,\ \text{DNA}_{i+1}^{x} \rightarrow_{e_1} \text{DNA}_i^{\text{nil}}\ ,\ \text{RNAP}\cdot\text{DNA}_{i+1}^{x} \quad (1)$$

$$\forall i \in \{0,\ldots,n-1\}.$$
$$\text{RNAP}\cdot\text{DNA}_i^{\text{codon}},\text{DNA}_{i+1}^{\text{nil}} \rightarrow_{e_1} \text{DNA}_i^{\text{codon}},\text{RNAP}\cdot\text{DNA}_{i+1}^{\text{nil}},\text{mRNA}_{m_0-1+(i+1)/3} \quad (2)$$

$$\text{RNAP}\cdot\text{DNA}_n^{nil} \rightarrow_{\infty} \text{RNAP}\ ,\ \text{DNA}_n^{nil} \quad (3)$$

$$\forall i \in \{0,\ldots,m-1\}.$$
$$\text{Ribosome}\cdot\text{mRNA}_i\ ,\ \text{mRNA}_{i+1} \rightarrow_{e_2} \text{mRNA}_i\ ,\ \text{Ribosome}\cdot\text{mRNA}_{i+1} \quad (4)$$

$$\text{Ribosome}\cdot\text{mRNA}_m \rightarrow_{\infty} \text{Ribosome}\ ,\ \text{mRNA}_m \quad (5)$$

the rule set will lead to the dynamic supply of additional $m - m_0$ codons, such that the final solution shall contain $m + 1$ codons (allowing for m translation steps), and that $1 \leq m_0 \leq m < \frac{1}{3}n$. The ribosome's target codon mRNA_m corresponds to the DNA position $3 \cdot (m - m_0 + 1) - 1$. Beyond this, we assume that RNAP eventually reaches its own target, the nth position of DNA, without injecting additional codons to the solution.

Table 5 lists our rule schemas. The rule 1 for one step of the RNAP, in which no codon is produced, resembles rule 1 of Sect. 4.1. It applies when leaving nucleotides of the form $\text{DNA}_i^{\text{nil}}$, whether or not the step leaving the *next* nucleotide yields a codon. Hence the quantification over $x \in \{\text{codon}, \text{nil}\}$ for $\text{DNA}_{i+1}^{\text{x}}$.

The complementary rule 2 injects a new codon into the solution when the RNAP leaves a nucleotide of the form $\text{DNA}_i^{\text{codon}}$. The new codon's index is calculated from the current DNA position i and the initially available number of codons by the arithmetic expression $m_0 - 1 + (i + 1)/3$. By doing so, we ensure that DNA_2 yields mRNA_{m_0}, DNA_5 yields mRNA_{m_0+1}, etc. up to the ribosome's target mRNA_m.

The rule for the RNAP's dissociation from DNA (3) only marginally differs from that of the previous subsection (in that the nucleotide bears the second attribute nil), and the ribosome advance and release rules (4 and 5) remain just the same.

Simulation. We simulated our *concurrent elongation* model within the Kappa factory with several combinations of m_0, m and n. Figure 6 shows the outcome of the race distinguishing whether only one codon is initially present ($m_0 = 1$), or all ($m_0 = m$), for the same number of translation ($m = 15$) and transcription steps ($n = 50$).

When all codons are contained in the initial solution ($m_0 = 15$), the simulation results reduce to those of the basic model, whereas for $m_0 = 1$, the simulation curve shifts to the right, meaning that the probability that transcription wins

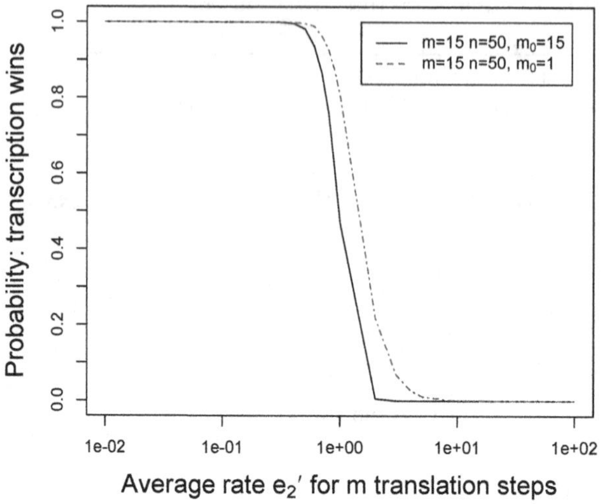

Fig. 6. Probability that transcription wins in the concurrent elongation model, as a function of the average translation rate e_2', for different numbers of initially available codons ($m_0 = 15$ vs $m_0 = 1$), but the same number of transcription steps ($n = 50$) and translation steps ($m = 15$)

the race increases. Indeed for each translation step, the ribosome's advance is potentially limited by the polymerase, that needs to add a further codon to the mRNA. Hence, even if translation is efficient, the polymerase wins more often than for $m_0 = 15$. In our simulations we observed a lesser shift for $m_0 = 10$, not included in the plot.

Our simulations underline that the outcome of the multi-step race is parameterized not only by the n transcription steps and the m translation steps, but also by the number m_0 of initially available codons. This last parameter only appears when the model integrates concurrent elongation. We can now summarize our analysis of the multi-step race parameterized by m, n and m_0, in terms of the shape of the curve that represents the probability that the polymerase wins the race:

- As pointed out by Elf and Ehrenberg [11], the ratio of m to n determines the curve's slope. They are the key parameters of the hyper-sensitivity of ribosome-dependent transcriptional attenuation.
- Varying m_0 shifts the curve. The polymerase's chance to win increases with m_0, when m and n remain fixed, because m_0 constrains the ribosome's advance along mRNA. As we observed, the shift increases with the difference between m and m_0.

Incorporating concurrent elongation into our model was facilitated by our rule-based approach with arithmetic. It would have been more difficult with probability functions. In a model that includes concurrent elongation, the positions of the ribosome and the RNAP are not independent. The advance of the former is limited by that of the latter. This point was not considered in [11].

5 Modeling Transcriptional Attenuation

This section presents our rule-based model of ribosome-mediated transcriptional attenuation at *E.coli*'s tryptophan operon. It refines our basic model of Sect. 4.1 in several points. The messenger RNA's representation dynamically grows while we simulate RNAP's advance, similarly as in the concurrent elongation model. But whereas in Sect. 4.2, we only used individual codons as building blocks of the transcript, the attenuation model also features mRNA segments as a whole. Explicit representations of S_1, S_2, S_3, and S_4 allow us to smoothly cover the dynamics of secondary structure formation, and incorporate the regulatory impact of hairpins on transcription. We make one notable exception to our all-in-one representation of mRNA segments. Regarding S_1, we switch between two different abstraction levels depending on the context, either representing it as a whole, or enumerating its codon sequence ($\oplus_{i=10}^{14}\mathrm{mRNA}_i$).

After introducing our attenuation model, we present simulation results in Sect. 5.2, and then explain the quantitative differences between our results and those of Elf and Ehrenberg [11] in Sect. 5.3.

5.1 Rule Schemas

Table 6 provides the rule schemas of our detailed attenuation model. The notational conventions are based on those of our basic model of Sect. 4.1. We use molecule names $\mathcal{N} = \{\mathrm{RNAP}, \mathrm{mRNA}, \mathrm{DNA}, \mathrm{Ribosome}, \mathrm{S}\}$, where DNA nucleotides are *unary*, attribute values $\mathcal{C} = \mathbb{N}_0 \cup \{\mathrm{fr}, \mathrm{bl}, \mathrm{hp}\}$, function symbols $\mathcal{F} = \{+\}$, and variables $\mathcal{V} = \{i, n, m, t, x\}$. Molecules with two attributes S_i^x represent segments of the mRNA leader. Their lower index $i \in \{1, 2, 3, 4\}$ denotes the segment's number, and the upper index x the segment's state which is among:

- *free* (fr): available for hairpin formation,
- *blocked* (bl): masked by the ribosome's footprint,
- *hairpin* (hp): complexed into a hairpin with a neighboring segment. For instance, $\mathrm{S}_1^{\mathrm{hp}} \cdot \mathrm{S}_2^{\mathrm{hp}}$ denotes the pause hairpin.

The initial solution for our simulations reflects the starting configuration for the attenuation race, depicted in Fig. 3 on page 205:

$$\begin{array}{l} \mathrm{RNAP} \cdot \mathrm{DNA}_0\ ,\ \oplus_{i=1}^{50}\mathrm{DNA}_i\ , \\ \mathrm{Ribosome} \cdot \mathrm{mRNA}_6\ ,\ \mathrm{mRNA}_7\ ,\ \mathrm{mRNA}_8\ ,\ \mathrm{mRNA}_9\ , \\ \mathrm{S}_1^{\mathrm{hp}} \cdot \mathrm{S}_2^{\mathrm{hp}}\ ,\ \mathrm{mRNA}_{15} \end{array} \qquad (0)$$

RNAP has transcribed the leader up to and including S_1 and S_2, that are paired into the pause hairpin, and is paused on the zero-th DNA nucleotide, that is followed by a sequence of 50. The ribosome has initiated translation and is located on the 6th codon of the transcript leader. We explicitly render the codons 6 to 9, which precede the segment S_1, and the stop codon 15, that is located between the segments S_1 and S_2. In contrast, we do not render the codons preceding 6, since they do not matter to the attenuation race, and for the same reason we will not provide rules for the initiation of transcription and translation.

Hairpin formation is covered by rule schema 1, be it for the pause hairpin, the anti-terminator or the terminator. Because hairpin formation occurs on a much faster time scale than any other reaction, we approximate it with an infinite rate constant.

Translation rules (schemas 2 to 7 in Table 6). Rule schema 2 covers the bulk of translation steps, that do not have side effects, nor depend on tryptophan availability or other side conditions. It bears the reaction rate constant $e_2 = 15s^{-1}$, i.e. the ribosome makes 15 steps over mRNA per second, in average. Rule schema 3 deals with the ribosome's step over the tryptophan codons within the leader, i.e. the control codons 10 and 11, where the distinct elongation rate constant e_3 holds. We will vary e_3 within $]0, 15]s^{-1}$ in our simulations, while e_2 remains fixed.

Starting from our initial solution (the above equation 0) the next important event is *melting the pause loop* $S_1^{hp} \cdot S_2^{hp}$, as the ribosome steps from $mRNA_6$ to $mRNA_7$. Two points are worthwhile noting in rule 4's right part. First, obviously since the pause loop is melt, S_2's state becomes *free* - and one could similarly expect a stage change at S_1. But second and more importantly, instead of switching S_1's state, we pass from the abstraction of the segment as a whole, to the enumeration of the codons $\oplus_{i=10}^{14} mRNA_i$ that make it up. The sequence enumeration remains part of the solution as long as the ribosome's footprint partially covers the first segment, i.e. until it dissociates from the stop codon. This implicitly sequesters S_1 from hairpin formation – which would instantaneously occur through schema 1 if both S_1 and S_2 were around and in their *free* state.

For the ribosome's step from $mRNA_{12}$ to $mRNA_{13}$, we introduce two rules with distinct preconditions. The common result of both is to reflect that the second segment gets masked by the ribosome's footprint, i.e. both rules produce S_2^{bl}. When S_2 is initially free, rule 5 applies. Otherwise, S_2 is paired into the anti-terminator hairpin, and rule 6 handles its melting through the ribosome's advance. When the ribosome dissociates from the stop codon (rule 7), S_1 re-assembles, and S_2 unblocks.

Transcription rules. Rules 8 to 13 in Table 6 represent transcription. Rule schema 8 represents one step of RNAP in the simplest possible fashion, that was already discussed in Sect. 4.1 with the rate constant $e_1 = 50s^{-1}$. It applies to all DNA positions from 0 to 50 with a few exceptions that we discuss in the order they are applied, starting from our initial solution.

Transcription *resumes* at position DNA_0 (rule schema 9) after the pause hairpin has been disrupted. This is witnessed by S_2 being either free or blocked.

Table 6. Rule schemas for hairpin formation (1), translation (2-7), and transcription (8-13)

$\forall i \in \{1,2,3\}.\mathrm{S}_i^{\mathrm{fr}}\ ,\ \mathrm{S}_{i+1}^{\mathrm{fr}} \rightarrow_{\infty} \mathrm{S}_i^{\mathrm{hp}} \cdot \mathrm{S}_{i+1}^{\mathrm{hp}}$	(1)
$\forall m \in \{7,8,9,13,14\}.\mathrm{Ribosome} \cdot \mathrm{mRNA}_m\ ,\ \mathrm{mRNA}_{m+1} \rightarrow_{e_2} \mathrm{mRNA}_m\ ,\ \mathrm{Ribosome} \cdot \mathrm{mRNA}_{m+1}$	(2)
$\forall t \in \{10,11\}.\mathrm{Ribosome} \cdot \mathrm{mRNA}_t\ ,\ \mathrm{mRNA}_{t+1} \rightarrow_{e_3} \mathrm{mRNA}_t\ ,\ \mathrm{Ribosome} \cdot \mathrm{mRNA}_{t+1}$	(3)
$\mathrm{Ribosome} \cdot \mathrm{mRNA}_6\ ,\ \mathrm{mRNA}_7\ ,\ \mathrm{S}_1^{\mathrm{hp}} \cdot \mathrm{S}_2^{\mathrm{hp}} \rightarrow_{e_2} \mathrm{mRNA}_6\ ,\ \mathrm{Ribosome} \cdot \mathrm{mRNA}_7\ ,\ \oplus_{i=10}^{14}\mathrm{mRNA}_i\ ,\ \mathrm{S}_2^{\mathrm{fr}}$	(4)
$\mathrm{Ribosome} \cdot \mathrm{mRNA}_{12}\ ,\ \mathrm{mRNA}_{13}\ ,\ \mathrm{S}_2^{\mathrm{fr}} \rightarrow_{e_2} \mathrm{mRNA}_{12}\ ,\ \mathrm{Ribosome} \cdot \mathrm{mRNA}_{13}\ ,\ \mathrm{S}_2^{\mathrm{bl}}$	(5)
$\mathrm{Ribosome} \cdot \mathrm{mRNA}_{12}\ ,\ \mathrm{mRNA}_{13}\ ,\ \mathrm{S}_2^{\mathrm{hp}} \cdot \mathrm{S}_3^{\mathrm{hp}} \rightarrow_{e_2} \mathrm{mRNA}_{12}\ ,\ \mathrm{Ribosome} \cdot \mathrm{mRNA}_{13}\ ,\ \mathrm{S}_2^{\mathrm{bl}}\ ,\ \mathrm{S}_3^{\mathrm{fr}}$	(6)
$\mathrm{Ribosome} \cdot \mathrm{mRNA}_{15}\ ,\ \oplus_{i=10}^{14}\mathrm{mRNA}_i\ ,\ \mathrm{S}_2^{\mathrm{bl}} \rightarrow_{d} \mathrm{Ribosome}\ ,\ \mathrm{mRNA}_{15}\ ,\ \mathrm{S}_1^{\mathrm{fr}}\ ,\ \mathrm{S}_2^{\mathrm{fr}}$	(7)
$\forall n \in \{1,\ldots,49\} \setminus \{35,46,47\}.\mathrm{RNAP} \cdot \mathrm{DNA}_n\ ,\ \mathrm{DNA}_{n+1} \rightarrow_{e_1} \mathrm{DNA}_n\ ,\ \mathrm{RNAP} \cdot \mathrm{DNA}_{n+1}$	(8)
$\forall x \in \{\mathrm{fr},\mathrm{bl}\}.\mathrm{RNAP} \cdot \mathrm{DNA}_0\ ,\ \mathrm{DNA}_1\ ,\ \mathrm{S}_2^{x} \rightarrow_{e_1} \mathrm{DNA}_0\ ,\ \mathrm{RNAP} \cdot \mathrm{DNA}_1\ ,\ \mathrm{S}_2^{x}$	(9)
$\mathrm{RNAP} \cdot \mathrm{DNA}_{35}\ ,\ \mathrm{DNA}_{36} \rightarrow_{e_1} \mathrm{DNA}_{35}\ ,\ \mathrm{RNAP} \cdot \mathrm{DNA}_{36}\ ,\ \mathrm{S}_3^{\mathrm{fr}}$	(10)
$\mathrm{RNAP} \cdot \mathrm{DNA}_{46}\ ,\ \mathrm{DNA}_{47} \rightarrow_{e_1} \mathrm{DNA}_{46}\ ,\ \mathrm{RNAP} \cdot \mathrm{DNA}_{47}\ ,\ \mathrm{S}_4^{\mathrm{fr}}$	(11)
$\mathrm{RNAP} \cdot \mathrm{DNA}_{47}\ ,\ \mathrm{S}_3^{\mathrm{hp}} \cdot \mathrm{S}_4^{\mathrm{hp}} \rightarrow_{e_1} \mathrm{DNA}_{47}\ ,\ \mathrm{RNAP}\ ,\ \mathrm{S}_3^{\mathrm{hp}} \cdot \mathrm{S}_4^{\mathrm{hp}}$	(12)
$\mathrm{RNAP} \cdot \mathrm{DNA}_{47}\ ,\ \mathrm{DNA}_{48}\ ,\ \mathrm{S}_2^{\mathrm{hp}} \cdot \mathrm{S}_3^{\mathrm{hp}} \rightarrow_{e_1} \mathrm{DNA}_{47}\ ,\ \mathrm{RNAP} \cdot \mathrm{DNA}_{48}\ ,\ \mathrm{S}_2^{\mathrm{hp}} \cdot \mathrm{S}_3^{\mathrm{hp}}$	(13)

Note that it would have been simpler to check the *absence* of the pause hairpin, but such negative tests are neither supported by the language used in this paper, nor by most current rule-based frameworks and tools, a notable exception to a certain extent is offered by [16].

The remaining rules deal with the creation of new mRNA segments, and (anti)termination of transcription. When RNAP steps over to DNA_{36}, the RNA segment S_3 is injected into the solution (reaction 10). S_4 follows at DNA_{47} (reaction 11). Transcription terminates on DNA_{47} provided there is a terminator hairpin, see $S_3^{hp} \cdot S_4^{hp}$ in reaction 12. If conversely the anti-terminator is present transcription proceeds to DNA_{48} (rule 13) and continues transcription into the operon.

Summary: global control by n-ary rules. Finally we summarize our use of n-ary rules, that separates into three categories. Such n-ary rules can not be rendered in an intelligible fashion within object-centric approaches limited to binary interactions, namely π-calculus based modeling languages.

In the *first category*, we check whether the current solution fulfills a certain *prerequisite*, e.g. contains a certain molecule, or a certain molecule in a specific state. The rules for abortion versus continuation of transcription (rules 12 and 13) depend on which hairpin is around, terminator or anti-terminator. Other prerequisites we check are unfortunately less intuitive: sometimes one would prefer to impose negative conditions on rule application, which however is neither supported by our language, nor most other current rule-based frameworks. For instance, rule 9 resumes transcription if the pause hairpin is absent, which is the case if S_2's state is free or blocked.

The *second category* are reactions that actually occur between two molecules, but entail the *modification* of a third. Examples are the rules for blocking S_2 as the ribosome proceeds to mRNA_{13}: rule 5 blocks S_2, rule 6 disrupts the anti-terminator hairpin at the same time, such that the states of both S_2 and S_3 change. The *third category* is the abstraction level switching for mRNA segments, that is assembling the first segment from the codons 10 to 14 in rule 7, versus splitting it in rule 4.

5.2 Simulation

Figure 7 plots the relative transcription frequency (y-axis) against the rate of *trp*-codon translation e_3 (x-axis). For each value of e_3 from 0 to 15 s^{-1} in steps of one, we performed 5000 Gillespie simulations of our model. Recall that two different pathways lead to the anti-terminator, each of them corresponds to a distinct curve, and a third curve sums up.

The curve *ribosome stalling* corresponds to anti-terminator formation while the ribosome remains stalled on the control codons. This predominates when *trp*-codon translation is slow, becomes rarer as the translation efficiency increases, and drops below 1% when $e_3 \geq 9s^{-1}$.

In a second pathway, the anti-terminator hairpin forms after the ribosome has released from the stop codon 15. This represents the basal read-through

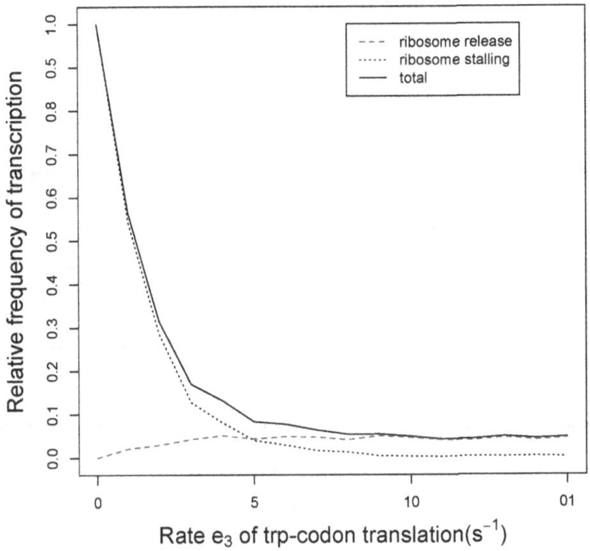

Fig. 7. Relative frequency of continued transcription as a function of the *trp*-codon translation rate. We distinguishing between anti-terminator formation during ribosome stalling, and after release of the ribosome from the stop codon.

level of the *trp* operon (see page 206). The corresponding curve *ribosome release* in our simulation plot starts from zero, steadily increases with the rate of *trp*-codon translation, reaches its maximal level around 4.5% at a rate of *trp*-codon translation of 4 s^{-1}, and remains stable henceforth.

Our results shown in Fig. 7 qualitatively confirm those of Elf and Ehrenberg [11]. However, even if the curves have the same shape, the asymptotic decrease of *ribosome stalling* toward 0 is less sharp in our case. While in their work the curves for ribosome release and ribosome stalling cross at a rate of $6s^{-1}$, in ours they already do so at $5s^{-1}$. Moreover our experiments predict a rate of basal transcription of slightly under 5% when *trp*-codon translation is efficient, where Elf and Ehrenberg predict 8%.

5.3 Discussion

We believe that the differences between our quantitative results and those of [11] are due to the greater level of detail rendered by our model. Elf and Ehrenberg reduce attenuation to a race between the ribosome and the polymerase, but their model does not make hairpins explicit. Instead, they infer transcription probabilities from the *relative positions* of both the RNAP and the ribosome. As opposed to this, our model explicitly renders hairpins.

Table 7 summarizes how the configuration of the system evolves, assuming simulations start from our initial solution (equation 0), and the ribosome has

Table 7. Transitions between configurations of the *trp* leader that our model yields, depending on the relative positions of the ribosome versus RNAP. We list the first segment in its blocked state for the sake of presentation, when the solution indeed enumerates its sequence.

RNAP \ Ribosome	[7,12]	[13,15]	OFF (A)	OFF (B)
[0,35]	S_1^{bl}, S_2^{fr}	S_1^{bl}, S_2^{bl}	100% $S_1^{hp} \cdot S_2^{hp}$	
[36,46]	$S_1^{bl}, S_2^{hp} \cdot S_3^{hp}$	$S_1^{bl}, S_2^{bl}, S_3^{fr}$ 50% / 50%	$S_1^{hp} \cdot S_2^{hp}, S_3^{fr}$	$S_1^{fr}, S_2^{hp} \cdot S_3^{hp}$
[47]	$S_1^{bl}, S_2^{hp} \cdot S_3^{hp}, S_4^{fr}$	$S_1^{bl}, S_2^{bl}, S_3^{hp} \cdot S_4^{hp}$	$S_1^{hp} \cdot S_2^{hp}, S_3^{hp} \cdot S_4^{hp}$	$S_1^{fr}, S_2^{hp} \cdot S_3^{hp}, S_4^{fr}$
OUTCOME	anti-termination *Ribosome stalling*	termination	termination	anti-termination *Ribosome release*

disrupted the pause hairpin. For each relative position of the ribosome on mRNA (columns) and the RNAP on DNA (rows), the table states which segments have been transcribed so far, and which hairpins have formed. Arrows between the table's cells corresponds to possible transitions of our solution. We distinguish the following positions of interest for the RNAP on DNA:

- **between DNA nucleotide 0 and 35:** only segments S_1 and S_2 are contained in the initial solution,
- **between 36 and 46:** RNAP has completed segment S_3,
- **on position 47:** RNAP has injected S_4 to the solution.

The *ribosome*'s positions of interest on the mRNA leader are:

- **between codon 7 and 12:** the ribosome's footprint has not yet reached S_2;
- **between codon 13 and 15:** the ribosome's footprint masks S_2 until dissociation from the mRNA.
- **off:** the ribosome has dissociated from the transcript, making segment S_2 newly available for hairpin formation.

Elf and Ehrenberg distinguish two cases of anti-termination (reproduced in our simulations in Fig. 7): anti-termination during *ribosome stalling* and anti-termination after *ribosome release.*

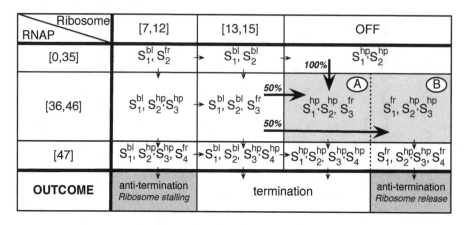

Stochastic trajectories leading to *ribosome stalling* descend our table's column [7, 12]. The ribosome then remains between the codons 7 and 12, most likely stalling on the control (*trp*) codons within the first segment. After the polymerase has transcribed segment S_3, the anti-terminator hairpin $S_2 \cdot S_3$ forms. This henceforth excludes the terminator hairpin, and transcription continues into the structural genes. We believe that our model and that of Elf and Ehrenberg show the same behavior for this first pathway.

We identified the following key difference between the models regarding the second pathway to anti-termination. Elf and Ehrenberg assume that the probability of anti-termination after *ribosome release* is half the probability of reaching the configuration (RNAP $\in [36, 46]$, ribosome OFF). The corresponding cell is highlighted in light gray in Table 7. It is important to note that we separate it into two sub-cells A and B. We refine Elf and Ehrenberg's assumption as follows: *stochastic trajectories leading to anti-termination after ribosome release pass through the sub-cell B, but* never *through the sub-cell A.*

Careful consideration of the table explains our refinement. The highlighted cell can be reached from either its top or left neighbor. Coming from the left neighbor (RNAP $\in [36, 46]$, ribosome $\in [13, 15]$), it is *equally likely* to reach sub-cell A that entails termination, as to reach B that entails anti-termination. However, coming from the top neighbor (RNAP $\in [0, 35]$, ribosome OFF) makes a difference. Because the ribosome has unbound the mRNA before segment S_3 was completed, the pause hairpin $S_1^{hp} \cdot S_2^{hp}$ is already part of the solution. Hence the anti-terminator does not appear once S_3 is completed. Thus, descending the column *ribosome off*, the system *always* reaches configuration A, and transcription always terminates.

Based on our detailed model, and contradicting Elf and Ehrenberg, we claim that the state (RNAP $\in [36, 46]$, ribosome off) does *not* lead with equal probabilities to anti-termination by ribosome release versus termination. The downward transitions into sub-cell A, that contains the pause hairpin $S_1^{hp} \cdot S_2^{hp}$ and hence excludes the anti-terminator, increase the termination probability[4]. This line of reasoning agrees with experimental knowledge on the impact of early ribosome release [29], and may explain why our experiments predict anti-termination less often than Elf and Ehrenberg's.

6 Conclusion

We have shown that rule-based modeling provides concise and elegant models for the fine-grained mechanism of transcriptional attenuation, a problem left open by previous work on discrete event modeling of the tryptophan operon [32]. The core ingredients for our model are rule schemas and n-ary chemical reactions. The importance of n-ary reactions renders representations of this case of genetic regulation in object-centered languages such as the stochastic pi-calculus [22,25,28] inappropriate, in practice. We used the Kappa factory for stochastic simulation [8], which provided us convenient analysis tools.

In our model we identified positions of individual nucleotides and codons within DNA and RNA sequences by numbers, abstracted over by variables, and addressed successors by simple arithmetic. This technique has its limitations when

[4] Another effect is due to the anti-terminator hairpin melting as the ribosome moves on to $m\text{RNA}_{13}$ that is included in our model, which lowers the *ribosome stalling* probability.

polymers become more complex than simple lists. Alternatively, we could assign names to molecular domains, and memorise those names in attribute values of adjacent molecules, similarly to Kappa. As shown in Sect. 3.4, the cost for such an alternative is that meaningless instances of rule schemas may be generated.

In future work, we plan to compute the exact probability that the ribosome dissociates before the segment S_3 is formed. This corresponds to the additional pathway that is not rendered by the model proposed by Elf and Ehrenberg (2005) [11]. This would formally prove what we conjectured to be the source of the quantitative difference between the two models. We can indeed compute such a probability because there are finitely many pathways and all pathways are terminating (i.e. the system always reaches a configuration in which no rule is applicable). In other words, one can exhaustively unfold the underlying CTMC and thus compute the probability associated to each pathway (i.e. each branch of the CTMC).

Acknowledgements. The Master's project with Valerio Passini at the Microsoft Research - University of Trento Center for Computational and Systems Biology sharpened our view of the system from a biological perspective and lead us to a rule-based approach. The previous Master's project of Gil Payet (co-supervised with Denys Duchier) had confronted us with the limitations of object-based approaches to the representation of the complex dependencies of transcriptional attenuation. We thank Maude Pupin, who was the first to point us at the intricate regulatory mechanisms at *E.coli*'s tryptophan operon, and Joachim Niehren for his valuable coaching. Plectix BioSystems kindly made the Kappa Factory available to us, and gave us useful support while we carried out the experiments reported in this paper. Finally, we thank the CNRS for a sabbatical to Cédric Lhoussaine, and the Agence Nationale de Recherche for funding this work through a *Jeunes Chercheurs* grant (ANR BioSpace, 2009-2011).

References

1. Arkin, A., Ross, J., McAdams, H.H.: Stochastic kinetic analysis of developmental pathway bifurcation in phage λ-infected Escherichia coli cells. Genetics 149, 1633–1648 (1998)
2. Baader, F., Nipkow, T.: Term rewriting and all that. Cambridge University Press, New York (1998)
3. Barboric, M., Peterlin, B.M.: A new paradigm in eukaryotic biology: HIV Tat and the control of transcriptional elongation. PLoS Biology 3(2), 0200–2003 (2005)
4. Beisel, C.L., Smolke, C.D.: Design principles for riboswitch function. PLoS Computational Biology 5(4), e1000363, 04 (2009)
5. Cardelli, L., Zavattaro, G.: On the computational power of biochemistry. In: Horimoto, K., Regensburger, G., Rosenkranz, M., Yoshida, H. (eds.) AB 2008. LNCS, vol. 5147, pp. 65–80. Springer, Heidelberg (2008)
6. Chabrier-Rivier, N., Fages, F., Soliman, S.: The biochemical abstract machine BioCham. In: Danos, V., Schachter, V. (eds.) CMSB 2004. LNCS (LNBI), vol. 3082, pp. 172–191. Springer, Heidelberg (2005)
7. Ciocchetta, F., Hillston, J.: Bio-PEPA: a framework for modelling and analysis of biological systems. Theoretical Computer Science (to apppear)

8. Danos, V., Feret, J., Fontana, W., Krivine, J.: Scalable simulation of cellular signaling networks. In: Shao, Z. (ed.) APLAS 2007. LNCS, vol. 4807, pp. 139–157. Springer, Heidelberg (2007)
9. Danos, V., Feret, J., Fontana, W., Harmer, R., Krivine, J.: Rule-based modelling of cellular signalling. In: Caires, L., Vasconcelos, V.T. (eds.) CONCUR 2007. LNCS, vol. 4703, pp. 17–41. Springer, Heidelberg (2007)
10. Dematté, L., Priami, C., Romanel, A.: The beta workbench: A tool to study the dynamics of biological systems. Briefings in Bioinformatics 9(5), 437–449 (2008)
11. Elf, J., Ehrenberg, M.: What makes ribosome-mediated trascriptional attenuation sensitive to amino acid limitation? PLoS Computational Biology 1(1), 14–23 (2005)
12. Gillespie, D.T.: A general method for numerically simulating the stochastic time evolution of coupled chemical reactions. Journal of Computational Physics 22, 403–434 (1976)
13. Gollnick, P.: Trp operon and attenuation. In: Lennarz, W.J., Lane, M.D. (eds.) Encyclopedia of Biological Chemistry, pp. 267–271. Elsevier, New York (2004)
14. Gollnick, P., Babitzke, P., Antson, A., Yanofsky, C.: Complexity in regulation of tryptophan biosynthesis in Bacillus subtilis. Annual Review of Genetics 39(1), 47–68 (2005)
15. Gutierrez-Preciado, A., Jensen, R.A., Yanofsky, C., Merino, E.: New insights into regulation of the tryptophan biosynthetic operon in Gram-positive bacteria. Trends in Genetics 21(8), 432–436 (2005)
16. Blinov, M.L., Faeder, J.R., Hlavacek, W.S.: Rule-Based Modeling of Biochemical Systems with BioNetGen. In: Systems Biology. Methods in Molecular Biology, vol. 500, pp. 1–55. Humana Press (2009)
17. John, M., Lhoussaine, C., Niehren, J., Uhrmacher, A.M.: The attributed pi-calculus with priorities. Transactions on Computational Systems Biology (to appear, 2009)
18. Nakamura, Y., Roesser, J.R., Yanofsky, C.: Regulation of basal level expression of the tryptophan operon of Escherichia coli. J. Biol. Chem. 264(21), 12284–12288 (1989)
19. Kasai, T.: Regulation of the expression of the histidine operon in Salmonella typhimurium. Nature 249, 523–527 (1974)
20. Konan, K.V., Yanofsky, C.: Role of ribosome release in regulation of tna operon expression in Escherichia coli. J. Bacteriol. 181, 1530–1536 (1999)
21. Krivine, J., Milner, R., Troina, A.: Stochastic bigraphs. In: 24th Conference on the Mathematical Foundations of Programming Semantics. Electronical notes in theoretical computer science, vol. 218, pp. 73–96. Elsevier, Amsterdam (2008)
22. Kuttler, C., Lhoussaine, C., Niehren, J.: A stochastic pi calculus for concurrent objects. In: Anai, H., Horimoto, K., Kutsia, T. (eds.) AB 2007. LNCS, vol. 4545, pp. 232–246. Springer, Heidelberg (2007)
23. Lloyd, J.W.: Foundations of Logic Programming, 2nd edn. Springer, Heidelberg (1987)
24. Pedersen, M., Plotkin, G.: A language for biochemical systems. In: Priami, C., et al. (eds.) Trans. on Comput. Syst. Biol. XII. LNCS (LNBI), vol. 5945, pp. 77–145. Springer, Heidelberg (2010)
25. Phillips, A., Cardelli, L.: Efficient, correct simulation of biological processes in the stochastic pi-calculus. In: Calder, M., Gilmore, S. (eds.) CMSB 2007. LNCS (LNBI), vol. 4695, pp. 184–199. Springer, Heidelberg (2007)
26. Pradalier, S., Credi, A., Garavelli, M., Laneve, C., Zavattaro, G.: Modelization and simulation of nano devices in the nano-kappa calculus. In: Calder, M., Gilmore, S. (eds.) CMSB 2007. LNCS (LNBI), vol. 4695, pp. 168–183. Springer, Heidelberg (2007)

27. Ramsey, S., Orrell, D., Bolouri, H.: Dizzy: stochastic simulation of large-scale genetic regulatory networks. Journal of Bioinformatics and Computational Biology 3(2), 415–436 (2005)
28. Regev, A.: Computational Systems Biology: A Calculus for Biomolecular Knowledge. Tel Aviv University, PhD thesis (2002)
29. Roesser, J.R., Yanofsky, C.: Ribosome release modulates basal level expression of the trp operon of Escherichia coli. Journal of Biological Chemistry 263(28), 14251–14255 (1988)
30. Santillan, M., Zeron, E.S.: Dynamic influence of feedback enzyme inhibition and transcription attenuation on the tryptophan operon response to nutritional shifts. Journal of Theoretical Biology 231(2), 287–298 (2004)
31. Shieber, S.M.: An Introduction to Unification-Based Approaches to Grammar, vol. 4. CLSI Publications (1986)
32. Simão, E., Remy, E., Thieffry, D., Chaouiya, C.: Qualitative modelling of regulated metabolic pathways: application to the tryptophan biosynthesis in E.coli. In: ECCB/JBI, pp. 190–196 (2005)
33. Trun, N., Trempy, J.: Gene expression and regulation. In: Fundamental bacterial genetics, pp. 191–212. Blackwell, Malden (2003)
34. von Heijne, G., Nilsson, L., Blomberg, C.: Translation and messenger RNA secondary structure. Journal of Theoretical Biology 68, 321–329 (1977)
35. Yang, J., Monine, M.I., Faeder, J.R., Hlavacek, W.S.: Kinetic monte carlo method for rule-based modeling of biochemical networks. Physical Review E 78(3), 7 (2008)
36. Yanofsky, C.: Attenuation in the control of expression of bacterial operons. Nature 289, 751–758 (1981)
37. Yanofsky, C.: Transcription attenuation: once viewed as a novel regulatory strategy. J. Bacteriology 182(1), 1–8 (2000)
38. Yanofsky, C.: RNA-based regulation of genes of tryptophan synthesis and degradation, in bacteria. RNA - A publication of the RNA Society 13(8), 1141–1154 (2007)

A Kappa Rules for Section 5

In the following we list the Kappa code of our detailed attenuation model in Section 5, as we implemented it in Kappa Factory version 12.2.0. in order to run stochastic simulations. This encoding remains as close as possible to our rule-based model, notably it does not represent DNA or mRNA as interconnected chains, hence does not use *name sharing.*

Some comments on the syntax of Kappa seem appropriate. For instance consider the following reaction produced by rule schema 1:

```
'LoopS2S3 (anti-terminator)' S2(s~fr),S3(s~fr)
      -> S2(s~hp!1),S3(s~hp!1) @ $INF
```

This reaction is given the name `'LoopS2S3(anti-terminator)'` and an infinite rate `@ $INF`. The reactants `S2(s~fr)` and `S3(s~fr)` have an attribute `s` both with values `fr`. The reaction produces the molecules `S2(s~hp!1)` and `S3(s~hp!1)`, where the attribute `s` has the value `hp`. The modifier `!1` indicates that the molecules form a complex, which is linked by the edge `1`.

```
# rule schema 1
'LoopS2S3 (anti-terminator)' S2(s~fr),S3(s~fr)
    -> S2(s~hp!1),S3(s~hp!1) @ $INF
'LoopS3S4 (terminator)' S4(s~fr),S3(s~fr)
    -> S4(s~hp!1),S3(s~hp!1) @ $INF
'LoopS1S2(pause)' S1(s~fr),S2(s~fr)->S1(s~hp!1),S2(s~hp!1) @ $INF
# rule schema 2
'RiboTo8' Ribo(m!1),mRNA7(t!1),mRNA8(t)
    -> Ribo(m!2),mRNA7(t),mRNA8(t!2) @ 15.0
...
'RiboTo15' Ribo(m!1),mRNA14(t!1),mRNA15(t)
    -> Ribo(m!2),mRNA14(t),mRNA15(t!2) @ 15.0
# rule scheOma 3
'RiboTo11_Trp' Ribo(m!1),mRNA10(t!1),mRNA11(t)
         -> Ribo(m!2),mRNA10(t),mRNA11(t!2) @ 1.0
'RiboTo12_Trp' Ribo(m!1),mRNA11(t!1),mRNA12(t)
         -> Ribo(m!2),mRNA11(t),mRNA12(t!2) @ 1.0
# rule schema 4
'RiboTo7_MeltS1S2' Ribo(m!1),mRNA6(t!1),mRNA7(t),
    S1(s~hp!2),S2(s~hp!2)
    -> Ribo(m!2),mRNA6(t),mRNA7(t!2),mRNA10(t),mRNA15(t),
    mRNA11(t),S2(s~fr),mRNA14(t),mRNA13(t),mRNA12(t) @ 15.0
# rule schema 5
'blockS2_ribo@13' Ribo(m!1),mRNA13(t!1),S2(s~fr)
  -> Ribo(m!1),mRNA13(t!1),S2(s~bl) @ $INF
# rule schema 6
'meltS2S3_ribo@13' Ribo(m!1),mRNA13(t!1),S2(s~hp!1),S3(s~hp!1)
  -> Ribo(m!1),mRNA13(t!1),S2(s~bl), S3(s~fr) @ $INF
# rule schema 7
'RiboRelease@15_joinS1' mRNA15(t!1),Ribo(m!1),mRNA10(t),
 mRNA11(t),mRNA12(t),mRNA13(t),mRNA14(t),S2(s~bl)
 -> Ribo(m),S1(s~fr),S2(s~fr) @ 1.0
# rule schema 8
'RNAPto2' RNAP(d!1),DNA1(t!1),DNA2(t)
  -> RNAP(d!2),DNA1(t),DNA2(t!2) @ 50.0
...
'RNAPto35' RNAP(d!1),DNA34(t!1),DNA35(t)
  -> RNAP(d!2),DNA34(t),DNA35(t!2) @ 50.0
'RNAPto37' RNAP(d!1),DNA36(t!1),DNA37(t)
  -> RNAP(d!2),DNA36(t),DNA37(t!2) @ 50.0
...
'RNAPto46' RNAP(d!1),DNA45(t!1),DNA46(t)
  -> RNAP(d!2),DNA45(t),DNA46(t!2) @ 50.0
'RNAPto49' RNAP(d!1),DNA48(t!1),DNA49(t)
  -> RNAP(d!2),DNA48(t),DNA49(t!2) @ 50.0
```

```
'RNAPto50' RNAP(d!1),DNA49(t!1),DNA50(t)
  -> RNAP(d!2),DNA49(t),DNA50(t!2) @ 50.0
# rule schema 9
'RNAPresumes_S1S2broken_a' RNAP(d!1),DNA0(t!1),DNA1(t),S2(s~bl) ->
 RNAP(d!2),DNA0(t),DNA1(t!2),S2(s~bl) @ 50.0
'RNAPresumes_S1S2broken_b' RNAP(d!1),DNA0(t!1),DNA1(t),S2(s~fr) ->
 RNAP(d!2),DNA0(t),DNA1(t!2),S2(s~fr) @ 50.0
# rule schema 10
'RNAPto36_spawnS3' RNAP(d!1),DNA35(t!1),DNA36(t)
   -> RNAP(d!2),DNA35(t),DNA36(t!2),S3(s~fr) @ 50.0
# rule schema 11
'RNAPto47_spawnS4' RNAP(d!1),DNA46(t!1),DNA47(t)
   -> RNAP(d!2),DNA46(t),DNA47(t!2),S4(s~fr) @ 50.0
# rule schema 12
'RNAP_dissociate' RNAP(d!1),DNA47(t!1),S3(s~hp!2),S4(s~hp,2!)
      -> RNAP(d),DNA47(t),DNA48(t),S3(s~hp,!2),S4(s~hp,!2) @ 50.0
# rule schema 13
'RNAP_antiTerm' RNAP(d!1),DNA47(t!1),DNA48(t),
   S2(s~hp,!2),S3(s~hp,!2) -> DNA47(t),RNAP(d!3),
   DNA48(t!3),S2(s~hp,!2),S3(s~hp,!2)  @ 50.0
```

All quantitative information reported in this article is indeed obtained from Kappa *stories*, which required additional control molecules that do not represent any actual biological actor. We refrain from showing this version of the code here, but it is available from the authors upon request.

Modelling and Analysis of the NF-κB Pathway in Bio-PEPA

Federica Ciocchetta[1], Andrea Degasperi[2], John K. Heath[3], and Jane Hillston[4]

[1] Microsoft Research - University of Trento Centre for Computational and Systems Biology, Trento, Italy
ciocchetta@cosbi.eu

[2] Department of Computing Science, University of Glasgow, G12 8QQ, UK
andrea@dcs.gla.ac.uk

[3] School of Biosciences, University of Birmingham, B15 2TT, UK
J.K.Heath@bham.ac.uk

[4] Laboratory for Foundations of Computer Science, University of Edinburgh, EH8 9AB, UK and Centre for Systems Biology at Edinburgh (CSBE)*, Edinburgh, Scotland
jeh@inf.ed.ac.uk

Abstract. In this work we present a Bio-PEPA model describing the Nuclear Factor κB (NF-κB) signalling pathway. In particular our model focuses on the dynamic response of NF-κB to an external stimulus. Each biochemical species in the pathway is represented by a specific Bio-PEPA component and the external stimulus is abstracted by time-dependent Bio-PEPA events describing the start and the end of the signal.

The Bio-PEPA model is a formal intermediate representation of the pathway on which various kinds of analysis can be performed. Both stochastic and deterministic simulations are carried out to validate our model against the experimental data and *in-silico* experiments in the literature and to verify some properties, such as, the impact of the duration of the external stimulus and of the total NF-κB initial amount on the behaviour of some species of interest. Furthermore we use stochastic simulation to compare the behaviour of the single cell against the average behaviour of a population of cells. Finally, sensitivity analysis is considered to investigate the most influential parameters of the model. Importantly, the approach taken suggests that the sensitivity of some parameters alters with the time evolution of the pathway.

Keywords: Process algebras, NF-κB pathway, modelling, analysis.

1 Introduction

Nuclear Factor κB (NF-κB) is a protein complex that regulates numerous genes that play important roles in inter- and intra-cellular signalling, cellular stress response, cell growth, survival and apoptosis [1,2]. The investigation of the specific mechanisms that

* The Centre for Systems Biology at Edinburgh is a Centre for Integrative Systems Biology (CISB) funded by the BBSRC and EPSRC in 2006.

C. Priami et al. (Eds.): Trans. on Comput. Syst. Biol. XII, LNBI 5945, pp. 229–262, 2010.

govern NF-κB activities is essential for the understanding of various biological processes and for the potential use of NF-κB as a drug target. In the literature there are numerous models describing different aspects of the NF-κB pathway [3,4,5,6,7,8]. These models are based on various assumptions about the biochemical mechanisms involved and describe specific subsets of species and interactions. Most of them are defined in terms of Ordinary Differential Equations (ODEs) and the validation and analysis are based on numerical integration of ODEs. Furthermore, in [9,10] Ihekwaba *et al.* proposed a *Gepasi* [11][1] model of the pathway and analysed it using deterministic (i.e. ODE-based) sensitivity analysis.

Recently, there have been various applications of process algebras for the modelling and analysis of biochemical networks [12,13,14,15,16,17]. These formalisms were originally defined in the context of concurrent systems in computer science and are useful in the field of systems biology too. In particular, among their various advantages and properties, they offer a formal model of a system in terms of interacting components, and support a compositional approach to model construction. They exemplify *algorithmic* or *executable* systems biology [18,19], an approach in which the intention is to construct models which are more than simply a mathematical function which recreates the mapping from system input to system output.

In this work we consider the process algebra Bio-PEPA [20,21] and define a Bio-PEPA model for the NF-κB pathway earlier described by Lipniacki *et al.* [5] in terms of a system of ODEs (we will refer to this as the *Lipniacki model*). Our choice is motivated by the fact that, at the time of our study, the Lipniacki model was the most recent concerning the NF-κB pathway and it made more realistic assumptions than previous models. Specifically, the model is characterized by explicit handling of compartments, with sizes derived from biological knowledge [22], and by the transport of species between compartments. Furthermore, the model takes into account some experimental constraints on important species at the steady state (for instance, the level of free IκBα is less than 15% of the total IκBα [22]). The pathway considered, and the data used for the validation, derive from mouse fibroblasts [23,3].

The aim of this work is twofold. Firstly, we demonstrate the power of Bio-PEPA as a modelling language for biochemical networks. In particular, we show how to capture some features of these networks in Bio-PEPA, such as static compartments and the presence of external stimuli which cause the activation of some reactions, using locations and temporal events, respectively. Furthermore, the model also benefits from Bio-PEPA's support for generic kinetic laws by means of functional rates, the explicit definition of stoichiometry and recording the role of each species in a reaction.

Secondly, we use some of the analysis techniques defined for Bio-PEPA in order to extend the existing analysis of the model. Whilst previous work focused on a set of ODEs which were subject to numerical integration and deterministic sensitivity analysis, we perform stochastic simulation to verify the possible variability across several

[1] Gepasi is a software package for modelling biochemical systems. It provides a number of tools to fit models to data, optimize any function of the model, perform metabolic control analysis and linear stability analysis. Gepasi translates the language of chemistry (reactions) to mathematics (ODEs).

runs of the model[2]. We first consider the average of several runs in order to obtain the average behaviour of a population of cells and validate our results against the experimental data, derived from a population of mouse fibroblasts [23,3]. Then we focus on single simulation runs and compare them with the average of several simulation runs. We are particularly interested in the oscillatory behaviour of nuclear NF-κB as it seems to have an important role in essential activities of the cell [7,8]. With this in mind, we perform some in silico experiments to investigate the impact of the duration of the signal and the influence of two inhibitors of the NF-κB, the proteins A20 and IκBα, on the oscillations. Finally, we use sensitivity analysis on that stochastic model to isolate the most influential parameters. This sensitivity analysis is applied using a novel algorithm, based on the definition of histogram distance over the simulation runs [24], implemented in the version of the Dizzy simulator developed at the University of Edinburgh [25]. Our analysis is complementary to the previous work in [9]; in particular the model properties are analysed from a different point of view and the differences between the two approaches are discussed.

A preliminary study of the pathway in Bio-PEPA was presented at the Dagstuhl seminar "Formal Methods in Molecular Biology" [26]. Here we give more details about the pathway and the approach used and we report new analysis results and experiments, especially with respect to the sensitivity analysis.

The rest of the paper is structured as follows. The NF-κB pathway and the Bio-PEPA model of the pathway are described in Sect. 2 and Sect. 5, respectively. Sect. 3 report an overview of related work. Bio-PEPA is introduced in Sect. 4. In Sect. 6 the validation of our model and some analysis results are presented. Finally, in Sect. 7 we report some concluding remarks.

2 The NF-κB Pathway

In the following we describe the pathway as captured by the Lipniacki model since it is our reference for this work. A general schema of the pathway is reported in Fig. 1. The main species involved in the pathway are the IκB kinase (IKK), NF-κB, the protein A20, the protein IκBα, their complexes, mRNA transcripts of A20, IκBα and a hypothetical control gene (cgen). The species cgen represents a control gene, regulated by NF-κB, distinct from the genes corresponding to A20 and IκBα. In the absence of an external stimulus (i.e. normal condition), NF-κB is bound to the inhibitor protein IκBα and remains in the cytoplasm. When an upstream stimulus (SIGNAL), such as the Tumor Necrosis Factor (TNF) or the interleukin-1α (IL-1α), is received, the IKK protein in the neutral form[3](IKKn) is transformed into its active phosphorylated form (IKKa) and then it is modified, under the influence of the stimulus and the protein A20, into another inactive form (IKKi). The inactive form IKKi is different from IKKn as it is overphosphorylated. The activation of IKK is enabled only when the stimulus is

[2] Note that each run can represent the behaviour of a single cell. If we assume that each cell is independent of the others, multiple simulation runs can approximate the average behaviour of a population of cells.

[3] Neutral refers to IKK in absence of any extracellular stimuli. Neutral IKK does not interact with IκBα and therefore does not trigger the cascade of the NF-κB pathway.

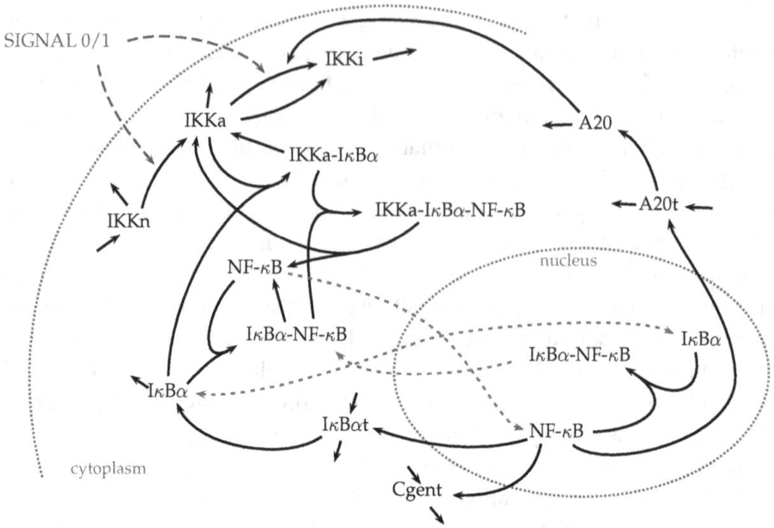

Fig. 1. Schematic depiction of the NF-κB signalling pathway considered in the paper. Signal 0 corresponds to the absence of the external stimulus, signal 1 corresponds to the presence of the stimulus. The red long dashed arrows are the interactions triggered by the signal, the brown short dashed arrows the transport reactions between compartments, black continuous arrows represent all the other kinds of interaction (association and dissociation reactions, translation of mRNAs into proteins). Compartments are delimited by green dotted lines.

present whereas its inactivation is possible both in the presence of the stimulus (in this case it is activated by A20) and also in the normal condition. When activated, IKKa can trigger the degradation of IκBα, which has the effect of releasing free cytoplasmic NF-κB. This enters the nucleus and upregulates the transcription of the two inhibitors, A20 and IκBα, and a large number of other genes (represented by cgen in Fig. 1). The newly synthesized IκBα again inhibits NF-κB while A20 can inhibit IKKa by catalysing its transformation into IKKi, which is no longer able to trigger the degradation of IκBα.

The pathway is characterized by the following main features.

1. There are *two compartments*, the nucleus and the cytoplasm. Realistic compartment sizes have been obtained from experiments and this information is taken into account in the derivation of rates and concentrations. These data relate to mouse fibroblasts [22]. Cytoplasmatic NF-κB, the complex nuclear IκBα-NF-κB, nuclear and cytoplasmatic IκBα, can move from one compartment to the other.
2. mRNA transcripts move from the nucleus to the cytoplasm as soon as they are created. Therefore, the translation of mRNA in the associated proteins happens in the cytoplasm. Whilst transcription is stimulated by NF-κB (the *inducible term* in the ODE representation), a basic rate of transcription will be observed at all times (the *steady term* in the ODE representation).
3. The *external stimulus* is represented by a *signal*. In [5] Lipniacki *et al.* assume that the stimulus is of long duration (persistent); indeed it is active for 6 hours,

starting after 1 hour. Short pulse-like stimuli can also be considered to describe specific kinds of inflammation [3,8]. The effect of the signal is to enable some reactions (i.e. the rate becomes non-zero); specifically, the activation of IKKn and the transformation of IKKa into IKKi.

4. There are two *regulatory feedback loops*: the former involves IκBα and the latter A20. Indeed nuclear NF-κB upregulates the transcription of both proteins and these, in turn, inhibit the activity of NF-κB. In the latter feedback loop the action of A20 on the regulation of NF-κB is not direct: A20 inactivates IKKa, this stops the degradation of IκBα and, consequently, there is an increase in the inhibition of NF-κB. We do not consider the third feedback loop in Ashall *et al.* [8] as in our model we consider just one of the IκB isoforms.

Some simplifications are made in the construction of the model (the most important ones are reported below). In particular, it involves a restricted number of species and reactions. As explained above, the model closely follows the Lipniacki model, reflecting the available information and making a number of assumptions in order to simplify the analysis [5].

- First, NF-κB and IKK are protein complexes, but the details of their structure and the complicated kinetics leading to their formation are neglected. Specifically, NF-κB proteins are small groups of dimeric transcription factors which consist of different members (for instance, in mammals these are RelA, Rel, RelB, p50, p52).
- Second, the inhibitory proteins A20 and IκBα mimic the common activity of groups of inhibitors. For instance, IκBα is just one of the possible IκB isoforms involved in the pathway. The choice to consider only IκBα and not all the other isoforms (see for instance Hoffmann's model [3]) reflects the fact that this isoform is the most active and abundant in the cell and its absence, in contrast to the other isoforms, is lethal [27].
- Third, all the other proteins which are not considered in the model remain at their normal (i.e. in the absence of signal) levels.
- Fourth, IKK has three different forms. Each of them undergoes degradation with the same rate and the normal form IKKn is the only one that can be synthesised. We can obtain the inactive form only from the active form IKKa and this is in part independent from the external stimulation.

The Lipniacki model describes the pathway reported above in terms of a system of ODEs, where variables stands for species concentrations (in micro molars, μM). All the kinetic laws associated with the various interactions are simply mass-action.

In order to validate the model, Lipniacki *et al.* analyse its ability to reproduce the data from experiments on mouse fibroblasts. Hoffmann *et al.* report measurements involving wild type cells (i.e. without external stimulation) in response to persistent and pulse-like TNF activation [3]. On the other hand, Lee *et al.* measure the response of wild type and A20-deficient cells to a persistent TNF signal [23]. The experimental data concern the average behaviour of a cell population (10^6 cells in [3]). We consider these experimental data to validate our model as well.

Due to the large number of unknown parameters, Lipniacki *et al.* proposed the following parameter estimation approach. They started from a reasonable set of parameters obtained from the literature (for instance from the experiments in [22,23]) able to produce a correct steady state in the absence of a TNF signal. Then they proceeded with the signal initiated by TNF and iterated until the fit to all the available data is satisfactory. In the definition of the parameters, the difference between the nuclear and the cytoplasmatic volumes and constraints following from experimental data are taken into account. In particular, as the original system of ODEs is in terms of concentrations, Lipniacki *et al.* took the proportion factor between the two compartments $k = V_c/V_n = 5$ into account in the kinetic laws of the reactions involving reactants and products in two different compartments, such as the transport of a species from the nucleus to the cytoplasm. Indeed, for a given number of molecules, the corresponding concentration of the species in the cytoplasm is k times less than the corresponding concentration in the nucleus and both the concentrations are present in the differential equations in the terms representing transport and interactions between compartments. The explicit and realistic use of compartments leads to more realistic constant parameters than in the previous model by Hoffmann *et al.* [3], especially with respect to the translation and the transport rates.

3 Related Work

There is a vast literature of models for the NF-κB signalling pathway [3,4,5,6,7,8], and each model focuses on particular aspects of the pathway and refers to different sets of available experimental data.

The first attempt to model the IκB–NF-κB signalling module was made by Hoffmann *et al.* [3]. They defined an ODE model describing the interplay between three isoforms of the inhibitor proteins IκB (IκBα, IκBβ and IκBϵ) and NF-κB, under both a persistent and short pulse-like TNF stimulus. They proved that IκBα is responsible for a strong negative feedback loop that allows a fast turn-off of the NF-κB whereas the other two isoforms reduce the system's oscillations and stabilize NF-κB responses during longer stimulation. In contrast to the Lipniacki model, they assumed that the two compartments (i.e. the nucleus and the cytoplasm) have the same volume. This assumption simplified the definition of the system of ODEs, but leads to unrealistic transport rates. Their model is fitted against the experimental data reported in the paper and was able to reproduce the expected behaviour. However, the model does not satisfy all the experimental constraints; for instance constraints on the quantity of free IκBα are not considered. Moreover it does not consider some keys species and mechanisms, such as the protein A20 and the IKK activation/disactivation. These two proteins and the related processes have an essential role in the pathway; indeed the knockout of A20 in mice dramatically alters the cell response to TNF stimulation due to persistent IKK activity and causes A20-deficient cells to die prematurely [23]. These shortcomings are addressed in the Lipniacki model and, therefore, in our work too.

Subsequent work by Ihekwaba *et al.* conducts a sensitivity analysis of the model presented by *Hoffmann et al.* [9,10]. In this case the model was defined in the notation of Gepasi [11], but it is a close representation of Hoffmann *et al.*'s ODE model. Thus as with that earlier model, Ihekwaba *et al*'s model fails A20 and IKK activation/inactivation, and supposed equal compartment sizes for the cytoplasm and the nucleus. In translating the model, the authors found some discrepancies in the Hoffmann's supplementary material and proposed a new set of parameters for ten of the reactions of the pathway in order to fit the available experimental data. These parameters concern the synthesis of the various IκB isoforms and association/dissociation of complexes involving NF-κB, IKK and IκB isoforms[4]. The Gepasi model was then mapped to ODEs in order to obtain the temporal evolution of the species and analysed using the parameter scan capability in order to identify those parameters in the IκB–NF-κB system (containing only the IκBα isoform) that most affect the oscillatory concentration of nuclear NF-κB, in terms of *period* (time taken for one oscillation), *phase* (the timing of the beginning of the period) and *amplitude* (the range values attained during an oscillation). Parametric sensitivity analysis was performed on all the system's parameters: each parameter was considered in isolation and varied in order to see the impact of its variation on the behaviour of nuclear NF-κB. Of the 64 parameters in the model, just nine exerted significant influence and these mainly involved IκBα and IKK. A more advanced analysis of the parameters was reported in [10], where pairwise modulation of the nine parameters found in the previous study was carried out. Synergistic effects were observed: the effect of one of the parameters was strongly dependent on the values of another parameter, proving a very strong non-linearity in the system, as expected.

A hybrid variant of the Lipniacki model is proposed by the same authors in [6]: ordinary differential equations, used for description of fast reaction channels of processes involving a large number of molecules, are combined with a stochastic switch to account for the activity of the genes involved. In this way the authors improved simulation efficiency and defined a model able to appropriately handle the small numbers of transcripts. This model was defined in MATLAB and an *ad hoc* algorithm for the simulation was implemented. In the other models defined in terms of ODEs these species are approximated by continuous variables with very low concentrations. This simplification is not realistic. However, for this pathway, it gives correct average behaviours.

Recently a new model for the NF-κB pathway has been proposed by Ashall *et al.* [8]. The authors define a hybrid model, similar to the one proposed in [6] and an associated simulation algorithm, able to capture the behaviour of the pathway under repeated short pulses of TNF at various intervals, mimicking pulsatile inflammatory signals. In order to reproduce the new experimental results based on pulsatile stimulation they modified the Lipniacki model. Specifically they proposed a new set of reactions to describe the IKK activation/disactivation. Differently from the Lipniacki model, inactive IKK (IKKi) can be transformed into neutral IKK (IKKn) and A20 is an inhibitor of this interaction. Furthermore, a third feedback loop is added describing the influence of the isoform IκBϵ on the pathway. Ashall *et al.* used their model to investigate the possible role of the NF-κB oscillations in the systems and how these oscillations are affected, for instance, by the various TNF pulse frequencies and the presence of feedback loops. In the Lipniacki model this third loop is not considered because the focus is on the most abundant and important isoform of IκB.

Cho *et al.* [4] focused on just the first part of the signalling cascade up to the activation of NF-κB, without considering the translocation into the nucleus and the activation of the transcription of the various genes by NF-κB. The authors proposed an ODE model describing in detail the various interactions that leads to the activation of NF-κB.

[4] Note that most of these reactions are not present in the Lipniacki model.

All the models reported above are defined directly in terms of systems of ODEs, as hybrid models or as Gepasi models (and then translated into ODEs). In the literature there are just a few applications of process algebras for the modelling and study of this pathway [28,29]. In both these papers the focus was on the modelling of the pathway using process algebras and just the validation of the model against the literature was reported. Larcher *et al.* [29] represented the pathway previously described by Hoffmann *et al.* [3] in *BetaWB*, a language based on Beta binders [15]. A clear mapping exists from biochemical entities, such as species and reactions, into the BetaWB language. Each species is abstracted by a bio-process, a box with an interface representing its interaction capabilities. The interactions among species are represented by various kinds of actions such as the formation of complexes and decomplexation, the communication between species and the change of the interface, the deletion or creation of a box, the join and split of boxes. The analysis is based on stochastic simulation and on the comparison with the results in the literature. Note that in [3] the two compartments are assumed to have the same size. The compartments are not considered explicitly in the BetaBW model but their equal size is reflected in the derivation of the number of molecules and stochastic rates. In [28] Hillston and Duguid defined a PEPA [30] model for the pathway presented in Cho *et al.* [4]. The level of abstraction is higher than the one in the work by Larcher *et al.*, in particular, species are abstracted by processes and all the reactions by interactions between processes. Hillston and Duguid considered the reagent-centric style of PEPA, on which Bio-PEPA is based. However, in PEPA it is not possible to defined some features of biochemical systems, such as stoichiometry and events. Furthermore, in the standard version, PEPA does not support functional rates. In [28], the map from the PEPA model into the associated ODE model is presented and ODE numerical integration results are shown.

4 Bio-PEPA

In this section we give a short description of Bio-PEPA [20,21], a language that has been developed for the modelling and analysis of biological systems. Recently Bio-PEPA has been extended to incorporate events [31] and to support biological locations [32], two features which will be useful in developing our model of the NF -κB pathway.

The main descriptive components of a Bio-PEPA system are the *species (or sequential) components*, describing the behaviour of each species, and the *model component*, describing the interactions between the various species. The species initial amounts are given in the model component.

The syntax of the Bio-PEPA components is defined as:

$$S ::= (\alpha, \kappa)\ \mathsf{op}\ S \mid S + S \mid C \mid C@L \qquad \text{with } \mathsf{op} = \downarrow \mid \uparrow \mid \oplus \mid \ominus \mid \odot$$

$$P ::= P \underset{\mathcal{H}}{\bowtie} P \mid S[x]$$

where S is the *species component* and P is the *model component*. We assume a countable set of model components C, a a countable set of locations $\mathcal{L}$ and a countable set of action types $\mathcal{A}$. These three sets are disjoint. In the prefix term $(\alpha, \kappa)\ \mathsf{op}\ S$, the action type α is ranged over by the set $\mathcal{A}$ and abstracts a reaction of the network, $\kappa \in \mathbb{N}$ is the *stoichiometry coefficient* of species S in reaction α and the *prefix combinator* "op" represents the role of S in the reaction. Specifically, $\downarrow$ indicates a *reactant* (i.e. the amount of the species decreases), $\uparrow$ a *product* (i.e. the amount of the species increases), $\oplus$ an *activator* (i.e. the species activates the reaction without modifying its amount), $\ominus$ an *inhibitor* (i.e. the species inhibits the reaction without modifying its amount) and $\odot$ a generic *modifier*. The general modifier operator is useful to indicate species that are involved in a reaction without changing their concentration but which cannot be classified as activators or inhibitors (e.g. a gene during transcription). We can use "(α, κ) op " as an abbreviation for "(α, κ) op S". The operator "+" expresses the choice between possible actions, and the constant C is defined by an equation $C \stackrel{def}{=} S$. The notation $C@L$ indicates that the species represented by the component C is in the location L.

The process $P \underset{\mathcal{H}}{\bowtie} Q$ denotes synchronisation between components P and Q, the set $\mathcal{H} \in 2^{\mathcal{A}}$ determines those action types on which the operands are forced to synchronise, with $\underset{*}{\bowtie}$ denoting a synchronisation on all common action types. Note that the synchronization of components on a given action type α represents the participation of the corresponding species in the same reaction (abstracted by the action type α). In the model component $S[x]$, the parameter $x \in \mathbb{R}^+$ represents the initial value. The reader is referred to [21] for further details of the language and its semantics.

In addition to species and model components, a Bio-PEPA system is characterised by a context containing the constant parameters, the functional rates, the locations, the possible events and auxiliary information about the species.

The parameters are defined in the model by means of a set of parameter definitions $\mathcal{K}$. Each parameter is defined by "k_{name} = value unit", where "$k_{name} \notin C$" is the parameter name, "value" denotes a positive real number and the (optional) "unit" denotes the unit associated with the parameter.

In order to collect the information about the dynamics of the system, we associate a functional rate f_α with each action type α. The set of functional rates is denoted $\mathcal{F}_R$. The function f_α can depend on parameters, names of species components and possibly on simulation time and represents the kinetic law of the associated reaction as a mathematical expression. The mathematical expressions are defined in terms of mathematical operators or predefined functions, expressing well-known kinetic laws such as mass-action, Hill kinetics and Michaelis-Menten. In the former case the names of the parameters and the names of the species components involved in the reaction must be given whereas with the predefined kinetic laws the components/species are derived from the context.

Locations represent both biological compartments (such as nucleus, cytoplasm, $\cdots$) and membranes. Membranes represent the boundaries of compartments and may or may not be explicitly included. Every model must have at least one compartment. Each location is described by "L : *s unit, kind*", where L is the (unique) location name, "*s*" expresses the size and can be either a positive real number or a more complex mathematical expression depending on time t; the (optional) "*unit*" denotes the unit of measure associated with the location size, and "*kind*" $\in \{\mathbf{M}, \mathbf{C}\}$ expresses if it is a membrane or a compartment, respectively. Although the relative position of locations is assumed to be static, their size may change with time by expressing the volume or area as a function of time. In this latter case only specific kinds of analysis are supported (i.e. numerical integration of ODEs).

A key reaction involving location is the translocation of a species S from one location L_i to the location L_j. This is simply abstracted by a reaction $S@L_i \rightarrow S@L_j$, where $S@L_i$ is the reactant and $S@L_j$ is the product.

Events are constructs that represent changes in the system due to some triggering conditions. This allows biochemical perturbations to the system to be represented, such as the timed introduction of reagents or the modulation of system components by external stimuli. A Bio-PEPA event has the form *(id, trigger, event_assignment, delay)*, where:

- *event_id* is the event identifier,
- *trigger* is a mathematical expression that, when it evaluates to true, makes the event fire. It can be composed of one or more conditions involving the components of the Bio-PEPA model and/or time;
- *event_assignment_list* is a list of changes (assignments) to elements of the system in response to the event;
- *delay* is the length of time between when the event fires and when the event assignments are executed. *delay* is either 0 (*immediate events*) or a positive real value (*delayed events*). In the model we consider in this paper we consider just immediate temporal events, i.e., they are not delayed and their trigger involves time.

Note that events are added to the language as a distinct set of elements and the rest of the syntax is unchanged in order to keep the specification of the model as simple as possible. This approach is particularly useful when the same biochemical system is studied under different experimental regimes as the list of events can be modified without any changes to the rest of the system. Details of analysis supporting events are reported in [31].

Finally, a set $\mathcal{N}$ is defined in order to collect some auxiliary information about the species used in some kinds of analysis supported by Bio-PEPA.

In order to illustrate Bio-PEPA syntax, we show how a simple network can be specified in Bio-PEPA. This network is composed of the following two reactions (in chemical reaction form):

$$S + E \longrightarrow P + E\,; \qquad P \longrightarrow$$

The former is an enzymatic reaction describing the transformation of a substrate S into the product P with the help of the enzyme E and the latter is the degradation of the product P. The kinetic law for the enzymatic reaction is $f_E = \frac{v_M \cdot E \cdot S}{(K_M + S)}$ whereas for the degradation it is $f_{deg} = k \cdot P$. All the species are in the same location L.

The three species can be specified in Bio-PEPA by the following components:

$$S@L \stackrel{def}{=} (\alpha, 1)\downarrow \qquad P@L \stackrel{def}{=} (\alpha, 1)\uparrow + (\beta, 1)\downarrow \qquad E@L \stackrel{def}{=} (\alpha, 1)\oplus$$

The action type α abstracts the enzymatic reaction and β the degradation. The species S is only involved in the reaction α as a reactant (i.e. it is consumed). The enzyme E is only an activator for α. The product P can take part in both reactions. In the case of α it is a product (i.e. it is created) and in the case of β is a reactant (i.e. it is consumed). In all cases the stoichiometry is one.

The system is described by

$$S@L[x_{S,0}] \underset{\{\alpha\}}{\bowtie} E@L[x_{E,0}] \underset{\{\alpha\}}{\bowtie} P@L[x_{P,0}]$$

where $x_{S,0}$, $x_{E,0}$ and $x_{P,0}$ are the initial values of the three species and the functional rates are $f_\alpha = fMM(v_M, K_M)$, i.e. Michaelis-Menten kinetics with parameters v_M, K_M, and $f_\beta = fMA(k)$.

Bio-PEPA offers a formal intermediate compositional representation of biochemical systems, on which different kinds of analysis can be carried out, through defined mappings into continuous-deterministic and discrete-stochastic mathematical models. The Bio-PEPA language is supported by software tools (for instance the Bio-PEPA Workbench [33]) which automatically process Bio-PEPA models and generate other representations in forms suitable for different kinds of analysis [21,34]. In particular, the generated simulation model can be executed using MATLAB [35] and the Dizzy simulation tool [36], in which both stochastic simulation algorithms and differential equation solvers are implemented. Here we use a version of the Dizzy simulator developed at the University of Edinburgh [25], which extends the original tool with sensitivity analysis techniques and additional simulation methods. Some events, such as, for instance, time-dependent events, can be translated into time-dependent reaction rates in the Dizzy model (defined in terms of the step function *theta*, which is predefined in Dizzy). Most stochastic algorithms and ODE solvers in Dizzy support this function.

5 A Bio-PEPA Model for the NF-κB Pathway

In the following we illustrate the Bio-PEPA model describing the NF-κB pathway presented in Sect. 2. We show the mapping from each biochemical entity (species, reaction, ...) to Bio-PEPA. We report just the main ideas of the abstraction, the full model is reported in Appendix A and available from the Bio-PEPA web page [34].

The pathway is characterised by the presence of two compartments (and the transport of some species between them) and by the influence of an external signal. These features can be easily represented in Bio-PEPA using locations and events.

Compartments. The *nucleus* and the *cytoplasm* are abstracted by *locations* in Bio-PEPA:

$$\begin{aligned} &\textit{location nuc} : \textit{kind} = \mathbf{C},\ \textit{size} = 3.33 \cdot 10^{-13}\, l;\\ &\textit{location cyt} : \textit{kind} = \mathbf{C},\ \textit{size} = 1.65 \cdot 10^{-12}\, l \end{aligned}$$

They are both of kind **C** (i.e. compartments) and their sizes are as given in [5].

Reactions. Each *reaction* is associated with an *action type* and with a *functional rate*. For instance, in the case of degradation of the protein A20 we have the action type *a20_degradation* and the associated functional rate:

$$f_{a20\_degradation} = fMA(c5)$$

where $fMA(r)$ stands for mass-action with rate constant r and $c5$ is the constant degradation rate for $A20$.

Species. Each *biochemical species* in the pathway is abstracted by a *species component*, describing its behaviour in terms of the interactions in which it is involved. For instance the protein A20 is represented as:

$$\begin{aligned} A20@cyt \stackrel{def}{=} \; & (a20\_translation, 1) \uparrow + \\ & (a20\_degradation, 1) \downarrow + \\ & (transformation\_IKKa\_into\_IKKi\_by\_A20, 1) \oplus \end{aligned}$$

This species is in the cytoplasm and it is involved in three interactions: its translation, its degradation and as an activator for the transformation of IKKa into the inactive form IKKi. The abbreviation presented in Section 4 is used. In all the three cases mass-action kinetic laws are considered [5].

Model Component. The species and their possible interactions are represented by the *model component*:

$$I\kappa B\alpha\text{–}NF\text{-}\kappa B@cyt[60000] \underset{*}{\bowtie} IKKn@cyt[0] \underset{*}{\bowtie} IKKa@cyt[0] \underset{*}{\bowtie} \cdots A20@cyt[0]$$

where the number between square brackets represents the initial number of molecules of each species.

Signal. The *signal* (TNF stimulus) is abstracted in Bio-PEPA by two *time-dependent events*, representing the start and the end of the signal.

$$\begin{array}{c} (begin\_signal,\; t = T_1,\; signal = 1, 0); \\ (end\_signal,\; t = T_2,\; signal = 0, 0) \end{array}$$

In our case $T_1 = 3600$ and $T_2 = 7 \cdot 3600 = 25200$ seconds (corresponding to 1 hour and 7 hours, respectively) and both events are immediate.

To simplify the handling of compartments for Bio-PEPA with locations we prefer to have the model expressed in terms of numbers of molecules, rather than concentration, for each species. Otherwise concentrations have to be converted to account for different volumes whenever molecules move between compartments. Considering the number of molecules is thus more convenient. Note that the rates and initial values for species in Lipniacki's model [5] are expressed in concentrations (μM). In order to derive a model in terms of molecule numbers for Bio-PEPA, the continuous concentration values are translated into discrete numbers of molecules and the rates are modified in order to take this transformation into account (see [37] for details).

In our model we have two compartments with different volume sizes, so we have to define two scaling factors for the transformation from concentrations to molecules and use them according to the location of each species and reaction. Specifically, the two scaling factors are:

$$nscale \;=\; V_n \cdot N_A \cdot 10^{-6} \;=\; 2 \cdot 10^5 \quad \text{molecules}/\mu\text{mol} \cdot \text{L}$$

[5] Note that in the reaction describing the transformation of IKKa into IKKi, the protein A20 remains constant and it is a sort of activator of the reaction. The reaction is described as $IKKa + A20 \rightarrow IKKi + A20$ with kinetic law $k_2 \cdot IKKa \cdot A20$.

$$cscale = V_c \cdot N_A \cdot 10^{-6} \approx 10^6 \quad \text{molecules}/\mu\text{mol} \cdot \text{L}$$

where *nscale* and *cscale* are the scaling factors, $N_A = 6.022 \cdot 10^{23}\,\text{mol}^{-1}$ is the Avogadro number and V_n and V_c, the volume sizes, for the nucleus and cytoplasm, respectively. As the concentrations are in terms of μM instead of M we have to multiply by 10^{-6}. As we consider numbers of molecules, we do not need to use the factor k, used to take into account the compartment volumes in the reactions involving species in different compartments when concentrations are used [5].

A list of all parameter values used in our model is reported in Table 1.

Table 1. Parameters and reactions of the model. For more details see [5,6].

Parameter	Value	Unit	Description
nscale	$2 \cdot 10^5$	molecules/μmol · L	scaling factor for the nucleus
cscale	10^6	molecules/μmol · L	scaling factor for the cytoplasm
k	5	-	cytoplasmic to nuclear volume
k_{prod}	0.000025·cscale= 25	molecules · s^{-1}	IKKn production rate
k_{deg}	0.000125	s^{-1}	IKKa, IKKn and IKKi degradation
k_1	0.0025	s^{-1}	IKK activation rate caused by TNF
k_2	0.1/cscale= 10^7	molecules^{-1} s^{-1}	IKK inactivation rate caused by A20
k_3	0.0015	s^{-1}	IKK spontaneous inactivation rate
c_1	$5 \cdot 10^{-7} \cdot k = 2.5 \cdot 10^{-6}$	s^{-1}	A20-inducible mRNA transcription
c_2	0	molecules · s^{-1}	A20-constitutive mRNA transcription
c_3	0.0004	s^{-1}	A20 mRNA degradation
c_4	0.5	s^{-1}	A20 translation rate
c_5	0.0003	s^{-1}	A20 protein degradation
t_1	0.1	s^{-1}	IKKa-IκBα catalysis
t_2	0.1	s^{-1}	IKKa-IκBα-NF-κB catalysis
c_{1_a}	$5 \cdot 10^{-7} \cdot k = 2.5 \cdot 10^{-6}$	s^{-1}	IκBα-inducible mRNA transcription
c_{2_a}	0	molecules · s^{-1}	IκBα-constitutive mRNA transcription
c_{3_a}	0.0004	s^{-1}	IκBα mRNA degradation
c_{4_a}	0.5	s^{-1}	IκBα translation rate
c_{5_a}	0.0001	s^{-1}	spontaneous, free IκBα degradation
c_{6_a}	0.00002	s^{-1}	IκBα degradation (complexed to NF-κB)
a_1	0.5/cscale= $5 \cdot 10^{-7}$	molecules^{-1} s^{-1}	IκBα-NF-κB association
a_2	0.2/cscale= $2 \cdot 10^{-7}$	molecules^{-1} s^{-1}	IKKa-IκBα association
a_3	1/cscale= $1 \cdot 10^{-6}$	molecules^{-1} s^{-1}	IKKa IκBα-NF-κB association
a_{1_n}	0.5/nscale= $2.5 \cdot 10^{-6}$	molecules^{-1} s^{-1}	IκBα-NF-κB association (nucleus)
c_{1_c}	$5 \cdot 10^{-7} \cdot k = 2.5 \cdot 10^{-6}$	s^{-1}	cgen inducible mRNA transcription
c_{2_c}	$0 \cdot k = 0$	molecules · s^{-1}	cgen constitutive mRNA transcription
c_{3_c}	0.0004	s^{-1}	cgen mRNA degradation
i_{1_a}	0.001	s^{-1}	IκBα nuclear import
e_{1_a}	$0.0005 \cdot k = 0.0025$	s^{-1}	IκBα nuclear export
e_{2_a}	$0.01 \cdot k = 0.05$	s^{-1}	IκBα-NF-κB nuclear export
i_1	0.0025	s^{-1}	NF-κB nuclear import

6 Validation and Analysis

The Bio-PEPA model of the NF-κB pathway was implemented in the Bio-PEPA Workbench [33] which supports a variety of analyses (see Fig. 2). We consider the ODE MATLAB model to provide validation against the results reported in [5] and the Dizzy model for stochastic simulation and sensitivity analysis. The simulations are carried out using Gillespie's direct method [37]. This choice was made on the basis of the efficiency of that algorithm with respect to our model and the fact that its implementation within Dizzy supports time-dependent events by means of the *theta* function, as discussed at the end of Sect. 4. Nevertheless we verified that the same results are obtained if other stochastic algorithms in Dizzy which support events are used (results not shown). Throughout the presented analyses the unit of time is one second.

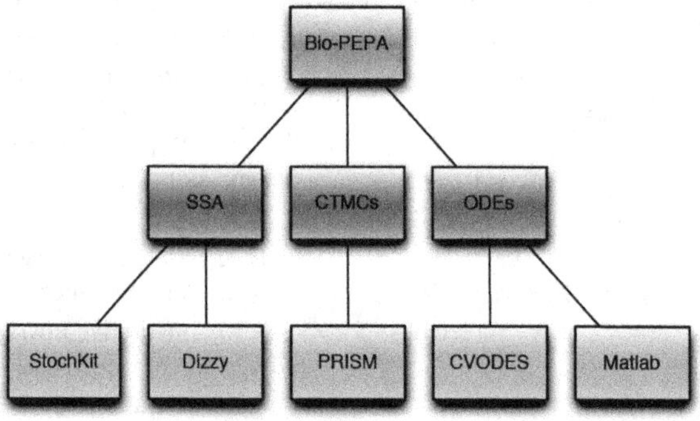

Fig. 2. Schematic view of the analyses available within the Bio-PEPA Workbench, from the Bio-PEPA description of the system (top level) three distinct classes of analysis are accessible (middle level): Stochastic Simulation (SSA), explicit state CTMC techniques such as numerical solution and stochastic model checking, and ordinary differential equations (ODEs); these analyses are supported using existing tools (bottom level)

6.1 Validation

As explained above, in order to validate our model against the original ODE model [5] and the available experimental data in [23,3] showing the behaviour of a cell population, we consider the ODE MATLAB model obtained from the Bio-PEPA Workbench [33].

In [5], Lipniacki *et al.* take the following approach to time series evolution of species. At time 0 just the complex cytoplasmic IκBα–NF-κB is present and all other species are zero. The simulation is run until the resting cell equilibrium state is reached (100 hours). The simulation is then run for a further seven hours, with the external signal enabled after one hour.

The ODE solver used is the MATLAB *ode23tb* [38], designed for stiff systems. Events are handled "by hand": stopping the simulation when the events occur and resetting the system of ODEs with the initial species values set to the final values obtained from the previous simulation and changing the rates of the two reactions enabled by the external stimulus (i.e. activation of IKK and the transformation of active IKK into inactive IKK by means of A20) in the appropriate way. Specifically, the rate constants of the two reactions are initially zero. At the start of the stimulus at time 1 hour these rates are set to values different from zero and, at the end of the stimulus at time 7 hours, they are reset to zero again.

In Fig. 3 we report the results obtained running our MATLAB ODE model with the approach described above. The figure corresponds to the Fig. 3 in [5], obtained using the ODE MATLAB Lipniacki model. The persistent TNF signal (subfigure [A]) causes a pulse activation of IKK (subfigure [C]). The pulse of active IKK (IKKa) initiates the cascade. First IκBα (subfigure [E]) and the complex IκBα–NF-κB (subfigure [F]) are degraded; the released NF-κB moved in the nucleus and, due to the low quantity of nuclear IκBα (subfigure [G]), it rapidly increases. Nuclear NF-κB (subfigure [H]) up-regulates mRNA expression of both IκBα and A20 (subfigures [I] and [H]); the peaks of the associated transcripts are followed by peaks in the corresponding free cytoplasmatic proteins. The new synthesised IκBα binds to NF-κB and leads it out of the nucleus while A20 triggers IKK inactivation. The two negative loops and the movement of species between compartments are fundamental for NF-κB's oscillations and therefore have a large impact of the cell's activities.

The species in our MATLAB model are expressed in terms of number of molecules and, therefore, the species values in Fig. 3 are rescaled in terms of concentration in order to have a more direct comparison with the figures reported in [5]. The results obtained

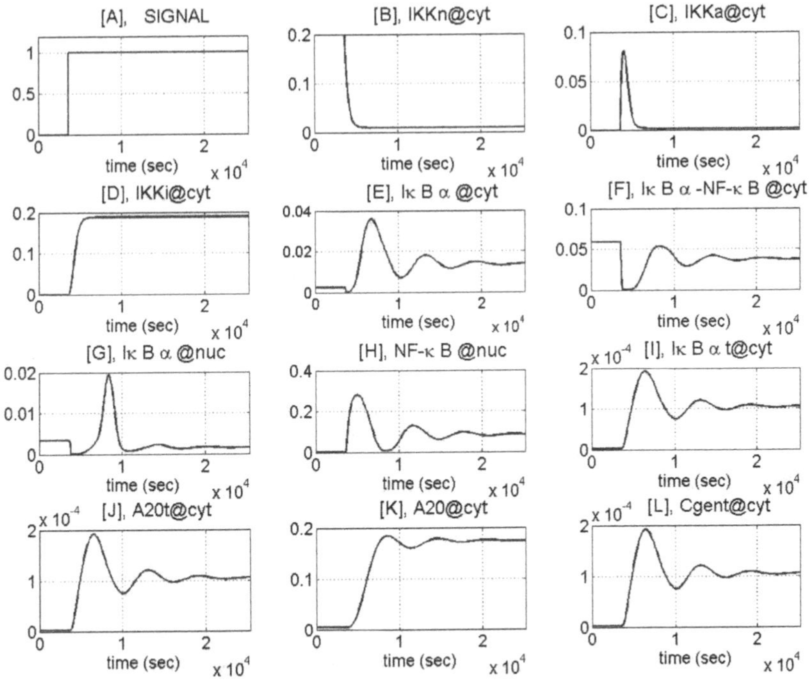

Fig. 3. Validation of the ODE MATLAB model obtained from Bio-PEPA (concentrations). The time (x value) is in seconds and the y values are concentrations (in μM).

are identical to the ones shown in the paper. This also confirms that our conversion from concentration to number of molecules is appropriate. The same results are obtained when the Dizzy tool is used for the simulation.

6.2 Stochastic Simulation: Population vs. Single Cell Behaviour

In this section we consider the Dizzy model derived from our Bio-PEPA and we perform stochastic simulation in order to investigate the effect of stochasticity on the behaviour of the species in the pathway, in particular of nuclear NF-κB.

While deterministic models are good approximations of real biochemical systems when the number of molecules is sufficiently high, at low copy numbers the effect of random fluctuations becomes significant and so stochasticity needs to be taken into account to obtain a faithful representation of the real biochemical system [39]. This is particularly true when the activation of genes is involved, as generally there are few copies of each gene in the cell. For this reason, we decide to consider stochastic simulation for the following analyses of the model.

Note that a single simulation of the model is a specific realization of the underlying stochastic process and therefore can abstract the behaviour of a single cell. If we assume that the behaviour of the cells are independent (at least with respect to the reactions of the pathway under consideration) the comphrensive behaviour of a population of

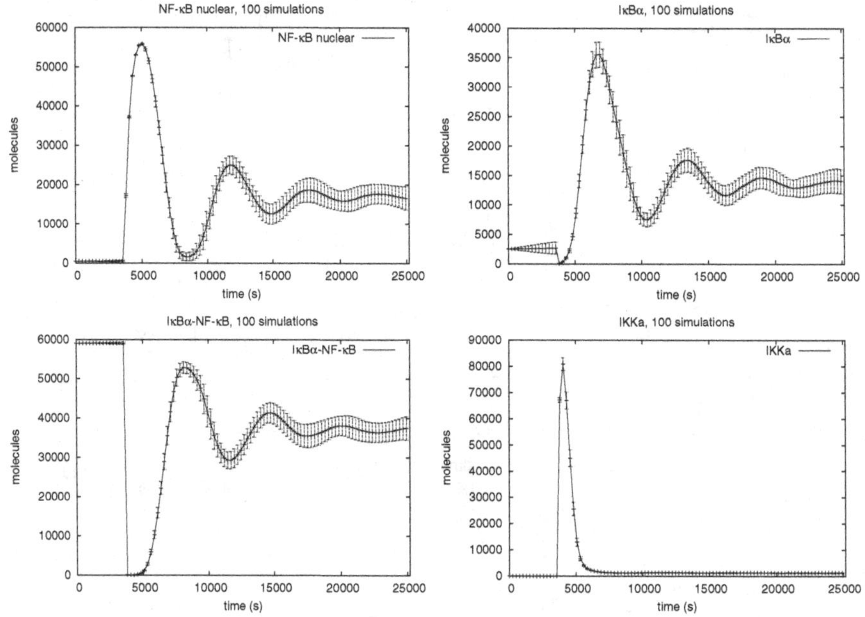

Fig. 4. Stochastic simulation (Gillespie's direct method) for nuclear NF-κB (top-left), cytoplasmic IκBα (top-right), cytoplasmic IκBα–NF-κB (bottom-left) and IKKa (bottom-right). For each of them the average value and the standard deviation are shown. 100 simulation runs are considered. These species correspond to the species in the subgraphs H, E, F and C of Fig. 3, respectively.

cells can be captured by the average of a number of simulation runs. Therefore, in the following we report the graphs showing the average of several stochastic simulation runs as we want to compare them with the experimental data in [23,3], which refer to comphrensive behaviour of a population of cells. On the other hand, we consider the single simulation runs too, in order to show the possible differences between the behaviour of a single cell and a population of cells.

Particular attention must be given to species that are characterized by oscillatory behaviour, indeed the study of oscillations for NF-κB is fundamental as this mechanism seems to have an essential role on important activities of the cell, such as death and immunity. The oscillations need to be largely synchronous to be observed in a cell population, as, if they are out of phase, in the average they are damped, often to the point of invisibility [40,8]. From ODE models in the literature [3,5] (including the ones derived from Gepasi) we can only observe the average behaviour that, for some experiments, is not representative of the behaviour of a single cell, especially with respect to oscillations.

Fig. 4 reports the average of 100 simulation runs and the standard deviation of the species nuclear NF-κB (top-left), cytoplasmic IκBα (top-right), cytoplasmic IκBα–NF-κB (bottom-left) and IKKa (bottom-right). The average behaviour for these species is very close to the deterministic solution. This is unsurprising since the amount of these species is indeed quite high. For nuclear NF-κB, IκBα and IκBα–NF-κB we can observe an average oscillatory behaviour, that is less evident after the first oscillations, when the variability between the different runs is greater. If we observe a single simulation run for the species (in Fig. 5 the results for two species are reported), the species present a persistent oscillatory behaviour, at least for the time interval considered. The oscillations in the various simulation runs are not completely in phase and, therefore, in the average they tend to be damped.

These results are in agreement with the literature [8]. Indeed experimental data concerning the pathway suggest a small degree of synchronization amongst the cell population when we have a persistent or long pulse TNF signal, with more evident synchronization in the first oscillations. However, in the case of shorter stimuli, the experimental results show that, under certain assumptions, a strong synchronization and the cell population behaviour is representative of the behaviour of the single cell [40,8].

Similar results, but with smaller standard deviations, are obtained with a larger number of simulation runs (results not presented). In this case, the simulation time increases.

6.3 In Silico Experiments

We use our model to study the behaviour of the system, particularly nuclear NF-κB, under various assumptions. In the following, we study how the duration of the TNF stimulus affects the behaviour of the nuclear NF-κB and we investigate the regulatory mechanisms, varying the initial total amount of the NF-κB and the inhibitory activity of A20 and IκBα. After that, we explore the behaviour of IκBα mRNA, IKKa and nuclear NF-κB in A20-deficient cells. Finally, we study the effect of varying both A20 and IκBα transcription rates. This last experiment is not present in [5].

The use of stochastic simulation within our analysis means that in addition to the cell population results, previously presented by [5], we can also analyse the behaviour

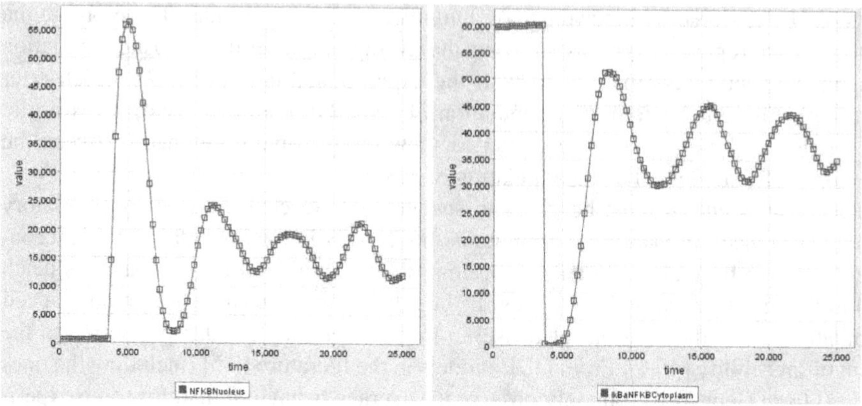

Fig. 5. Single stochastic simulation runs (Gillespie's direct method) for nuclear NF-κB (left) and cytoplasmic IκBα–NF-κB (right). These species correspond to the species in the subgraphs H and F of Fig. 3, and in the graphs top-left and bottom-left in Fig. 4, respectively.

of single cells. In particular, we compare the results obtained from single simulation runs (a single cell) to the average behaviour over many runs (a population of cells). The results of these single runs have yet to be validated against appropriate experiments but the data produced is ideally suited to live cell imaging techniques. This is in contrast to the population case (ODEs and multiple runs of stochastic simulations) which correspond to population approaches such as western blotting. However the results presented here, whilst not yet validated, do show agreement with present knowledge of the system.

Effect of the duration of TNF stimulus on nuclear NF-κB. In the previous sections we have assumed that the TNF stimulus is persistent and lasts for 6 out of 7 simulation hours. It is interesting to investigate how a shorter duration stimulus can affect the nuclear NF-κB activity, in particular the oscillatory behaviour. Different durations describe specific kinds of inflammation [3,8].

Fig. 6 shows the average over 100 stochastic runs for IKKa and nuclear NF-κB when the stimulus lasts for 15 minutes (left), 60 minutes (middle) and 6 hours (original value, right). The three graphs show the same behaviour for both IKKa and nuclear NF-κB for the first three hours. In particular, the pulse of nuclear NF-κB starts after one hour, has a peak at 90 minutes and lasts about one hour. This similarity between the three situations described above may be due to the fact that the pulse of NF-κB is strictly influenced by IKKa and the behaviour of IKKa is the same for all the three cases. Indeed IKK activation seems, at least in part, independent of the duration of the TNF stimulus. This can be explained by the fact that the inactivation rate under the stimulus is very high. Therefore, the inactivation of IKK is very fast and all IKKa is consumed soon after the beggining of the stimulus. However, the duration of the stimulus has an impact on the behaviour of NF-κB after the initial pulse: when the stimulus lasts for 15 minutes or 1 hour nuclear NF-κB drops to a very small amount (2 or 4 molecules, respectively). In contrast, with the longer stimulus we see a pronounced oscillation. The same results

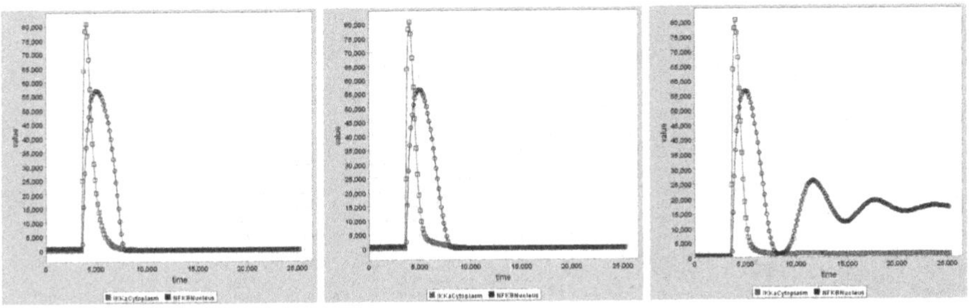

Fig. 6. IKKa (red line) and nuclear NF-κB (blue line) at and after 15 minute-long (left), 60 minute-long TNF stimulation (centre) and at and after 6 hours (right). The graphs show the average amount over 100 simulation runs.

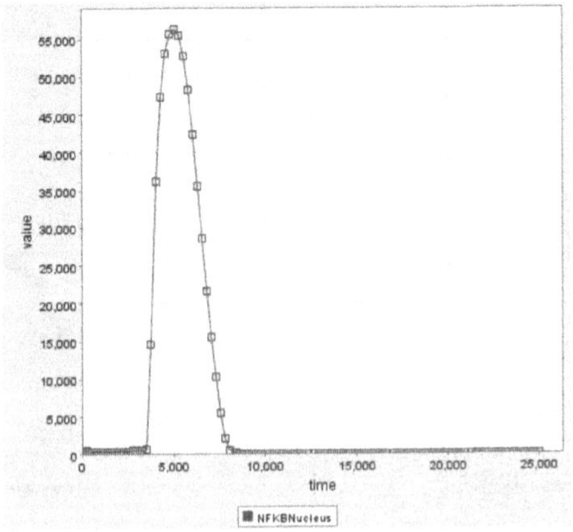

Fig. 7. A single stochastic simulation run for nuclear NF-κB with one-hour stimulus

are obtained when we consider single simulation runs, as we can see in Fig. 7, where a single run for NF-κB with one-hour stimulus is reported. Therefore we can deduce that shorter stimuli do not induce oscillations for nuclear NF-κB either at the level of a single cell nor the level of a population of cells.

These results are in full agreement with the experimental data shown in [3] for wild-type cell under TNF stimuli of short duration.

Study of the regulatory mechanisms. First, we consider various initial amounts for the total NF-κB. In Fig. 8 we report the average behaviour over 100 runs of the nuclear NF-κB when the initial total NF-κB is the original value, three times the original value and one third of the original value. These values have been chosen to reproduce the experimental data and in silico experiments in [5]. At the level of a population of cells, an increase in the amount of total NF-κB makes the oscillation more pronounced whereas, taken in the average, a decrease smooths out the oscillations.

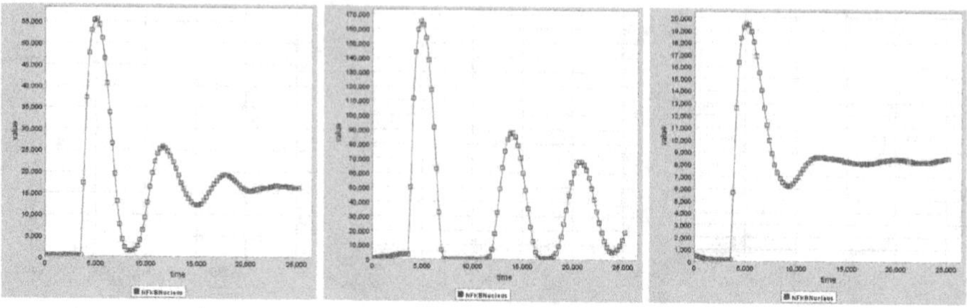

Fig. 8. Nuclear NF-κB when the initial total NF-κB is $6 \cdot 10^4$ (original value, left), is $1.8 \cdot 10^5$ (three times the original value, centre), and $2 \cdot 10^4$ (one third of the original value, right). The graphs show the average amount over 100 simulation runs.

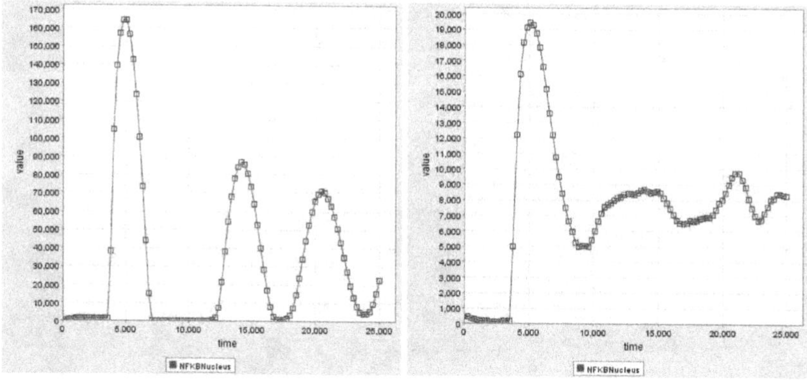

Fig. 9. Single simulation runs for nuclear NF-κB when the initial total NF-κB is $1.8 \cdot 10^5$ (three times the original value, left) and $2 \cdot 10^4$ (one third of the original value, right)

In Fig. 9 we report single simulation runs (single cells) under the same assumptions. For the larger quantity of initial NF-κB the behaviour at the level of the population of cells and at the level of a single cell is identical. On the other hand, when smaller initial quantities are assumed, in the single cell we can observe an oscillatory trend, even though less regular than before, not observable when the average is considered. Therefore, the initial amount of NF-κB seems essential to obtain an oscillatory behaviour, especially when the population of cells is considered.

In order to see how the system behaves at different inhibitor levels, we elevate the A20 and IκBα mRNA transcription rates (c_1 and c_{1a}, respectively) three-fold with respect to the original value whereas the other parameters remain unchanged. As before, these values have been selected in order to compare our results with the experiments in [5]. The results are reported in Fig. 10. The higher level of A20 mRNA transcription leads to a higher level of the protein A20 and therefore to a lower level of IKKa. The resulting nuclear NF-κB has less damped oscillations. A similar result for nuclear NF-κB

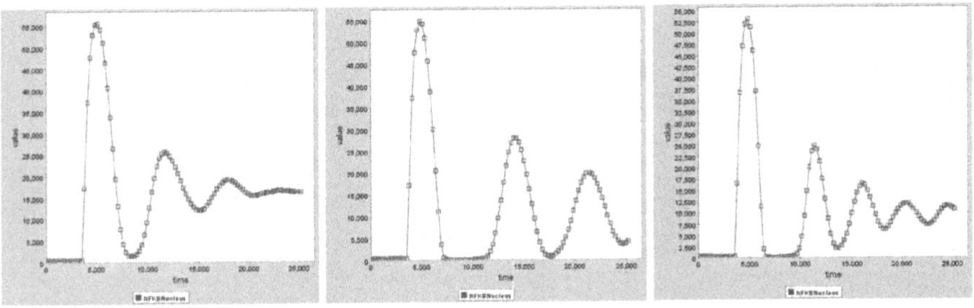

Fig. 10. Nuclear NF-κB when A20 and IκBα mRNA transcription rates have the original values (left), when A20 mRNA transcription rate is three-fold the original value (centre) and when IκBα mRNA transcription rate is three-fold the original value (right). The graphs show the average amount over 100 simulation runs.

is obtained when we elevate the IκBα mRNA transcription rate three-fold with respect to the original value.

These results are in agreement with the ones in [5]. Concerning single simulation runs, also for these experiments we have more persistent and sustained oscillations.

A20 deficient cells. The protein A20 is strongly NF-κB responsive and has an important inhibitory effect on the activation of NF-κB. The knockout of A20[6] in mouse dramatically alters the cell response to TNF stimulation due to persistent IKK activities and causes A20 deficient mice to die prematurely.

In Fig. 11 we report the stochastic simulation (average of 100 runs) for some species of the pathway. A20 deficient cells are simulated by setting A20 mRNA transcription rate to zero ($c1 = 0$). IKKa presents a level of activation in the tail after the peak. This is different from what happens in wild-type cells (see for instance Fig. 3) and it is mainly due to the fact that A20 is responsible for the inactivation of IKKa. This difference in the IKK activity influences the rest of the pathway response. In particular, this disregulates NF-κB, which then accumulates in the nucleus during TNF stimulation. As a consequence of this, the amount of IκBα mRNA increases and at the end of the stimulation we have twice the quantity with respect to the wild-type cells. These results are in agreement with the published experiments.

Fig. 12 reports a single simulation run for nuclear NF-κB. For A20 deficient cells nuclear NF-κB does not have an oscillatory behaviour also at the level of a single cell. This confirms the importance of the A20 feedback loop on oscillations.

Study of varying both c_1 and c_{1a}. In addition to the experiments reported above, where we vary each mRNA transcription rate (c_1 and c_{1a}) in isolation, in this work we use our model to investigate how the nuclear NF-κB is affected by varying both c_1 and c_{1a}. Indeed, from previous investigations [5,10], each of these two parameters influences the behaviour of nuclear NF-κB. Thus it is interesting to study their synergistic effect on the system. This experiment is not present in [5]. In Fig. 13 we report the results

[6] We call such cells A20 deficient.

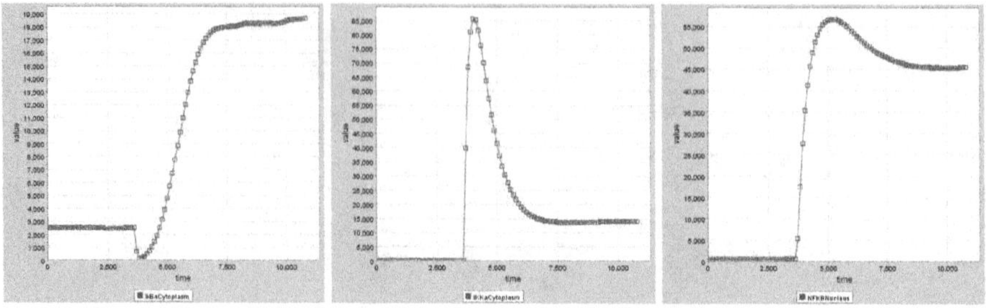

Fig. 11. A20 deficient cells. Stochastic simulation (average of 100 runs) for IκBα mRNA (left), IKKA (centre) and nuclear NF-κB (right).

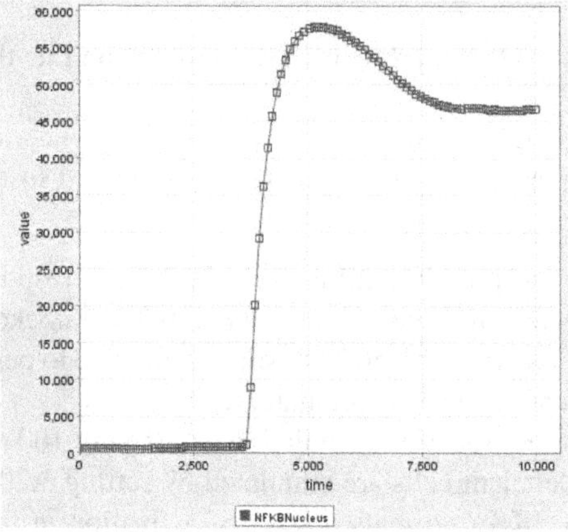

Fig. 12. A20 deficient cells. Single stochastic simulation for nuclear NF-κB.

for some choices of the two parameter values. Specifically, we focus on the following cases: both c_1 and c_{1a} are three-fold the original values (top-left), c_{1a} is three-fold its original value and c_1 is one third of its original value (top-right), c_1 is three-fold its original value and c_{1a} is one third of its original value (bottom-left) and, finally, both c_1 and c_{1a} are one third of the original values (bottom-right). Sustained oscillations (in the average) are obtained for the first and the third cases, when c_1 has a higher value. On the other hand, when c_1 has lower and c_{1a} has higher than original values the oscillations are damped and disappear immediately after the first oscillation when both the parameters are low. The effect of varying c_{1a} when c_1 is high is on the amplitude and the period of the oscillations: the higher the c_{1a} value, the greater the amplitude and narrower the oscillations.

As observed for the previous experiments, for single simulation runs the oscillations are more sustained and persistent (see Fig. 14), especially with respect to the smaller values of the parameter c_{1a}. Indeed, when c_{1a} is a third of its original value, the oscillations observable in a single run disappear at the population level (graphs on the right in Fig. 14 and Fig. 13, respectively).

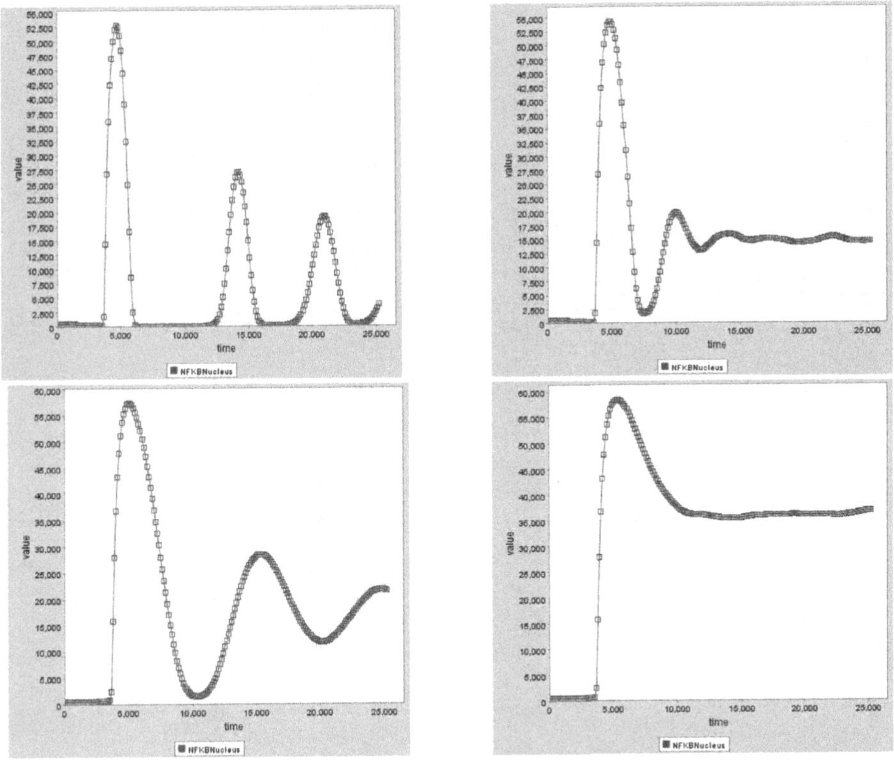

Fig. 13. Nuclear NF-κB under various assumptions about the A20 mRNA transcription rate c_1 and the IκBα mRNA transcription rate c_{1a}. Both c_1 and c_{1a} three-fold the original values (top, left), c_{1a} three-fold the original value and c_1 one third of the original value (top, right), c_1 three-fold the original value and c_{1a} one third of the original value (bottom, left) and both c_1 and c_{1a} one third of the original values (bottom, right). The graphs show the average amount over 100 simulation runs.

These results have yet to be validated against appropriate experiments but they are in agreement with the present knowledge of the system.

6.4 Sensitivity Analysis

In order to improve our understanding of the influence of parameter perturbation on species concentration, we employed *Sensitivity Analysis* (SA) [41]. In general this is performed as follows:

1. An initial set of parameter values (*nominal parameters*) is identified. Usually these are the values that are considered the most likely or are the result of parameter fitting. The analysis takes place in the neighbourhood of this configuration. We will call the model with nominal parameters the *nominal model*.
2. A measure or index of the sensitivity is defined, based on the aspect of the model that is the subject of the analysis. An aspect might be the amount of a species S

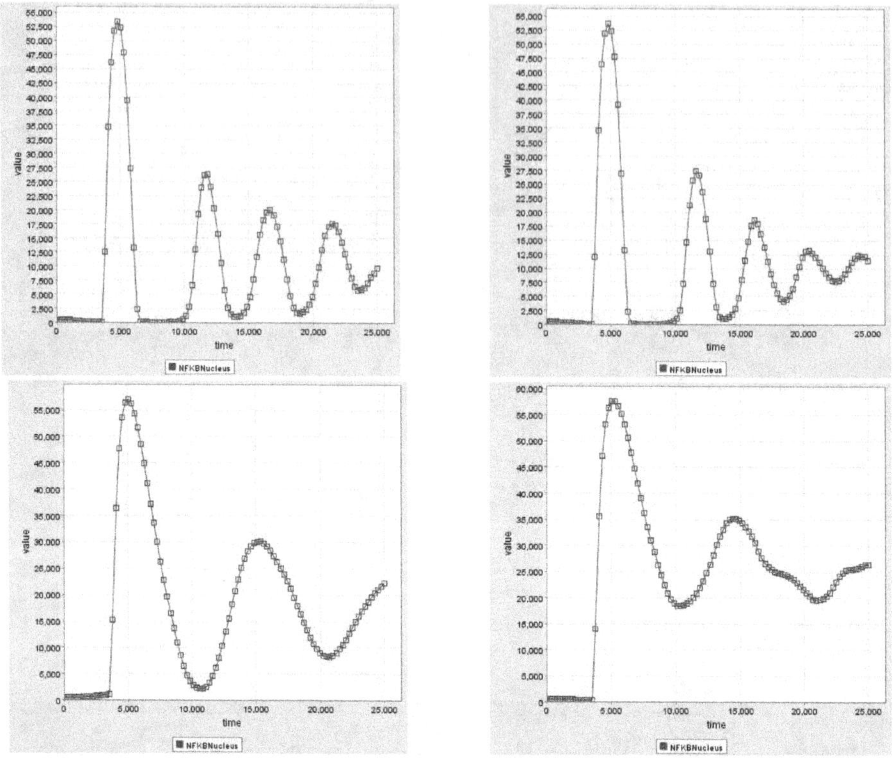

Fig. 14. Single simulation runs for nuclear NF-κB under various assumptions about the A20 mRNA transcription rate c_1 and the IκBα mRNA transcription rate c_{1a}. Both c_1 and c_{1a} three-fold the original values (top, left), c_{1a} three-fold the original value and c_1 one third of the original value (top, right), c_1 three-fold the original value and c_{1a} one third of the original value (bottom, left) and both c_1 and c_{1a} one third of the original values (bottom, right). These graphs correspond to the ones reported in Fig. 13 with a single simulation run instead of the average of 100 simulation runs.

at a specific time or the amplitude of an oscillation, while the index might be the difference between the amount of S computed by the model with nominal parameters and the amount of S computed by the model with perturbed parameters. This difference is an example of a *sensitivity index* (SI), which quantifies the influence the parameters have on a particular aspect.

3. One or more parameters are modified by a fixed value or a percentage, and SIs are obtained. Reactions governed by parameters with high sensitivity indicate where the model is most susceptible to variations. On the other hand, low sensitivities indicate robustness.

A previous sensitivity analysis of the NF-κB pathway can be found in [9], where an extensive investigation of the influences of the parameters on the amplitude and period of the oscillations of nuclear NF-κB was presented. As described earlier, this analysis was based on a slightly improved version of the system of ODEs presented by Hoffmann *et al.* [3], described in terms of a Gepasi model. The "one-at-a-time" (*OAT*) method, which consists of perturbing only one parameter of the model at a time, was used and nine parameters out of a total of 64 where identified as the most important. This is based on numerical integration of ODEs. Since our stochastic version of the Lipniacki model has a lot of interactions in common with Hoffmann's, these nine parameters find a match in our model, allowing us to compare the results.

In our sensitivity analysis we also used an OAT approach, but three main differences with respect to [9] should be highlighted. First, we perform the analysis on a stochastic model, which permits a more informative analysis than in the case of ODE models, as we shall see shortly. Second, the model we consider includes only one IκB inhibitor (IκBα but not IκBβ and IκBϵ) and the additional A20 inhibitory effect on IKK. Third, we do not use the difference in amplitude or period of oscillations of nuclear NF-κB as a measure of sensitivity. Instead we consider the difference in amount of nuclear NF-κB and cytoplasmic IκBα at regular time intervals.

The parameters subjected to analysis are the already mentioned nine parameters identified in [9] (c_{3_a}, a_3, a_2, i_{1_a}, t_1, t_2, k_{deg}, c_{4_a} and c_{1_a}), with the addition of three parameters related to A20 activity (c_1, c_3 and c_4). For the details of these parameters see Table 1.

As sensitivity indices we use two complementary measures: the *average distance* and the *histogram distance* of stochastic simulations. The former consists of the simple difference in the amount of a species at a selected time, after the perturbation of one parameter. Since we use stochastic simulations, such an amount is the average over a certain number of runs. The latter has been introduced as a sensitivity measure in [24]. The idea behind this distance is that an estimated probability density function (edf) of the amount of a species at a time point can be constructed using a suitable number of simulation runs. Such an edf will have area equal to one, like a probability density function. The overlapping area of two edfs obtained from two different models is then an estimate of the likelihood that the two models will reproduce the same output (see Fig. 15). This measure was introduced originally in [42] to quantify the ability of an approximated version of Gillespie's SSA to replicate the original. We will call the results of the analysis *average sensitivities* if obtained using average distance, or *stochastic sensitivities* if obtained using histogram distance.

The complementarity of the two measures is evident. The average distance does not contain any notion of the distribution of the runs, but it becomes necessary if the distributions of the runs do not overlap at all (in which case the histogram distance is always two). Therefore, the stochastic OAT sensitivity analysis applies when one is interested in observing the change in the distribution of the amount of a particular species at a given time. This method has been implemented in the version of the Dizzy simulator developed at the University of Edinburgh [25].

In our analysis we observed the influence of a perturbation of 10% of the nominal value of the parameters (as in [9]) on the amount of nuclear NF-κB and cytoplasmic IκBα, using 200 simulation runs. The analysis is performed every 15 minutes, as the sensitivity indices might be time dependent. Results have been grouped in four heat diagrams, shown in Figure 16, with average distance at the top and histogram distance below. Each diagram can be read in two ways: horizontally, for the sensitivity of a

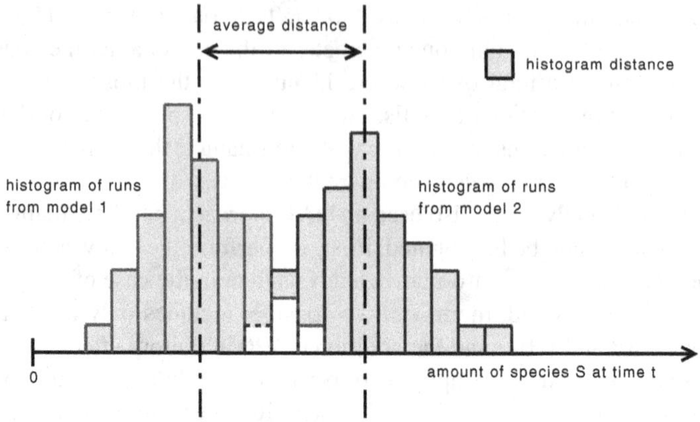

Fig. 15. Average distance and histogram distance between two sets of simulation runs at time t. The former represents the distance of the mean while the latter quantifies the likelihood of run overlapping.

parameter through time and vertically, to compare sensitivity of the parameters at a specific time. A scale of colours is used to represent sensitivity indices, from dark red (low SI), passing through yellow, to white (high SI).

The sensitivity analysis gives rise to the following observations:

- In contrast with [9], four of the nine originally identified parameters, a_2, t_1, t_2 and k_{deg}, present a low sensitivity and are clearly the least influential as they show very little variation through time with respect to both nuclear NF-κB and cytoplasmic IκBα. This is probably due to the introduction of the A20 feedback loop. High sensitivity of the three parameters connected with A20 activity sustains this hypothesis, suggesting a key role for this molecule.
- The stochastic sensitivities reveal that with a 10% perturbation of the considered rates there is still significant overlap in the distributions of values obtained from the simulations. Moreover, the trend of the stochastic sensitivities seems to agree with the average sensitivities, suggesting a distribution of the runs around the mean.
- c_{1a} and c_{4a} (bottom two rows of the heat diagrams) exhibit a similar trend, with high sensitivities when nuclear NF-κB peaks (compare with Fig. 4), suggesting that the rates of transcription and translation of IκBα play a major role in the determination of peak intensity. In particular, the sensitivities of c_{1a} and c_{4a} oscillate in time, suggesting agreement with the nominal model (low sensitivity) alternated to disagreement (high sensitivity) at the times when NF-κB peaks. Moreover, this trend is complementary to the one of c_{3a}, that presents low sensitivity when nuclear NF-κB peaks;
- Similarly, c_1 and c_4, rates of A20 transcription and translation, share a similar trend, and this appears to be dependent on the target species.
- The sensitivity of i_{1a}, the rate of IκBα nuclear import, shows a very clear oscillatory trend only when targeting cytoplasmic IκBα. In particular, a perturbation of i_{1a} yields major changes in the cytoplasmic IκBα peaks.

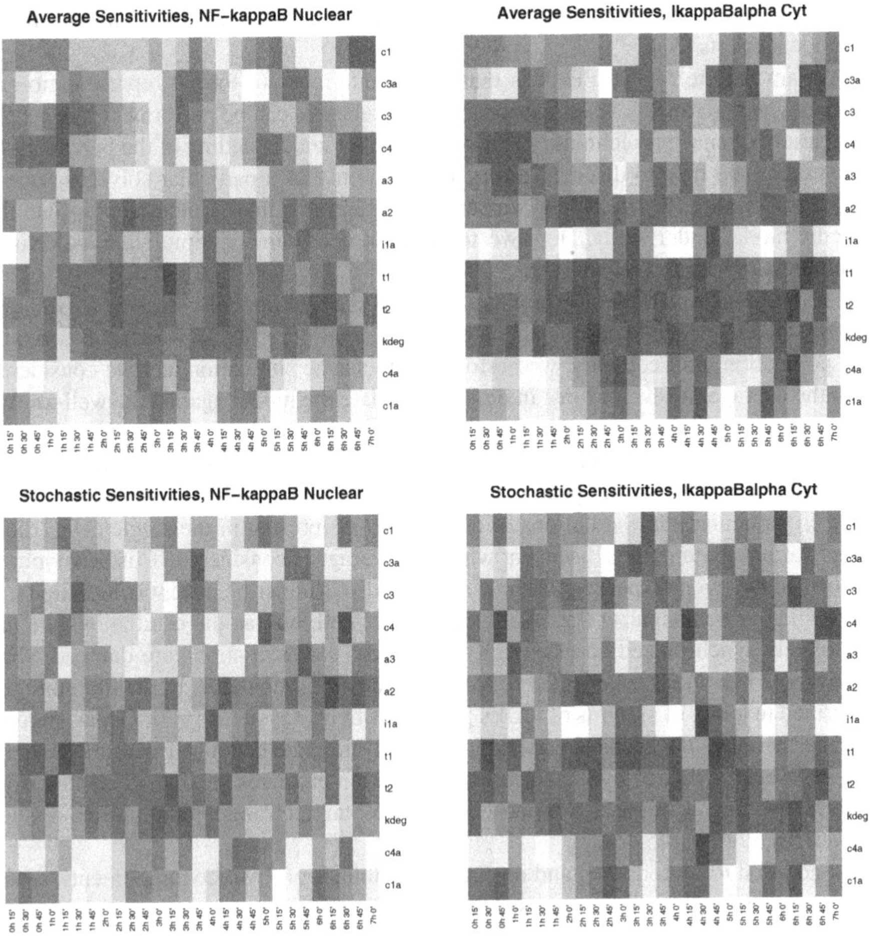

Fig. 16. Heat diagrams of the sensitivity analysis performed on selected parameters of the NF-κB model. Red colours indicate low sensitivity, while yellow or white indicate high sensitivity. The two diagrams above, refer to sensitivity indexes computed using average of simulations, while the two diagrams at the bottom refer to the stochastic approach, that considers a density distance.

7 Conclusions

In this work we have presented a Bio-PEPA model for the NF-κB signalling pathway. This is the first algorithmic model [18] of the pathway, which captures how and why the system changes state: a much more mechanistic account of the pathway than that given by previous representations as systems of ODEs, or even as chemical reactions since the role of compartments and events are fully captured in the model. We have studied it using a selection of the analysis techniques supported by the language and implemented in the Bio-PEPA Workbench [43]. With our model we were able to describe in detail important features of the system, such as compartments and the activation of NF-κB by an external stimulus, using a recent extension of the Bio-PEPA language which incorporates explicit representation of locations [32] and time-dependent events [31].

A main feature of Bio-PEPA is that it is a intermediate formal representation of biochemical systems, on which various kinds of analysis can be performed. The access to a variety of analysis techniques can foster a better understanding of the behaviour of the system, and help to discover errors due to the use of a particular solver/simulator [44]. Furthermore, the modeller can select the approach that is most appropriate for specific model under study. Here we focused on deterministic simulation, stochastic simulation and sensitivity analysis.

The system of ODEs corresponding to the Bio-PEPA model was used to validate the system against the experimental data from the literature [23,3], which is based on a cell population. Subsequently, we performed stochastic simulation and we considered both the average of several runs, in order to validate the model against the well-known population behaviour, and single runs, to analyse the behaviour of a single cell. Even though the average behaviour is in agreement with the experimental data and the results obtained from ODEs, some species are characterized by variability between the various runs. In particular, the oscillations of nuclear NF-κB present in the single cells (runs) under various assumptions disappear when the average is considered. This behaviour is confirmed by recent experiments showing that NF-κB under persistent stimulation has an oscillatory behaviour at the level of the single cell, but, as the oscillations are not completely synchronized across a cell population, the oscillations are damped when the average behaviour is considered. Note that most of the models in the literature are described in terms of systems of ODEs [3,5] or as hybrid systems [6] or Gepasi models [9] and therefore the focus is generally on the average behaviour of cells. The single cell approach facilitated by the stochastic simulation emulates the data generated from live cell imaging techniques, as opposed to those obtained from population studies such as western blotting [45,46].

In addition to reproducing and discussing a number of in-silico experiments, sensitivity analysis was applied to investigate the most influential parameters of the model. Our results are complementary to the previous work [9]. Our model has some significant differences, such as the inclusion of a negative feedback loop, and is analysed stochastically. As a result we observe some interesting new phenomena. Moreover, we investigated model properties from a different point of view, using alternative sensitivity measures with respect to [9]. In the presented analysis, we considered a local approach, i.e. one focused around a specific point in the parameter space. This can be informative, giving an idea of the impact of parameter changes on the behaviour of the system. Moreover, this strongly suggested that parameters have specific dependences at particular points in the time evolution of the pathway. This prediction would be amenable to experimental test by single cell imaging methods. In the future, we plan to apply global methods in order to explore the full parameter space (or a meaningful subset of it) and to quantify the relationships between different parameters.

Another route to analysis of a Bio-PEPA model is via the mapping to continuous time Markov chains (CTMC). In particular, it is possible to derive a PRISM [47,48] model in order to verify some properties expressed as a logical formula, by model checking. We have yet to explore the possibilities offered by this route for our NF-κB model. Two main challenges for the use of model checking with this model are the dimension of the state space (it is extremely large) and the presence of temporal events. For the former, one possibility is to apply an abstract-view for the CTMC in terms of concentration levels [49], which can substantially reduce the state space.

A final biological point is that the growing body of evidence from modelling biological pathways is that they exhibit parameter dependences that characterise the structure of the pathway. Most exhibit a small number of vulnerable nodes and are insensitive to perturbation of many values. This is important for considering the development of therapeutic interventions and may well reflect the evolution of the pathway via quantitative changes in key interactions [50].

Acknowledgements

This work was completed while Federica Ciocchetta was a research fellow at the University of Edinburgh, supported by the EPSRC grant EP/c54370x/01. Andrea Degasperi is supported by a University of Glasgow Lord Kelvin/Adam Smith scholarship. Jane Hillston is supported by the EPSRC ARF EP/c543696/01. John K. Heath is supported by Cancer Research UK.

References

1. Cheong, R., Levchenko, A.: Wires in the soup: quantitative models of cell signalling. Trends in Cell Biology 18, 112–118 (2008)
2. Hayden, M., Ghosh, S.: Shared principle in NF-κB Signalling. Cell 132, 344–362 (2008)
3. Hoffmann, A., Levchenko, A., Scott, M., Baltimore, D.: The IκB–NF-κB Signaling Module: Temporal Control and Selective Gene Activation. Science 298, 1241–1245 (2002)
4. Cho, K.H., Shin, S.Y., Lee, H.W., et al.: Investigations Into the Analysis and the Modeling of the TNF-mediated NFκB Signaling Pathway. Genome Res. 13, 2413–2422 (2003)
5. Lipniacki, T., Paszek, P., Brasier, A., Luxon, B., Kimmel, M.: Mathematical model of NF-κB regulatory module. Journal of Theoretical Biology 228, 195–215 (2004)
6. Lipniacki, T., Paszek, P., Brasier, A., Luxon, B., Kimmel, M.: Stochastic Regulation in Early Immune Response. Biophysical Journal 90, 725–742 (2006)
7. Nelson, D., Ihekwaba, A., Elliott, M., Johnson, J., Gibney, C., Foreman, B., Nelson, G., See, V., Horton, C., Spiller, D.G., Edwards, S., McDowell, H., Unitt, J.F., Sullivan, E., Grimley, R., Benson, N., Broomhead, D.S., Kell, D., White, M.: Oscillations in NF-κB Signalling Control the Dynamics of Gene Expression. Science 306, 704–708 (2004)
8. Ashall, L., Horton, C., Nelson, D., Paszek, P., Harper, C., Sillitoe, K., Ryan, S., Spiller, D.G., Unitt, J.F., Broomhead, D., Kell, D., Rand, A., Sée, V., White, M.: Pulsatile Stimulation Determines Timing and Specificity of NF-κB–Dependent Transcription. Science 324, 242–246 (2009)
9. Ihekwaba, A., Broomhead, D., Grimley, R., Kell, D.: Sensitivity analysis of parameters controlling oscillatory signalling in the NF-κB pathway: the roles of IKK and IκBα. Systems Biology 1, 93–103 (2004)
10. Ihekwaba, A., Broomhead, D., Grimley, R., Kell, D.: Synergistic control of oscillations in the NF-κB signalling pathway. IEE Proc.-Syst. biol. 152, 153–160 (2005)
11. Mendes, P.: GEPASI: a software package for modelling the dynamics, steady states and control of biochemical and other systems. Comput. Appl. Biosci. 9, 563–571 (1993)
12. Priami, C., Regev, A., Silverman, W., Shapiro, E.: Application of a stochastic name-passing calculus to representation and simulation of molecular processes. Information Processing Letters 80, 25–31 (2001)

13. Curti, M., Degano, P., Priami, C., Baldari, C.: Modelling biochemical pathways through enhanced π-calculus. Theoretical Computer Science 325, 111–140 (2004)
14. Regev, A., Panina, E., Silverman, W., Cardelli, L., Shapiro, E.: BioAmbients: an Abstraction for Biological Compartments. Theoretical Computer Science 325, 141–167 (2004)
15. Priami, C., Quaglia, P.: Beta binders for biological interactions. In: Danos, V., Schachter, V. (eds.) CMSB 2004. LNCS (LNBI), vol. 3082, pp. 20–33. Springer, Heidelberg (2005)
16. Calder, M., Gilmore, S., Hillston, J.: Modelling the Influence of RKIP on the ERK Signalling Pathway Using the Stochastic Process Algebra PEPA. In: Priami, C., Ingólfsdóttir, A., Mishra, B., Riis Nielson, H. (eds.) Transactions on Computational Systems Biology VII. LNCS (LNBI), vol. 4230, pp. 1–23. Springer, Heidelberg (2006)
17. John, M., Lhoussaine, C., Niehren, J., Uhrmacher, A.: The Attributed Pi-Calculus with Priorities. In: Priami, C., et al. (eds.) Transactions on Computational Systems Biology XII. LNCS (LNBI), vol. 5945, pp. 13–76. Springer, Heidelberg (2010)
18. Priami, C.: Algorithmic systems biology. Communications of the ACM 52 (2009)
19. Fisher, J., Henzinger, T.: Executable cell biology. Nature Biotechnology 25, 1239–1249 (2007)
20. Ciocchetta, F., Hillston, J.: Bio-PEPA: a framework for the modelling and analysis of biological systems. School of Informatics University of Edinburgh Technical Report EDI-INF-RR-1231 (2008)
21. Ciocchetta, F., Hillston, J.: Bio-PEPA: a Framework for the Modelling and Analysis of Biochemical Networks. Theoretical Computer Science 410, 3065–3084 (2009)
22. Carlotti, F., Dower, S., Qwarnstrom, E.: Dynamic shuttling of nuclear factor kappa B between the nucleus and cytoplasm as a consequence of inhibitor dissociation. J. Biol. Chem. 275, 41028–41034 (2000)
23. Lee, E., Boone, D., Chai, S., Libby, S., Chien, M., Lodolce, J., Ma, A.: Failure to regulate TNF-induced NF-κB and cell death responses in A20-deficient mice. Science 289, 2350–2354 (2000)
24. Degasperi, A., Gilmore, S.: Sensitivity Analysis of Stochastic Models of Bistable Biochemical Reactions. In: Bernardo, M., Degano, P., Zavattaro, G. (eds.) SFM 2008. LNCS, vol. 5016, pp. 1–20. Springer, Heidelberg (2008)
25. Dizzy Edinburgh version (2009), `http://homepages.inf.ed.ac.uk/stg/software/Dizzy/`
26. Ciocchetta, F., Degasperi, A., Heath, J., Hillston, J.: Modelling and analysis of the NF-κB pathway in Bio-PEPA. In: Breitling, R., Gilbert, D.R., Heiner, M., Priami, C. (eds.) Formal Methods in Molecular Biology, Dagstuhl, Germany. Dagstuhl Seminar Proceedings, vol. 09091. Schloss Dagstuhl - Leibniz-Zentrum fuer Informatik, Germany (2009)
27. Gerondakis, S., Grossmann, M., Nakamura, Y., Pohl, T., Grumont, R.: Genetic approaches in mice to understand Rel/NF-κB and IκB function: transgenics and knockouts. Oncogene 18, 6888–6895 (1999)
28. Hillston, J., Duguid, A.: Deriving Differential Equations from Process Algebra Models in Reagent-Centric Style. In: Algorithmic Bioprocesses. Natural Computing Series. LNCS (2009)
29. Larcher, R., Ihekwaba, A., Priami, C.: A BetaBW model for the NF-κB pathway. Technical Report TR 25/2007, The Microsoft Research-University of Trento Centre for Computational and Systems Biology (2007)
30. Hillston, J.: A Compositional Approach to Performance Modelling. Cambridge University Press, Cambridge (1996)
31. Ciocchetta, F.: Bio-PEPA with events. In: Priami, C., Back, R.-J., Petre, I. (eds.) Transactions on Computational Systems Biology XI. LNCS (LNBI), vol. 5750, pp. 45–68. Springer, Heidelberg (2009)

32. Ciocchetta, F., Guerriero, M.: Modelling Biological Compartments in Bio-PEPA. In: Proc. of MeCBIC 2008. ENTCS, vol. 227, pp. 77–95 (2009)
33. The Bio-PEPA Workbench (2009), http://www.dcs.ed.ac.uk/home/stg/software/biopepa/about.html
34. Bio-PEPA (2008), http://www.biopepa.org/
35. MATLAB (2009), http://www.mathworks.com/products/matlab/
36. Dizzy (2008), http://magnet.systemsbiology.net/software/Dizzy
37. Gillespie, D.: Exact stochastic simulation of coupled chemical reactions. J. Phys. Chem. 81, 2340–2361 (1977)
38. Bank, R., et al.: Transient simulation of silicon devices and circuits. IEEE Transactions on Electron Devices 32, 1992–2007 (1985)
39. McAdams, H., Arkin, A.: Stochastic mechanisms in gene expression. Proc. Natl. Acad. Sci. USA 94, 814–819 (1997)
40. Ihekwaba, A., Wilkinson, S., Waithe, D., Broomhead, D., Li, P., Grimley, R., Benson, N.: Bridging the gap between in silico and cell-based analysis of the nuclear factor κB signalling pathway by in vitro studies of IKK2. FEBS Journal 274, 1678–1690 (2007)
41. Saltelli, A., Chan, K., Scott, E.: Sensitivity Analysis. Wiley, Chichester (2000)
42. Cao, Y., Petzold, L.: Accuracy limitations and the measurements of errors in the stochastic simulation of chemically reacting systems. J. Comput. Phys. 212, 6–24 (2006)
43. Ciocchetta, F., Duguid, A., Gilmore, S., Guerriero, M., Hillston, J.: The Bio-PEPA Tool Suite. In: Proceedings of the 6th International Conference on Quantitative Evaluation of SysTems (QEST 2009), Budapest, Hungary, pp. 309–310 (2009)
44. Calder, M., Duguid, A., Gilmore, S., Hillston, J.: Stronger computational modelling of signalling pathways using both continuous and discrete-state methods. In: Priami, C. (ed.) CMSB 2006. LNCS (LNBI), vol. 4210, pp. 63–77. Springer, Heidelberg (2006)
45. Ankers, J., Spiller, D., White, M., Harper, C.: Spatio-temporal protein dynamics in single living cells. Curr. Opin. Biotechnology 19, 375–380 (2008)
46. Sillitoe, K., Horton, C., Spiller, D., White, M.: Single-cell time-lapse imaging of the dynamic control of nfκb signalling. Biochem. Soc. Transactions 35, 263–266 (2007)
47. Kwiatkowska, M., Norman, G., Parker, D.: Prism: Probabilistic model checking for performance and reliability analysis. ACM SIGMETRICS Performance Evaluation Review (2009)
48. PRISM (2009), http://www.prismmodelchecker.org
49. Ciocchetta, F., Degasperi, A., Hillston, J., Calder, M.: Some Investigations Concerning the CTMC and the ODE Model Derived from Bio-PEPA. In: Proc. of FBTC 2008. ENTCS, vol. 209, pp. 145–163 (2009)
50. Gerhart, J., Kirshner, M.: The theory of facilitated variation. PNAS (2007)

A The Bio-PEPA Model for the NF-κB Pathway

In this Appendix we report the full Bio-PEPA model of the NF-κB pathway studied in this paper. First, the set of locations is considered. Then the set of functional rates and the set of parameters are reported. The name of each action type describes the function of the associated reaction. The notation $fMA(r)$ indicates that the kinetic law is mass-action with constant rate r. After that, there is the definition of species components and of the model component. Finally, the events describing the TNF stimulus are defined. The effect of events is to reset the value of the parameter signal, from 0 (signal inactive) to 1 (signal active) and vicerversa. These events enable /unable two reactions, the transformation of IKKn into IKKa and the transformation of IKKa into IKKi with the support of A20, represented by the action types activation_IKKn and transformation_IKKa_into_IKKi_by_A20, respectively (the corresponding kinetic laws depend on the parameter signal and are 0 if the signal is 0). Here we do not report the set $\mathcal{N}$ with auxiliary information for species as this information is not considered in our study. Note that species and parameters are given in terms of number of molecules.

In order to derive a model in terms of number of molecules from the model in terms of concentration [5], all the initial species concentrations are rescaled by the factors *nscale*= $2 \cdot 10^5$ (species in the nucleus) and *cscale*= 10^6 (species in the cytoplasm) and the parameters are modified accordingly. The initial concentration of the cytoplasmic complex IκBα–NF-κB is therefore $0.06 \cdot cscale = 60000$ molecules and all the other species are zero.

$$location\ nuc : kind = \mathbf{C},\ size = 3.33 \cdot 10^{-13}\ l;$$
$$location\ cyt : kind = \mathbf{C},\ size = 1.65 \cdot 10^{-12}\ l$$

$activation\_IKKn = [fMA(signal \cdot k_1)]$;
$transformation\_IKKa\_into\_IKKi\_by\_A20 = [fMA(signal \cdot k_2)]$;
$production\_IKKn = [k_{prod}]$;
$transformation\_IKKa\_into\_IKKi = [fMA(k_3)]$;
$degradation\_IKKn = [fMA(k_{deg})]$;
$a20t\_transcription\_by\_NFkBn = [fMA(c_1)]$;
$degradation\_IKKa = [fMA(k_{deg})]$;
$cgent\_transcription\_by\_NFkBn = [fMA(c_{1c})]$;
$degradation\_IKKi = [fMA(k_{deg})]$;
$association\_IkBa\_IKKa = [fMA(a_2)]$;
$dissociation\_IkBaIKKa = [fMA(t_1)]$;
$a20\_translation = [fMA(c_4)]$;
$association\_IKKa\_IkBaNFkB = [fMA(a_3)]$;
$a20\_degradation = [fMA(c_5)]$;
$dissociation\_IKKaIkBaNFkB = [fMA(t_2)]$;
$a20t\_transcription = [c_2]$;
$association\_IkBa\_NFkB = [fMA(a_1)]$;
$a20t\_degradation = [fMA(c_3)]$;
$association\_IkBa\_NFkBn = [fMA(a_{1n})]$;
$cgent\_transcription = [c_{2c}]$;
$transport\_IkBaNFkBn\_nucl\_cyt = [fMA(e_{2a})]$;
$cgent\_degradation = [fMA(c_{3c})]$;
$ikBat\_transcription\_by\_NFkBn = [fMA(c_{1a})]$;

$$
\begin{aligned}
&dissociation\_IKKaIkBaNFkB = [fMA(c_{6a})];\\
&transport\_IkBa\_cyt\_nucl = [fMA(i_{1a})];\\
&transport\_IkBa\_nucl\_cyt = [fMA(e_{1a})];\\
&transport\_NFkB\_cyt\_nucl = [fMA(i_{1})];\\
&ikBa\_translation = [fMA(c_{4a})];\\
&ikBa\_degradation = [fMA(c_{5a})];\\
&ikBat\_transcription = [c_{2a}];\\
&ikBat\_degradation = [fMA(c_{3a})]
\end{aligned}
$$

$$
\begin{aligned}
IKKn@cyt &\stackrel{def}{=} (production\_IKKn, 1)\uparrow + (degradation\_IKKn, 1)\downarrow +\\
&\quad (activation\_IKKn, 1)\downarrow\\
IKKa@cyt &\stackrel{def}{=} (activation\_IKKn, 1)\uparrow + (transformation\_IKKa\_into\_IKKi, 1)\downarrow +\\
&\quad (transformation\_IKKa\_into\_IKKi\_by\_A20, 1)\downarrow +\\
&\quad (degradation\_IKKa, 1)\downarrow +\\
&\quad (association\_IkBa\_IKKa, 1)\downarrow +\\
&\quad (dissociation\_IkBaIKKa, 1)\uparrow +\\
&\quad (association\_IKKa\_IkBaNFkB, 1)\downarrow +\\
&\quad (dissociation\_IKKaIkBaNFkB, 1)\uparrow\\
IKKi@cyt &\stackrel{def}{=} (transformation\_IKKa\_into\_IKKi, 1)\uparrow +\\
&\quad (transformation\_IKKa\_into\_IKKi\_byA20, 1)\uparrow +\\
&\quad (degradation\_IKKi, 1)\downarrow\\
A20@cyt &\stackrel{def}{=} (transformation\_IKKa\_into\_IKKi\_by\_A20, 1) \odot +\\
&\quad (a20\_translation, 1)\uparrow + (a20\_degradation, 1)\downarrow\\
I\kappa B\alpha@cyt &\stackrel{def}{=} (IkBa\_translation, 1)\uparrow + (IkBa\_degradation, 1)\downarrow +\\
&\quad (association\_IkBa\_IKKa, 1)\downarrow + (association\_IkBa\_NFkB, 1)\downarrow +\\
&\quad (transport\_IkBa\_cyt\_nucl, 1)\downarrow + (transportIkBa\_nucl\_cyt, 1)\uparrow\\
NF\text{-}\kappa B@cyt &\stackrel{def}{=} (dissociation\_IkBaNFkB, 1)\uparrow + (association\_IkBa\_NFkB, 1)\downarrow +\\
&\quad (dissociation\_IKKaIkBaNFkB, 1)\uparrow +\\
&\quad (transport\_NFKB\_cyt\_nucl, 1)\downarrow\\
complex1 &\stackrel{def}{=} (association\_IkBa\_NFkB, 1)\uparrow + (dissociation\_IkBaNFkB, 1)\downarrow +\\
&\quad (association\_IKKa\_IkBaNFkB, 1)\downarrow +\\
&\quad (transport\_IkBaNFkBn\_nucl\_cyt, 1)\uparrow\\
complex2 &\stackrel{def}{=} (association\_IkBa\_IKKa, 1)\uparrow + (dissociation\_IkBaIKKa, 1)\downarrow\\
complex3 &\stackrel{def}{=} (association\_IKKa\_IkBaNFkB, 1)\uparrow +\\
&\quad (dissociation\_IKKaIkBaNFkB, 1)\downarrow\\
I\kappa B\alpha@nuc &\stackrel{def}{=} (association\_IkBan\_NFkBn, 1)\downarrow +\\
&\quad (transport\_IkBa\_IkBa\_cyt\_nucl, 1)\uparrow + (transport\_IkBa\_nucl\_cyt, 1)\downarrow
\end{aligned}
$$

$$
\begin{aligned}
complex4 &\stackrel{def}{=} (association\_IkBa\_NFkBn, 1)\uparrow + \\
&\quad (transport\_IkBaNFKBn\_nucl\_cyt, 1)\downarrow \\
A20t@cyt &\stackrel{def}{=} (a20\_translation, 1)\odot + (a20t\_transcription, 1)\uparrow + \\
&\quad (a20t\_transcription\_by\_NFkBn, 1)\uparrow + (a20t\_degradation, 1)\downarrow \\
I\kappa B\alpha t@cyt &\stackrel{def}{=} (ikBat\_transcription, 1)\uparrow + (ikBat\_transcription\_by\_NFkBn, 1)\uparrow + \\
&\quad (ikBat\_degradation, 1)\downarrow + (ikBa\_translation, 1)\odot \\
Cgent@cyt &\stackrel{def}{=} (gent\_transcription, 1)\uparrow + (cgent\_transcription\_by\_NFkBn, 1)\downarrow + \\
&\quad (cgent\_degradation, 1)\downarrow
\end{aligned}
$$

where the names *complex*1, *complex*2, *complex*3 and *complex*4 stand for *IκBα–NF-κB@cyt*, *IKKa–IκBα@cyt*, *IKKa–IκBα–NF-κB@cyt* and *IκBα–NF-κB@nuc*, respectively. The species components *A20t*, *IκBαt* and *cgent* are the mRNA transcripts of the proteins *A20*, *IκBα* and *cgen* and are assumed in the cytoplasm as in the original model.

$$
\begin{gathered}
IKKn@cyt[0] \underset{*}{\bowtie} IKKa@cyt[0] \underset{*}{\bowtie} IKKi@cyt[0] \underset{*}{\bowtie} IKKa\text{–}I\kappa B\alpha@cyt[0] \underset{*}{\bowtie} \\
A20@cyt[0] \underset{*}{\bowtie} I\kappa B\alpha@cyt[0] \underset{*}{\bowtie} NF\text{-}\kappa B@cyt[0] \underset{*}{\bowtie} \\
I\kappa B\alpha\text{–}NF\text{-}\kappa B@cyt[60000] \underset{*}{\bowtie} IKKa\text{–}I\kappa B\alpha\text{–}NF\text{-}\kappa B@cyt[0] \underset{*}{\bowtie} \\
I\kappa B\alpha@nuc[0] \underset{*}{\bowtie} NF\text{-}\kappa B@nuc[0] \underset{*}{\bowtie} I\kappa B\alpha\text{–}NF\text{-}\kappa B@nuc[0] \underset{*}{\bowtie} \\
A20t@cyt[0] \underset{*}{\bowtie} I\kappa B\alpha t@cyt[0] \underset{*}{\bowtie} Cgent@cyt[0]
\end{gathered}
$$

$$
\begin{aligned}
Events = [&(begin\_signal,\ time = T_1,\ signal = 1,\ 0); \\
&(end\_signal,\ time = T_2,\ signal = 0,\ 0)]
\end{aligned}
$$

In our case $T_1 = 3600$ and $T_2 = 25200$ (time expressed in *seconds*).

Author Index